A First Course on PARAMETRIC INFERENCE

A First Course on PARAMETRIC INFERENCE

B.K. Kale

Narosa Publishing House
New Delhi Madras Bombay Calcutta
London

Dr. B.K. Kale
Department of Statistics
University of Pune
Pune-411 007, INDIA

NAROSA PUBLISHING HOUSE

6 Community Centre, Panchsheel Park, New Delhi 110 017
22 Daryaganj, Prakash Deep, Delhi Medical Association Road, New Delhi 110 002
35-36 Greams Road, Thousand Lights, Madras 600 006
306 Shiv Centre, D.B.C. Sector 17, K.U. Bazar P.O., New Mumbai 400 705
2F-2G Shivam Chambers, 53 Syed Amir Ali Avenue, Calcutta 700 019
3 Henrietta Street, Covent Garden, London WC2E 8LU, UK

ISBN 81-7319-196-4

Published by N.K. Mehra for Narosa Publishing House, 6 Community Centre, Panchsheel Park, New Delhi 110 017 and printed at Replika Press Pvt Ltd, Delhi 110 040 (India).

हें सारस्वताचें गोड
तुम्हींचि लाविलें जी झाड
तरी आता अवधानामृते वाड
सिंपौनि की जो ।।

संत ज्ञानेश्वर, शालिवाहन शके १२१२

This fruitful tree of knowledge
Which you yourself have planted
Please nurture it with the nectar
Of your full attention it deserves

Saint Dnyaneshwar—1290 A.D.

Perface

This book is intended to be used as a text for an introductory course on model based parametric inference in senior undergraduate (honours) or in the first year graduate program. The course can be covered well in one academic year or two semesters or three quarters as is normally done in most U.S./Canadian and Indian Universities where it is regarded as a core course in the Statistics Master's degree program. A major (principal) in Statistics and minor (subsidiary) in Mathematics at Bachelor's level would be adequate prerequisites for this text. Therefore it is assumed that the student knows single and multivariable calculus, elementary matrix algebra, standard distribution theory and elementary forms of Weak Laws of Large Number and Central Limit Theorems.

In Chapter 1 a historical perspective is presented which led to the current frequentist approach where the evalution of the performance of any inference procedure is based on its sampling distribution. Such a perspective is necessarily subjective and other teachers may want to view the evolution of ideas somewhat differently. However it is necessary to present such a historical perspective emphasizing that most of the current inference procedures have been develped to provide answers to important questions in other sciences. It is necessary that the students and teachers should be aware of this evolution so that they can respond effectively in a similar way to provide statistical methods as needed by other scientists facing important questions in their fields.

It is assumed here that statistical inference is an integral part of rational empiricism proposed by Descartes—Bacon leading to deliberate experimentation as an important tool to gain knowledge about the external world around us. Thus the text starts with the concept of information in a sample about the unknown parameter of interest and starts with the problem of defining and obtaining a minimal sufficient statistic. There after the text follows more or less the well trodden path but also making some detours which it is hoped that the students and the teachers would find interesting and useful.

I have taught the material covered here for several years and there are many students and colleagues at Iowa State, Manitoba and Pune whose contributions, I wish to acknowledge and take this opportunity to thank them. I thank University Grants Commission of India for providing me a grant under their scheme of writing University level text book during May 1994-May 1996 and subsequently an Emeritus Fellowship. I also wish to express my thanks to the authorities of University of Pune and Department of Statistics for providing me necessary facilities to continue my research and complete the text after my retirement in November 1993.

Last but not least, I must thank Mrs. A.V. Sabane for her careful typing of the manuscript and M/s Narosa Publishing House, New Delhi for subsequent processing and bringing out the text so nicely in its present form.

Pune B.K. KALE

January 1999

Contents

A First Course on PARAMETRIC INFERENCE

1

Introduction

1.1 Basic Framework

In the standard framework of parameter estimation one starts with the data $(x_1, ..., x_n)$ which are observations on the characteristic X of n individual items and the data is regarded as a realization of random variables (r.v.) $(X_1, ..., X_n)$ which are assumed to be independent identically distributed (i.i.d.) as X the r.v. whose probability law describes stochastic behaviour of the character X in the population. This is also described as having a random sample of size n on X with probability distribution specified by probability density function (pdf) in case of continuous r.v. X or by probability mass function (pmf) in case of discrete X. Henceforth we will use the word pdf which would also include pmf when X is discrete. We further assume that the pdf has a known functional form but involves an unknown parameter a real or vector valued $\theta = (\theta_1, ..., \theta_m)$, which is a labelling or indexing parameter. That is if we know the value of θ then the pdf of X is fully specified and the stochastic behaviour in the population of the character X is completely known. The character X could be real or vector valued. The labelling or indexing parameter θ varies over a set of values, called as parameter space and is denoted by Ω. The object of inference therefore is the parameter θ or a function of the parameter say $\psi(\theta)$, which is of interest. We say that θ is an indexing or a labelling parameter when the probability distribution of X is uniquely specified by θ or when $F(x, \theta_1) = F(x, \theta_2)$ for all values of $x \in R_1$ implies that $\theta_1 = \theta_2$. We illustrate the above set up by a few examples.

EXAMPLE 1.1 Suppose that 100 seeds of brand A were planted one in each pot and let X_i equal one or zero according as the seed in the i-th pot germinates or not. The data consists of $(x_1, ..., x_{100})$ a sequence of ones and zeroes and is regarded as a realization of $(X_1, X_2, ..., X_{100})$ such that components are i.i.d.r.v.s with $P[X_i = 1] = \theta$ and $P[X_1 = 0] = 1 - \theta$, where θ represents the probability that a seed germinates. The object of estimation is θ itself or a function $\psi(\theta)$ that may be of interest. For example, consider $\psi_1(\theta) = \begin{bmatrix} 10 \\ 8 \end{bmatrix} \theta^8(1 - \theta)^2$, which is the probability that in a batch of 10 seeds exactly 8 seeds will germinate or $\psi_2(\theta) = \sum_{r=15}^{20} \begin{bmatrix} 20 \\ r \end{bmatrix} \theta^r(1 - \theta)^{20-r}$ which represents the probability that in a batch of 20 seeds at least 15 seeds will germinate. Another function of interest may be $\psi_3(\theta) = 1$ if $\theta \geq .90$ and zero otherwise. This corresponds to the situation where the brand A seeds would be recommended to the farmers provided the probability of germination of a seed is at least .90 or roughly speaking on an average at least 90% of the seeds sown would germinate.

It is worth pointing out that K. Pearson, one of the founding fathers of Statistics in Biometrika (1920) mentions that one of the fundamental problems of Statistics is to estimate probability of at least r_2 successes in future n_2 trials, on the basis of the data that r_1 successes have been observed in n_1 trials.

One can easily show that θ is infact a labelling parameter. Consider $F(u, \theta_1) = F(u, \theta_2)$ for all $u \in R_1$. In particular take any $u \in (0, 1)$ say $u = 1/2$. Then

$$F\left(\frac{1}{2}, \theta_1\right) = 1 - \theta_1 = F\left(\frac{1}{2}, \theta_2\right) = 1 - \theta_2$$

from which we conclude that $\theta_1 = \theta_2$.

EXAMPLE 1.2 It is quite well known that the Poisson distribution with pmf $P[X = x] = e^{-\lambda} \dfrac{\lambda^x}{x!}$, $x = 0, 1, 2, \ldots$ serves as a good model for the number of times a given event E occurs in a unit time interval, e.g. number of telephone calls received in a telephone exchange or the number of α particles hitting the Geiger counter. Instead of time interval one could have other type of "intervals", e.g. X could be the number of breakages per 100 m of a yarn or X_i could be number of bacteria of a specific type found in 1 cc of blood of the i-th patient. One can easily show again that the parameter λ is a labelling parameter. To prove this consider $F(u, \lambda_1) = F(u, \lambda_2)$ for $0 < u < 1$ then we have $e^{-\lambda_1} = e^{-\lambda_2}$ which implies that $\lambda_1 = \lambda_2$.

In a pathbreaking experiment Rutherford Chadwick and Elis (1920) observed 2608 time intervals of 7.5 seconds each and counted the number of time intervals N_r in which exactly r number of α particles hit the counter. They obtained the following table (W. Feller, 1957)

r	0	1	2	3	4	5	6	7	8	9	≥ 10
N_r	57	203	383	525	532	408	273	139	45	27	18

Here $n = 2608$ and $(X_1, \ldots, X_{2608})$ are i.i.d. Poisson r.v.s with mean λ where X_i denotes the number of α particles hitting the counter in the i-th time interval. Note that the data above is presented in a different way from the original observations $(x_1, x_2, \ldots, x_{2608})$.

EXAMPLE 1.3 Consider determination of an ideal physical constant such as gravity g. Usual way to estimate g is by the pendulum experiment and observe $X = 4\pi^2 l/T$, where l is the length of the pendulum and T the time required for a fixed number of oscillations. Due to variation which depends on several factors such as the skill of the experimentor and measurement errors, the i-th observation X_i can be represented as $X_i = g + \varepsilon_i$ where ε_i is the random error. Assuming error is normal with zero mean and variance σ^2 we have $(X_1, \ldots, X_n)$ are i.i.d. $N(g, \sigma^2)$. Here the parameter θ is, two dimentional vector, $\theta = (g, \sigma^2)'$. Is it a labelling parameter? The answer is yes and we can use the following argument. Let $F(x, g, \sigma^2)$ denote the d.f. of $N(g, \sigma^2)$. Then $F(x, g_1, \sigma_1^2) = F(x, g_2, \sigma_2^2)$ for all $x \in R_1$ implies that their characteristic functions are identical i.e.

$$\exp\left\{ig_1 t - \frac{t^2\sigma_1^2}{2}\right\} = \exp\left\{ig_2 t - \frac{t^2\sigma_2^2}{2}\right\} \text{ for all real } t$$

or $ig_1 t - \frac{t^2\sigma_1^2}{2} = ig_2 t - \frac{t^2\sigma_2^2}{2}$ for all t. Equating real and imiginary parts we have $g_1 = g_2$ and $\sigma_1^2 = \sigma_2^2$. If the student is not familiar with characteristic function one can use moment generating function of $N(g, \sigma^2)$ which is given by $\exp\left\{gt + \frac{t^2\sigma^2}{2}\right\}$ to claim that $g_1 t + \frac{t^2\sigma_1^2}{2} = g_2 t + \frac{t^2\sigma_2^2}{2}$ for all real t. Taking derivative at $t = 0$ we have $g_1 = g_2$ and then have $\sigma_1^2 = \sigma_2^2$. Here we can view estimation of g as estimating $\psi_1(g, \sigma^2) = g$. On the other hand one may be interested in estimating the error variance σ^2 through which we can estimate the ability of the experimenter. Thus we may want to estimate $\psi_2(g, \sigma^2) = \sigma^2$.

Exercise 1.1 Consider the repeated measurement model $X_i = \theta + \varepsilon_i$, where ε_i are i.i.d. (a) Uniform over $(-\sigma, \sigma)$, (b) double exponential with pdf $f(\varepsilon, \sigma) = \frac{1}{2\sigma} e^{-|\varepsilon|/\sigma}$ or (c) Cauchy with scale σ with pdf $f(\varepsilon, \sigma) = \frac{1}{\pi\sigma} \cdot \frac{1}{1 + (\varepsilon/\sigma)^2}$. Show that in each case the parameter (θ, σ) is a labelling parameter.

Exercise 1.2 Consider the indirect measurement problem, or standard regression problem where the response y_i at level d_i is given by $y_i = \alpha + \beta d_i + \varepsilon_i$, $i = 1, 2, \ldots, n$, where ε_i are i.i.d. $N(0, \sigma^2)$. Assuming that there are at least two distinct levels i.e. $d_i \neq d_j$ for some pair (i, j), show that $(\alpha, \beta, \sigma^2)$ is an indexing parameter.

We remark here that there are some practical situations in which the way data arises leads to the problem of non-identifiability. Consider the classical Signal plus Noise problem where observable r.v. $X_i = Y_i + \varepsilon_i$ where Y_i the 'signal' is $N(\theta, \sigma^2)$ and ε_i the 'noise' is $N(0, \phi^2)$. Then $(X_1, \ldots, X_n)$ are i.i.d. $N(\theta, \sigma^2 + \phi^2)$ and $(\theta, \sigma^2 + \phi^2)$ is a labelling parameter whereas the parameter $(\theta, \sigma^2, \phi^2)$ is not. Similarly consider a series system with two components (Y, Z) such that Y, Z are independent exponentials with failure rates θ and λ respectively. The observable random variable is $X = \text{Min}\ (Y, Z)$ the failure time of the i-th unit. Here $(X_1, \ldots, X_n)$ are i.i.d. exponentials with failure rate $(\theta + \lambda)$. The parameter (θ, λ) is not a labelling parameter as say $\theta = 1$, $\lambda = 3$ and $\theta = 2$ and $\lambda = 2$ would give rise to same distribution of X. The problem of non-identifiability arises quite often in many fields particularly in Econometrics. However, this being the first course we will not consider the problems of non-identifiability but instead concentrate on those situations in which the data $(X_1, \ldots, X_n)$ arises from a probability distribution with real or vector valued parameter θ a labelling parameter.

As mentioned earlier the object of obtaining data $(x_1, \ldots, x_n)$ is to estimate θ that determines the stochastic behaviour of the characteristic X which is under study. For example in the classical Rutherford Chadwick Elis experiment discussed in Ex. 1.2 the object is to estimate the parameter λ of the Poisson distribution which describes the stochastic behaviour of number of α-particles that hit the counter in a unit interval of time 7.5 seconds. On the other hand in Ex. 1.1, the object of experiment is to estimate the parameter θ the probability that the seed will germinate on the basis of observations on Bernoulli Series of trials $(X_1, \ldots, X_{100})$. In Ex. 1.1, an obvious estimate of θ is the observed proportion of seeds that have germinated out of the 100 which were planted. This estimate can be expressed as $\hat{\theta} = \frac{\Sigma x_i}{n} = \bar{x}$ the sample mean. In Ex. 1.2 we may suggest the estimate of λ as $\bar{x}$, the sample mean again since the parameter λ is also the population mean or expected value of X. With a similar logic we can propose $\bar{x}$ as an estimate of g in Ex. 1.3.

Two points emerge immediately, namely the fact that the estimate is a well defined function of the sample observations and therefore the estimate may change from sample to sample. For example in Ex. 1.1 the estimate $\hat{\theta} = \bar{x} = r/n$ for all those sequences, $\binom{n}{r}$ out of 2^n sequences, in which r seeds have germinated out of $n = 100$ seeds planted but would change as r, the total number of seeds germinated, would change from sample to sample. Thus an estimate is infact a random variable taking value $T(x_1, \ldots, x_n)$ depending on the observed sample values $(x_1, \ldots, x_n)$. To emphasize this aspect, we would now use the term estimator, i.e. a rule which specifies the function $T(X_1, \ldots, X_n)$ for all possible values of random vector $(X_1, \ldots, X_n)$, varying over the sample space. The estimate $T(x_1, \ldots, x_n)$ thus would be a realization of $T(X_1, \ldots, X_n)$ for the observed sample $(x_1, \ldots, x_n)$. By using the techniques of transformation or from basic principles, we could, at least theoretically obtain the sampling distribution of the estimator T. Thus in Ex. 1.1 estimator $T(X_1, \ldots, X_n) = \frac{1}{n} \Sigma X_i = \bar{X}$ would take $(n + 1)$ distinct values $(0, 1/n, \ldots, 1)$ and $P(\bar{X} = r/n) = P(\Sigma X_i = r) = \binom{n}{r} \theta^r (1 - \theta)^{n-r}$, $r = 0, 1, 2, \ldots, n$. Similarly, in repeated measurement model with normal errors considered in Ex. 1.3, the estimator $T(X_1, \ldots, X_n) = \bar{X}$ would itself be normal with mean g and variance σ^2/n. We thus note that in both cases, and in general also, the sampling distribution of an estimator T would also depend on θ. Thus to the basic frame work in which $(X_1, \ldots, X_n)$ are i.i.d. with pdf $\{f(x, \theta), \theta \in \Omega\}$ we add now the estimator $T(X_1, \ldots, X_n)$ with induced corresponding class of pdfs $\{g(t, \theta), \theta \in \Omega\}$. Of course one can suggest different estimators for the same parameter θ. For example, in the normal error model one could suggest using sample median $X_{([n/2]+1)}$, the $([n/2] + 1)$=th component of the order statistic of the sample or, in general, $T_l = \sum_{i=1}^{n} l_i X_{(i)}$, where l_i's are specified constants and $X_{(1)}, \ldots, X_{(n)}$ are the order statistics of the sample. The weights l_i's are chosen in such a way that extreme observation such as $X_{(1)}$, $X_{(n)}$ get less weight even zero and the components of order statistics around median get higher weightage. Similarly, for estimating variance σ^2 and the standard deviation σ in the normal model, apart from the classical estimators $\hat{\sigma}_2^2 = \frac{1}{n} \Sigma (X_i - \bar{X})^2$ and its square roots $\hat{\sigma}_2 = \left\{\frac{1}{n} \Sigma (X_i - \bar{X})^2\right\}^{1/2}$, many other estimators were suggested and used. For example, $\hat{\sigma}_1 = \sqrt{\pi/2}\, \frac{1}{n} \Sigma | X_i - \bar{X} |$ was recommended by Peter and $\hat{\sigma}_2$ was recommended by Bessel, one can use a multiple of the sample range $(X_{(n)} - X_{(1)})\, C_n$ as an estimate of σ, the standard deviation, where C_n is a suitable constant.

1.2 Historical Perspective—I

The problem of estimation arose in a very natural way in problems of Astronomy and Geodesy in the first half of the 18-th century. For example in Astronomy, the determination of interplanetary distances, determining the position of planets and their movements in time were some of the important problems. Whereas in Geodesy, determining the spheroidical shape of the earth was one of the most important problems. It was known that the figure of the earth is almost a sphere except for some flatness near the poles. Observations were obtained on the measurement of the length of one degree of a certain meredian and the problem was to determine the parameters α and β which specified the spheroid of the earth. Indirect observations on (α, β) were given by the relation

$$y_i = \alpha + \beta x_i, \quad i = 1, 2, \ldots, n$$

where x_i's are known fixed constants. Note that (α, β) are uniquely determined if only two observations on Y at different values of (x_1, x_2) are available. However, as is customary in science, several observations were made at different values $(x_1, \ldots, x_n)$ and this led to the theory of combination of observations with random error which directly or indirectly measured "magnitudes of interest" or parameters.

Most of the students today know that the estimates of α and β are obtained by the Method of Least Squares by minimizing $Q = \sum_{i=1}^{n} (y_i - \alpha - \beta x_i)^2$ where $y_i - \alpha - \beta x_i$ is called as error or residual. The method of least squares was proposed by Gauss and Legendre both very well known and distinguished mathematicians of 18-th century. There is some controversy regarding priority between them. Legendre published the method in 1805 studying it in great details but it appears that Gauss has been using the same since 1796. This is not the place to go into further details of this controversy but we note that about half a century earlier Boscovitzh (1757) had proposed a solution to this problem using different approach.

Boscovitch suggested that the estimates of α and β be determined such that

(i) The sum of positive and negative residuals or errors should balance i.e. $\Sigma (y_i - \alpha - \beta x_i) = 0$ and

(ii) Subject to the above constraint we determine (α, β) such that $R = \sum_{i=1}^{n} | y_i - \alpha - \beta x_i |$ the sum of absolute values of errors is as small as possible.

Using geometric argument Boscovitch solved the problem for the five observations that he had. Laplace (1789) gave a general algebric algorithm to obtain estimates of (α, β) on the above principles for any number of observations. It may be pointed out that Boscovitch method is the first instance of solving an optimization problem with constraint. Laplace (1789) in fact preferred Boscovitch method over that of Cotes (1722) which was then currently in use.

Boscovitch's contribution to the solution of the problem of "determination of magnitudes of interest" or the estimation of unknown parameters is a fundamental contribution in that it prescribed the method of estimation using some basic principles. Boscovitch suggested the fitting of straight line $y = \alpha + \beta x$ to the data $\{(x_1, y_1) \ldots (x_n, y_n)\}$ requiring that the fitted line should pass through the point $(\bar{x}, \bar{y})$ and the sum of absolute deviations of the points (x_i, y_i) from the fitted line in the direction of y-axis is minimized. Later in the method of Least Squares, Legendre and Gauss proposed fitting of the straight line using the principle that sums of squares of deviations in the direction of y-axis be minimized. Note that the line, fitted by method of Least Squares, passes through the point $(\bar{x}, \bar{y})$.

In the same spirit later in 19-th century K. Pearson proposed the method of moments and Fisher (1912) advocated the method of Maximum Likelihood. With several methods of estimation being proposed on different and possibly equally appealing grounds there arose the inevitable question of comparing estimates obtained by using different methods.

Prompted by a remark of Edington in his book "Stellar Movements" that $\hat{\sigma}_1 = \sqrt{\pi/2}\,\frac{1}{n}\Sigma | X_i - \bar{X} |$ has some theoretical advantage over

$\hat{\sigma}_2 = \left\{\frac{1}{n}\Sigma|X_i - \bar{X}|\right\}^{1/2}$, Fisher (1920) decided to investigate the performance of these two estimators of σ and laid down the basic principles as to how this comparison should be made. Fisher first obtained the sampling distribution of $\hat{\sigma}_1$ and $\hat{\sigma}_2$ and their means and variances and their mean squared errors (MSE). Comparing MSE ($\hat{\sigma}_1$) and MSE ($\hat{\sigma}_2$) he showed that the sampling distribution of $\hat{\sigma}_2$ is more concentrated around σ than that of $\hat{\sigma}_1$ assuming that errors are normally distributed. Thus Fisher laid down the foundation of the present practice that the performance of an estimator has to be judged on the basis of the sampling distribution of the estimator and a few characteristics of such a sampling distribution such as the mean, the variance or bias and the mean squared error.

The concept of the sampling distribution is a relatively new concept and the standard sampling distributions such as χ^2, t, F, and that of sample mean, variance, multiple and partial correlation coefficients etc. were derived during the second quarter of the 20-th century. Now many of these derivations are taught at the final year undergraduate level. Such sampling distributions can be obtained when one assumes that $(X_1, \ldots, X_n)$ has a given joint probability distribution. Once such a model for the data is assumed then using the calculus of probability and standard techniques such as transformations, one can at least theoretically obtain the distribution of an estimator. The distributions such as χ^2, t, F etc. were obtained by assuming that observations constitute a random sample from a normal distribution.

Primarily under the leadership of Fisher during the decade of 1920–30, foundations were laid down for the theory of estimation. Indeed Fisher (1922) in his pathbreaking paper entitled "On Mathematical Foundations of Theoretical Statistics" listed the basic problems of theoretical statistics as follows:

(1) *Problems of Specification*: Defining the distribution of the population.

(2) *Problems of Estimation*: Obtaining from the sample, the statistics or estimates of the parameters of the population.

(3) *Problems of Distribution*: Obtaining sampling distribution of statistics derived from the sample.

Following Fisher (1922) we adopt the attitude that the problems of specification have been satisfactorily resolved by the scientists and practising statisticians in the field and would thus assume that the stochastic behaviour of X is specified by the class of pdfs $\{f(x, \theta), \theta \in \Omega\}$, where f is known.

We will also assume that the problems of distribution have also been solved and we thus concentrate on the problem of estimation. Following Fisher we will compare the performance of various estimators on the basis of their sampling distributions under assumed model, namely $(X_1, \ldots, X_n)$ are independent identically distributed random variables each with pdf belonging to the class $\{f(x, \theta), \theta \in \Omega\}$ and an estimator T has the induced probability distribution given by the corresponding class of pdf $\{g(t, \theta), \theta \in \Omega\}$.

Following Section 1.3 presents a brief historical review of some of the models that arose in a natural way.

1.3 Historical Perspective—II

Recall that in the simplest model where Observation = True value + Error, the errors of overestimation and underestimation must balance out (Boscovitch assumption). Simpson (1776) translated this idea by assuming that errors are symmetrically uniformly distributed about zero or the model is given by $X_i = \theta + \varepsilon_i$ where the pdf of the error is given by $f(\varepsilon) = 1/2h, -h < \varepsilon < h$. Euler (1778) proposed the arc of a parabolic curve given by $f(\varepsilon) = \frac{3}{4r^3}(r^2 - \varepsilon^2), -r < \varepsilon < r$ as the pdf of the random error. Euler seems to be the first one to propose the distribution of errors with the basic property that large errors, i.e. large values of $|\varepsilon|$ are less likely than the small values of $|\varepsilon|$. Laplace proposed the pdf $f(\varepsilon) = \frac{1}{2h}\exp\{-|\varepsilon|/h\}$ for $-\infty < \varepsilon < \infty$ as the model for distribution of errors and Gauss proposed the normal distribution with pdf $f(\varepsilon) = \frac{1}{\sqrt{2\pi h^2}}\exp\{-\varepsilon^2/2h^2\}, -\infty < \varepsilon < \infty$. It is important to point out here that the double exponential distribution used by Laplace to represent error distribution led to the median of the sample as the "best" estimator of the "True value" whereas the normal distribution used by Gauss led to the mean of the sample as the "best" estimator of the "True value".

The popularity of the normal error model was mainly due to the ease with which Least Square Method could be applied as opposed to the method of least absolute deviations originally proposed by Boscovitch and followed by Laplace. Further the normal distribution arose also in the context of approximating the binomial probability of r successes in n trials given by De-Moivre Laplace limit theorem, namely

$$\binom{n}{r} p^r (1-p)^{n-r} \simeq \frac{1}{\sqrt{2\pi np(1-p)}} \exp\left\{-\frac{(r-np)^2}{2np(1-p)}\right\}.$$

The binomial distribution as a model of occurrence of an event E in a series of trials arose from the games of chance but it was also due to the interest in sex ratio, male to female births, in new born babies. The Bernoulli series of trials led to Poisson distribution to model the situation when n, the number of trials is large but p, (the probability of the event) is very small. Other problems related to Bernoulli series of trials led to such other well known distributions as geometric, and negative binomial among others. Now Poisson, geometric and negative binomial are generally asymmetric discrete distributions and differ very much from the symmetric error distribution models evolved earlier. Further K. Pearson observed that many data sets collected by him and other statisticians in connection with biometrical measurements indicated that the characteristic of interest (X) would have a distribution which could be skew on either side and very far from normal. Gaining experience from such collected data sets, Pearson proposed a very wide class of distributions now known as Pearsonian system of curves. As was customary then in applied mathematics, the pdf was defined by a differential equation

$$\frac{df(x)}{dx} = \frac{(x-a)f(x)}{b_0 + b_1x + b_2x^2}$$

where (a, b_0, b_1, b_2) are constants which determine the nature and shape of the pdf. Depending on the nature of the roots of the quadratic in the denominator the equation leads to variety of pdfs. [Kendall and Stuart, Vol. I (1958)]. Assuming that the first four moments exist, Pearson showed that the constants (a, b_0, b_1, b_2) are uniquely determined by the first four moments of the pdf or equivalently by the mean, variance, Skewness, Kurtosis. Since the distribution is uniquely determined by the first four moments, fitting of the frequency curve (pdf) by method of moments became a standard exercise which was further facilitated by detailed technique prescribed in Biometrika Tables for Statisticians (1954) prepared by Pearson and his colleagues and included in the text by Elderton (1954). However in this transition the "magnitudes of interest" or the parameters of the distribution were defined indirectly in terms of moments of r.v. X.

The next major advance was made by Fisher (1912) when he presented the method of maximum likelihood for determining the parameters occurring in the distribution of X given by the pdf $f(x, \theta)$ in which the function f is known and the constants $\theta = (\theta_1 \ldots, \theta_m)$, are determined on the basis of a random sample of size n on X with joint pdf $L(x, \theta) = \prod_{i=1}^{n} f(x_i, \theta)$. The function $L(x, \theta) = \prod_{i=1}^{n} f(x_i, \theta)$ for fixed $x = (x_1, \ldots, x_n)$ and variations in θ was defined by Fisher as the likelihood function. Note that for discrete r.v. X, $L(x, \theta)$ is the probability of observing $X = x$ for a given value of θ. As θ varies over the set of possible values Ω, one can establish an order relationship among the values of θ. We say that θ_1 is more likely than θ_2 if $L(x, \theta_1) > L(x, \theta_2)$ and θ_2 is more likely than θ_1 if $L(x, \theta_1) < L(x, \theta_2)$. If $L(x, \theta_1) = L(x, \theta_2)$ then on the basis of the data $X = x$, θ_1 and θ_2 are equally likely. In his method of maximum likelihood, Fisher proposed to estimate θ for a given value of $X = x$ by that value of θ for which $L(x, \theta)$ is maximum.

In all these approaches to estimation, it is assumed that it is possible to suggest an estimator of θ by making observations on X. This implies that we assume that observations on X are somehow providing information on θ. One can ask the question as to why we feel that observations provide some information about the unknown parameter. Note that the probability distribution changes as θ changes due to our assumption of θ being an indexing or lebelling parameter. Therefore the likelihood function $L(x, \theta)$ changes as θ varies over Ω and it is this change in the likelihood function that provides information about θ. If $L(x, \theta)$ were to remain constant for an observed x then such an observation x would be regarded as not providing any information about θ. At the base of above thinking is the assumption that if the probability distribution does not change with the parameter θ then observations from such a distribution would not provide any information about θ. Similarly, if $T(x)$ is a statistic then its probability distribution must effectively depend on θ if T is to provide information about θ. If the probability distribution of T does not depend on θ then it cannot provide any information about θ. For example, for a sample of size two from $N(\theta, 1)$, $T_1 = X_1 + X_2$ provides some information about θ since $T_1 \sim N(2\theta, 2)$ but $T_2 = X_1 - X_2$ will not provide any information about θ as $T_2 \sim N(0, 2)$.

Chapter 2 will consider information in detail and study the concept of sufficiency of a statistic T which plays a vital role in Statistical Inference.

2

Sufficient Statistic

2.1 Motivation

Let $(X_1, X_2, \ldots, X_n)$ be i.i.d.r.v. with pdf belonging to class $\{f(x, \theta), \theta \in \Omega\}$. Let $T(X_1, \ldots, X_n)$ be a statistic with corresponding pdf belonging to the class $\{g(t, \theta), \theta \in \Omega\}$. Following Fisher (1925) we call T a sufficient or an exhaustive statistic if it contains all the information about θ that is contained in the samle $(X_1, X_2, \ldots, X_n)$. Note that so far we have not quantified the amount of information either in the sample $(X_1, \ldots, X_n)$ or in a statistic T. However as observed earlier we have assumed that if a sample has its pdf not depending on the parameter θ then it does not contain any information about θ. Using this logic consider any other statistic T_1 and consider the conditional distribution of T_1 given T with pdf $h(t_1, \theta_1, t)$. If this does not depend on θ then the conditional distribution of T_1 given T does not contain any information about θ. If the statitic T is such that the above holds for every other statistic T_1 and for all possible values $T = t$ then Fisher defined T to be sufficient or exhaustive. Note that in this case the conditional distribution of any other statistic T_1 given T is independent of θ and therefore does not provide any additional information than that which is already contained in T. On the other hand if T is such that the conditional distribution of a statistic T_1 given T changes with θ then there is some information about θ in the conditional distribution of T_1 given T and T is not sufficient or exhaustive. Fisher (1920) noted this phenomenon first in connection with the comparison of the estimators $\hat{\sigma}_1$ and $\hat{\sigma}_2$ of the standard deviation of the $N(\mu, \sigma^2)$ model mentioned in Chapter I. He observed that the marginal pdf of $\hat{\sigma}_1$ as well as that of $\hat{\sigma}_2$ both depend on σ and as such contain information about σ. But whereas, the conditional distribution of $\hat{\sigma}_1$ given $\hat{\sigma}_2$ does not depend on σ, the conditional distribution of $\hat{\sigma}_2$ given $\hat{\sigma}_1$ does depend on σ. Fisher interpreted this as having observed $\hat{\sigma}_2$, no additional information about σ can be obtained by considering $\hat{\sigma}_1$. On the other hand if we observe $\hat{\sigma}_1$, the statistic $\hat{\sigma}_2$ provides still some additional information about σ. One must appreciate the genius of Fisher which from a single specific example formulated the property of sufficiency which later proved to be a fundamental concept in the theory of Statistical Inference. Fisher's initial work about $\hat{\sigma}_1$, $\hat{\sigma}_2$ is quite complex. We consider the following example which involves fairly simple calculations and illustrate the phenomenon discussed above.

EXAMPLE 2.1.1 Let (X_1, X_2) be i.i.d. $N(\theta, 1)$. Then $X_1 + X_2 \sim N(2\theta, 2)$ and $X_1 \sim N(\theta, 1)$ both have some information about θ. Now by standard techniques one can show that $(X_1, X_1 + X_2)'$ is bivariate normal (BVN) with mean vector $(\theta, 2\theta)'$ and covariance matrix $\begin{pmatrix} 1 & 1 \\ 1 & 2 \end{pmatrix}$ and the conditional distribution of X_1 given $X_1 + X_2 = t$ is $N(t/2, 1/2)$ which does not depend on θ. Thus having fixed $(X_1 + X_2)$, X_1 does not provide any additional information about θ. On the other hand the conditional distribution of $(X_1 + X_2)$ given $X_1 = x_1$ is normal with mean $(x_1 + \theta)$ and variance one. Thus conditional distribution of $(X_1 + X_2)$ given X_1 has some information about θ. In general consider $T_1 = lX_1 + mX_2$ and $T = (X_1 + X_2)$. Then $(T_1, T)'$ is BVN with mean $(\theta(l + m), 2\theta)'$ and covariance matrix $\begin{pmatrix} l^2 + m^2 & (l + m) \\ (l + m) & 2 \end{pmatrix}$. Then the conditional distribution of T_1 given $T = t$ is normal with mean $\frac{t}{2}$ and variance $\frac{(l - m)^2}{2}$. Therefore, conditional distribution of $lX_1 + mX_2$ given $X_1 + X_2 = t$ does not contain any information about θ. On the other hand the conditional distribution of T for $T_1 = t_1$ is normal with mean $2\theta + \frac{l + m}{(l^2 + m^2)}$ $[t_1 - (l + m)\theta]$ and variance $\frac{(l - m)^2}{(l^2 + m^2)}$. Thus distribution of T given $T_1 = t_1$ depends on θ and therefore contains some information about θ. Note that conditional distribution of $(X_1 + X_2)$ for fixed $lX_1 + mX_2 = t_1$ is independent of θ if and only if $2 - (l + m)^2/(l^2 + m^2) = 0$ or $l = m = k$ say i.e. $lX_1 + mX_2$ is a constant multiple of $(X_1 + X_2)$ and therefore fixing $lX_1 + mX_2 = t_1$ fixes $X_1 + X_2$ uniquely to be equal to t_1/k and the conditional distribution of $(X_1 + X_2)$ is a singular distribution with entire probability mass concentrated at the point t_1/k.

The above analysis shows that $lX_1 + mX_2$ where $l \neq m$ can not be sufficient as there exists a statistic $(X_1 + X_2)$ such that its conditional distribution, given $lX_1 + mX_2 = t_1$ depends on θ. Although conditional distribution of $lX_1 + mX_2$ given $X_1 + X_2 = t$ is independent of θ for any choice of $(l, m)'$ this does not prove the sufficiency of $(X_1 + X_2)$. To prove sufficiency of $T = (X_1 + X_2)$ we must show that for any statistic $T_2(X_1, X_2)$ the conditional distribution of T_1 given $T = (X_1 + X_2) = t$ is independent of θ. Let $T_2 = X_1^2 + X_2^2$ then for $X_1 + X_2 = t$, we have $X_2 = t - X_1$ and for fixed t, $T_2 = X_1^2 + (t - X_1)^2 = 2X_1^2 - 2tX_1 + t^2 = 2(X_1 - t/2)^2 + t^2/2$. Now conditional distribution of X_1 given $X_1 + X_2 = t$ is $N(t/2, 1/2)$ and therefore $2(X_1 - t/2)^2$ is χ_1^2 and as $t^2/2$ is fixed, the conditional distribution of T_2 given $(X_1 + X_2) = t$ is independent of θ. This exercise increases the plausibility of sufficiency of $(X_1 + X_2)$ and shows how we can prove sufficiency of $T = (X_1 + X_2)$. Consider any statistic $T_3(X_1, X_2)$ then for fixed $T = t$, we have $T_3(X_1, X_2)$

$= T_3(X_1, t - X_1)$ which is purely a function of X_1. Since conditional distribution of X_1 given $X_1 + X_2 = t$ is $N\left(\frac{t}{2}, \frac{1}{2}\right)$ the distribution of $U(X_1) = T_3(X_1, t - X_1)$ is independent of θ, and thus $T = (X_1 + X_2)$ is sufficient.

EXAMPLE 2.1.2 Let $(X_1, \ldots, X_n)$ be i.i.d. Poisson with mean θ and let $T = \sum_{i=1}^{n} X_1$. Then we know that T is Poisson with mean $n\theta$ and

$$P[X = x \mid T = t] = P(X_1 = x_1, \ldots, X_n = x_n \mid T = t)$$

$$= \frac{P(X_1 = x_1, \ldots, X_n = x_n, \sum_{k=1}^{n} X_i = t)}{P(\sum_{k=1}^{n} X_i = t)}$$

$$= \frac{P(X_1 = x_1, \ldots, X_{n-1} = x_{n-1}, X_n = t - \sum_{i=1}^{n-1} x_i)}{P(\sum_{k=1}^{n} X_i = t)}$$

$$= \prod_{i=1}^{n-1} \frac{e^{-\lambda}\lambda^{x_i}}{x_i!} \cdot \frac{e^{-\lambda}\lambda^{t - \sum_{i=1}^{n-1} x_i}}{(t - \sum_{i=1}^{n-1} x_i)!} \Bigg/ \frac{e^{-n\lambda}(n\lambda)^t}{t!}$$

$$= \frac{t!}{x_1! x_2! \ldots x_{n-1}! \, (t - \sum_{i=1}^{n-1} x_i!)} \frac{1}{n^t}$$

$$= \frac{t!}{\prod_{i=1}^{n} x_i!} \frac{1}{n^t} \text{ for } x_i \geq 0, \sum_{i=1}^{n} x_i = t$$

$$= 0 \quad \text{otherwise.}$$

This shows that the conditional distribution of $(X_1, X_2, \ldots, X_n)$ given $\sum_{i=1}^{n} X_i = t$ is a multinomial distribution in n equiprobable cells and frequencies $(X_1, \ldots, X_n)'$ such that $\sum_{i=1}^{n} X_i = t$. Since this conditional distribution on the sample space of $(X_1, \ldots, X_n)'$ given $\sum X_i = t$ does not depend on θ, it follows that for any subset A of sample space, $P_\theta(A \mid A_t)$, where $A_t = \{x \mid \sum x_i = t\}$ does not depend on θ, as $P_\theta(A \mid A_t) = \sum_{x \in A \cap A_t} P[X = x \mid T = t]$. Both the examples point out that the constratint $\sum_{i=1}^{n} X_i = t$ determines one variable say X_n in terms of remaining $(n - 1)$ variables $(X_1, X_2, \ldots, X_{n-1})$ and the conditional distribution of $(X_1, X_2, \ldots, X_{n-1})$ given $T = t$ does not depend on θ.

Above idea can be generalized to a vector valued statistic $T = (T_1, \ldots, T_k)'$ where $k \le n$. Fixing $T = t$ i.e. $T_1 = t_1, \ldots, T_k = t_k$ leaves $(n - k)$ of the X_i's say $(X_1, X_2, \ldots, X_{n-k})$ free and the remaining k variables $(X_{n-k+1}, \ldots, X_n)$ can be expressed as functions of $(X_1, X_2, \ldots, X_{n-k})$ and any real or vector valued statistic $U(X_1, \ldots, X_n)$ given $T = t$ can be expressed as a function of $(X_1, \ldots, X_{n-k})$ and one can work out conditional distribution of $U(X_1, \ldots, X_n)$ given $T = t$ from the conditional distribution of $(X_1, \ldots, X_{n-k})$ given $T = t$. If this distribution is independent of θ for each t, it follows that distribution of any statistic $U(X_1, \ldots, X_n)$ given $T = t$ would be independent of θ and T would be sufficient.

Remark 2.1.1 Note that for any two statistic T_1, T with joint pdf $g(t, t_1, \theta)$ the following factorization holds

$$g(t, t_1, \theta) = g_0(t, \theta) \cdot g_1(t_1, \theta \mid t) \tag{2.1.1}$$

where $g_0\,(t, \theta)$ is the marginal pdf of T and $g_1(t_1, \theta \mid t)$ is the conditional pdf of T_1 given $T = t$. If $g_1(t_1, \theta \mid t)$ does not depend on θ for all possible values of t then there is no information about θ in the conditional distribution of T_1 given $T = t$ for each t. If this holds for every T_1, then T is sufficient. Note that (2.1.1) holds for vector valued $\theta = (\theta_1, \ldots, \theta_m)'$ and vector valued statistic T_1, T of dimensions say k_1 and k respectively.

Remark 2.1.2 The observations themselves are always sufficient. For consider a random sample of size n, $(X_1, \ldots, X_n)'$ from $\{f(x, \theta), \theta \in \Omega\}$. Then consider T to be the n dimensional statistic $(X_1, \ldots, X_n)'$. Then given $T = t$ i.e. $X_1 = x_1, X_2 = x_2, \ldots, X_n = x_n$, for any real of vector valued statistic T_1, T_1 takes only one value $T_1(x_1, \ldots, n_n) = t_1$ and the conditional distribution of T_1 given $X_1 = x_1, \ldots, X_n = x_n$ is a singular distribution with entire probability mass concentrated at the point t_1. Since this conditional distribution does not depend on θ for any T_1, and every $(x_1, \ldots, x_n)$ it follow that $(X_1, \ldots, X_n)'$ is a sufficient statistic.

Remark 2.1.3 Taking T and $T_1 = (X_1, \ldots, X_{n-k})'$ in (2.1.1) we have

$$g(t, x_1, \ldots, x_{n-k}, \theta) = g_0(t, \theta)\, g_1(x_1, \ldots, x_{n-k} \mid t, \theta) \tag{2.1.2}$$

Now consider the transformation $y_i = x_i$, $i = 1, 2, \ldots, n - k$ and $y_{n-k+1} = t_1, \ldots, y_n = t_k$ such that $\left|\dfrac{\partial(y_1, \ldots, y_n)}{\partial(x_1, \ldots, x_n)}\right| \ne 0$. Then the joint pdf of $(Y_1, \ldots, Y_n)'$ is given by $g(y_1, \ldots, y_n, \theta) = L(h_1(y), \ldots, h_n(y), \theta) \mid J^{-1} \mid$ where J is the Jacobian of the transformation and $x_i = h_i(y)$, $i = 1, 2, \ldots, n$ is the inverse transformation. Referring back to the $(X_1, \ldots, X_n)$ co-ordinates and noting that when T is sufficient, g_1 in (2.1.2) does not depend on θ, we get

$$L(x_1, \ldots, x_n, \theta) = g_0(T_1(x), \ldots, T_k(x), \theta). h(x) \quad (2.1.3)$$

That is the joint pdf of $(X_1, \ldots, X_n)$ factorizes in two parts, one which is the pdf of $(T_1, \ldots, T_k)$ and the other which is purely a function of x.

In the above discussion the concept of information in a r.v. about a parameter θ has not been quantified. We consider this quantification and define Fisher Information in the next section.

2.2 Fisher Information

Let X be a real or vector valued r.v. whose pdf depends on a real parameter $\theta \in \Omega$. Then as indicated in Chapter 1 we assume that the variations in pdf of X for given $X = x$ as θ varies over Ω provides us some information about θ. For example let X be binomial with $P(X = x) = f(x, \theta) = \binom{n}{x} \theta^x (1 - \theta)^{n-x}$ and let $n = 10$ and $x = 2$ then we have $f(x, \theta)$ varies with θ as follows:

θ	.1	.2	.3	.4	.5	.6	.7	.8	.9
$f(2, \theta)$	.1937	.3020	.2335	.1209	.0439	.0106	.0014	.0001	.0000

It is this variation which provides some information about θ.

Now as in case of study of motion where change in position is studied by velocity and acceleration, (the derivatives of position w.r.t. time t), the information is studied through the derivatives of $f(x, \theta)$ w.r.t. θ for fixed x. Let $S_\theta = \{x \mid f(x, \theta) > 0\}$ denote the support of the pdf $f(x, \theta)$ under θ. Let $S = \bigcup_{\Omega} S_\theta$. Then we can ignore the points $x \notin S$, as such points will have zero probability under any $\theta \in \Omega$. For example in Binomial distribution with known n, $S_\theta = \{0, 1, 2, \ldots, n\}$ and we need not consider $x \notin S_\theta$. It may be pointed out here that although $f(2, .9)$ is given to be .0000, this is an approximation to $\binom{10}{2} (.9)^2 (.1)^8$ upto four places of decimals.

We now assume that (i) S_θ does not depend on θ, i.e. $S_\theta = S, \forall \theta \in \Omega$ and (ii) the pdf is such that differentiation under integral sign in the identity $\int_s f(x, \theta)\, dx = 1 \; \forall \; \theta \in \Omega$, is valid at least twice. We then can obtain

$$\int \frac{\partial \log f(x, \theta)}{\partial \theta} f(x, \theta)\, dx = 0 \quad (2.2.1)$$

and

$$\int \frac{\partial^2 \log f(x, \theta)}{\partial \theta^2} f(x, \theta)\, dx + \int \left[\frac{\partial \log f(x, \theta)}{\partial \theta}\right]^2 f(x, \theta)\, dx = 0 \quad (2.2.2)$$

Now $\frac{\partial \log f(x, \theta)}{\partial \theta}$ is the rate of change of the log-likelihood of $X = x$ at θ, say 'velocity' and similarly, $\frac{\partial^2 \log f(x, \theta)}{\partial \theta^2}$ represents the "acceleration".

The function $\dfrac{\partial \log f(x, \theta)}{\partial \theta}$ viewed as a function of x for fixed θ is called as a "score function" and for each fixed θ, is a random variable. Then (2.2.1) says that

$$E_\theta\left[\frac{\partial \log f(x, \theta)}{\partial \theta}\right] = 0 \tag{2.2.3}$$

and (2.2.2) gives

$$E_\theta\left[-\frac{\partial^2 \log f(x, \theta)}{\partial \theta^2}\right] = E_\theta\left[\left(\frac{\partial \log f(x, \theta)}{\partial \theta}\right)^2\right]$$

$$= \text{Var}_\theta\left[\frac{\partial \log f(x, \theta)}{\partial \theta}\right] \tag{2.2.4}$$

We define Fisher Information about θ in a r.v. X, by

$$I_X(\theta) = E_\theta\left(-\frac{\partial^2 \log f(x, \theta)}{\partial \theta^2}\right) = E_\theta\left[\left(\frac{\partial \log f(x, \theta)}{\partial \theta}\right)^2\right]$$

Observe that $I_X(\theta) \geq 0$ and $I_X(\theta) = 0$ if and only if

$E_\theta\left[\left(\dfrac{\partial \log f(x, \theta)}{\partial \theta}\right)^2\right] = 0$ or $\dfrac{\partial \log f(x, \theta)}{\partial \theta} = 0$ with probability one or the pdf of X, $f(x, \theta)$ does not depend on θ or the distribution of X does not change with θ.

If $(X_1, \ldots, X_n)$ is a random sample of size n with joint pdf $L(x, \theta) = \prod_{i=1}^{n} f(x_i, \theta)$, by taking logs and differentiating twice w.r.t. θ and taking expectations with negative sign we have Fisher information in the sample

$$I_{(X_1, X_2 \ldots X_n)}(\theta) = nI_X(\theta) \tag{2.2.4}$$

Let $T(X_1, \ldots, X_n)$ be a real or vector valued statistic with corresponding pdf $\{g(t, \theta), \theta \in \Omega\}$, then anologously one can define the Fisher information in a statistic T as

$$I_T(\theta) = E_\theta\left(\frac{-\partial^2 \log g(t, \theta)}{\partial \theta^2}\right) = E_\theta\left[\left(\frac{\partial \log g(t, \theta)}{\partial \theta}\right)^2\right] \tag{2.2.5}$$

We must however assume that the range of T say $\tau = \{t \mid g(t, \theta) > 0\}$ does not depend on θ and differentiation under integral sign in the identity $\int_\tau g(t, \theta)\, dt = 1$, $\forall\ \theta \in \Omega$ is valid at least twice.

Next we observe that if $(X_1, X_2, \ldots, X_n)'$ is a random sample of size n and if $(Y_1, Y_2, \ldots, Y_n)'$ is any one-to-one transformation with $\left|\dfrac{\partial(y_1, \ldots, y_n)}{\partial(x_1, \ldots, x_n)}\right| = |J| \neq 0$ then

$$g(y_1, \ldots, y_n, \theta) = L(x_1, \ldots, x_n, \theta) \mid J^{-1} \mid$$

where $x_i = h(y_1, \ldots, y_n)$, $i = 1, 2, \ldots, n$ is the unique inverse transformation. Noting that $|J^{-1}|$ does not depend on θ, taking logs and differentiating twice w.r.t. θ and then taking expectations with negative sign, it follows that

$$I_{(Y_1,\ldots,Y_n)}(\theta) = I_{(X_1,\ldots,X_n)}(\theta) = \mathrm{n}I_X(\theta)$$

Next suppose $T = (T_1, \ldots, T_k)'$ is a statistic then by (2.1.2)

$$L(x, \theta) = g_0(t, \theta)\, h(x_1, \ldots, x_{n-k}, \theta \mid t)$$

Hence $\log L(x, \theta) = \log g_0(t, \theta) + \log h(x_1, \ldots, x_{n-k}, \theta \mid t)$. Taking derivatives twice and taking expectations we have

$$I_{(X_1,\ldots,X_n)}(\theta) = I_T(\theta) + E\{I_{(X_1,\ldots,X_{n-k})|T=t)}(\theta)\}. \qquad (2.2.6)$$

where $I_{(X_1,X_2,\ldots,X_{n-k}|T=t)}(\theta)$ is the (conditional) Fisher information in $X_1, \ldots, X_{n-k}$ given $T = t$. As observed earlier we have $I_{(X_1,X_2,\ldots,X_{n-k}|T=t)}(\theta) \geq 0$ with probability one and therefore the second term on RHS of (2.2.6) is non-negative. Noting that $I_{(X_1,\ldots,X_n)}(\theta) = nI_X(\theta)$ we have for any k-dimensional statistic $(T_1, \ldots, T_k)$, with $k \leq n$

$$I_{(T_1,\ldots,T_k)}(\theta) \leq nI_X(\theta) \qquad (2.2.7)$$

This shows that information in a statistic T is always less than or at most equal to that in the original sample. Since constructing statistic is equivalent to a reduction of data, any reduction of data would generally involve a loss of information. There is no loss of information if the second term on RHS of (2.2.6) is zero which is possible only when $I_{(X_1,\ldots,X_{n-k}|T=t)}(\theta) = 0$ with probability one or $g_1(x_1, \ldots, x_{n-k}, \theta \mid t)$ for each t does not depend on θ. But as seen in Section 2.1 this implies that T is a sufficient statistic.

Therefore we have now a quantification of information about a paramenter θ and under certain regularity conditions a result that the information in any statistic T is always less than or at most equal to that in the sample. There is no loss of information if and only if $T = (T_1, \ldots, T_k)$ is a sufficient statistic, in the sense that conditional distribution of $(X_1, \ldots, X_{n-k})$ given $T = t$ does not depend on θ for each t and therefore conditional distribution of any other statistic T_1 given $T = t$ does not depend on θ. To fix the ideas we consider an example.

EXAMPLE 2.2.1 Let (X_1, X_2) be i.i.d. $N(\theta, 1)$ then a straightforward calculation gives $I_{(X_1,X_2)}(\theta) = 2$. Consider $T_1 = (lX_1 + mX_2) \sim N((l + m)\,\theta, l^2 + m^2)$ then

$$g(t_1, \theta) = \frac{1}{\sqrt{2\pi(l^2 + m^2)}} \exp\{-(t_1 - (l + m)\theta)^2/2(l^2 + m^2)\}$$

Taking logs and differentiating twice w.r.t. θ we have

$$\frac{-\partial^2 \log g}{\partial\theta^2} = \frac{(l+m)^2}{(l^2+m^2)} \text{ and } I_{T_1}(\theta) = \frac{(l+m)^2}{(l^2+m^2)}$$

Now $I_{(X_1,X_2)}(\theta) - I_{T_1}(\theta) = 2 - \dfrac{(l+m)^2}{(l^2+m^2)} = \dfrac{(l-m)^2}{(l^2+m^2)}$ represents the loss of information due to using T_1 instead of (X_1, X_2). Note that there is no loss of information iff $l = m$ i.e. $T = k(X_1 + X_2)$ which is sufficient. As already seen in Example 2.1.1, the conditional distribution of X_1 given T_1 is normal with mean $2\theta + \dfrac{(l+m)}{l^2+m^2}[t_1 - (l+m)\theta]$ and variance $\dfrac{(l-m)^2}{(l^2+m^2)}$.

$$\log g_1(x_1, \theta \mid t_1) = -\frac{1}{2}\log\left[2\pi\frac{(l-m)^2}{(l^2+m^2)}\right]$$

$$-\frac{\left\{x_1 - 2\theta - \dfrac{(l+m)}{l^2+m^2}[t_1 - (l+m)\theta]\right\}^2}{2(l-m)^2/(l^2+m^2)}$$

and

$$-\frac{\partial^2 \log g_1}{\partial\theta^2} = \frac{(l-m)^2}{(l^2+m^2)}$$

Thus the identity $I_{(X_1,X_2)}(\theta) = I_{T_1}(\theta) + E[I_{X_1|T=t}(\theta)]$takes the form $2 = \dfrac{(l+m)^2}{(l^2+m^2)} + \dfrac{(l-m)^2}{(l^2+m^2)}$ and $I_{X_1|T=t}(\theta) = 0$ iff $\dfrac{(l-m)^2}{(l^2+m^2)} = 0$ or $l = m$.

EXAMPLE 2.2.1 (*Contd.*) Now consider a sample of size n from $N(\theta, 1)$ then the information in the sample $I_{(X_1,\ldots,X_n)}(\theta) = n$ and let $T_1 = \sum_{i=1}^{n} c_iX_i$.

Then $T_1 \sim N(\Sigma\, c_i\theta, \Sigma\, c_i^2)$ and one can easily show that $I_{T_1}(\theta) = \dfrac{(\Sigma\, c_i)^2}{\Sigma\, c_i^2} \le n$ and equality holds iff $c_1 = c_2 = \ldots = c_n = l$. Note that if T_2 is such that $\Sigma\, c_i = 0$ then $T_2 \sim N(0, \Sigma\, c_i^2)$ and $I_{T_2}(\theta) = 0$.

We note that $I_X(\theta)$ is well defined under certain regularity conditions on the class of pdfs $\{f(x, \theta), \theta \in \Omega\}$, namely range of the r.v. X should not depend on θ or $S_\theta = S$ and further the operations of differentiation w.r.t. θ and integration w.r.t. x can be interchanged atleast twice.

For models such as Uniform disribution over $(0, \theta)$ or exponential distribution with $f(x, \theta) = \exp\{-(x - \theta)\}$, $x \geq \theta$, the support depends on θ as $S_\theta = (0, \theta)$ and (θ, ∞) respectively. Hence Fisher Information is not defined for these distributions. Even when S_θ does not depend on θ as in $N(\theta, 1)$ case we still require validity of interchange of operations of differentiation and integration. Many standard texts on Advanced Calculus would give sufficient conditions under which such an interchange of operations is valid. For validity of the interchange of operations of differentiation at θ_0 and integration w.r.t. x, it is sufficient that

(a) $$\left|\frac{f(x, \theta_0 + h) - f(x, \theta_0)}{h}\right| \leq G_0(x) \text{ for all } |h| < \delta$$

and $\int_s G_0(x)\, dx$ is finite and (b) $\dfrac{\partial f}{\partial \theta}$ exists for almost all values of x for each $\theta \in (\theta_0 - \delta, \theta_0 + \delta)$. If these two conditions are satisfied then differentiation under integral sign is valid once. For second derivative we require

(a′) $$\left|\frac{f(x, \theta_0 + 2h) - 2f(x, \theta_0 + h) + f(x, \theta_0)}{h^2}\right|$$

$$\leq G_1(x) \text{ for all } |h| < \delta$$

and (b′) $\dfrac{\partial^2 f}{\partial \theta^2}$ exists for almost all x for each $\theta = (\theta_0 - \delta, \theta_0 + \delta)$. The phrase 'almost all' in the conditions (b) or (b′) may be explained by way of an examle. Consider Double exponential distribution, or Laplace distribution with pdf $f(x, \theta) = \frac{1}{2} \exp\{-|x - \theta|\}$, $x \in R_1$, $\theta \in R_1$. Here $S_\theta = R_1$ and the range does not depend on θ. However $\dfrac{\partial f}{\partial \theta}$ does not exist at $\theta = x$ and therefore the set of exceptional points at which $\dfrac{\partial f}{\partial \theta}$ does not exist for $\theta \in (\theta_0 - \delta, \theta_0 + \delta)$ is the interval $(\theta_0 - \delta, \theta_0 + \delta)$ which has +ve probability under any θ. Now we say that a property Q holds for almost all x under the pdf $f(x, \theta)$ if the set of points Q^c where the property fails has probability zero, i.e. $\int_{Q^c} f(x, \theta)\, dx = 0$ under each $\theta \in \Omega$. Here $Q^c = (\theta_0 - \delta, \theta_0 + \delta)$ and $P_{\theta_0}(Q^c) > 0$. Therefore the condition (b) does not hold. We now show that for $N(\theta, 1)$ family, $\theta \in \Omega = R_1$, both conditions hold.

EXAMPLE 2.2.2 Let $X \sim N(\theta, 1)$ then $S_\theta = R_1$ and the range of X is independent of θ. Now $f(x, \theta) = \dfrac{1}{\sqrt{2\pi}} \exp\left\{-\dfrac{(x - \theta)^2}{2}\right\}$, $x \in R_1$, $\theta \in R_1$. Now consider

$$A_1 = \frac{f(x, \theta_0 + h) - f(x, \theta_0)}{h} = \frac{f(x, \theta_0)}{h}\left[\frac{f(x, \theta_0 + h)}{f(x, \theta_0)} - 1\right]$$

$$= \frac{f(x, \theta_0)}{h}\,[e^{-h^2/2}\, e^{h(x-\theta_0)} - 1]$$

Hence $\left|\frac{f(x, \theta_0 + h) - f(x, \theta_0)}{h}\right| \le f(x, \theta_0) \left|\frac{e^{h(x-\theta_0)} - 1}{h}\right|$ as $e^{-h^2/2} \le 1$ for $|h| \le \delta$. Now consider the function $\left|\frac{e^{ha} - 1}{h}\right|$ for $0 < |h| \le \delta$ then expanding $(e^{ha} - 1)$ in powers of (ha) and using $|\Sigma a_i| \le \Sigma |a_i|$ we have, for $|h| \le \delta$, $\left|\frac{e^{ha} - 1}{h}\right| \le \sum_{r=1}^{\infty} |\delta|^{r-1} |a|^r/r! = \frac{e^{\delta|a|} - 1}{\delta}$. Hence $\left|\frac{e^{ha} - 1}{h}\right| \le \frac{e^{\delta|a|}}{\delta}$ and therefore

$$|A_1| \le \frac{1}{|\delta|\sqrt{2\pi}} \exp\left\{-\frac{(x-\theta_0)^2}{2} + \delta|x - \theta_0|\right\} = G_0(x) \quad (2.2.8)$$

Now $I_1 = \int_{R_1} G_0(x)\, dx$ can be evaluated explicitly by using transformation $w = (x - \theta_0)$. Then

$$I_1 = \frac{1}{\delta}\int_{R_1} \frac{1}{\sqrt{2\pi}}\, e^{-w^2/2}\, e^{\delta|w|}\, dw \quad (2.2.9)$$

We note that integral in (2.2.9) is the moment generating function (mgf of $|W|$ evaluated at δ where $W \sim N(0, 1)$ and can be explicitly evaluated by completing the square in the exponent. Now observing that the integrand is an even function of w we have

$$I_1 = \frac{2e^{\delta^2/2}}{\sqrt{2\pi}\delta}\int_0^{\infty} \exp\left\{-\frac{(w-\delta)^2}{2}\right\} dw$$

$$= \frac{2e^{\delta^2/2}}{\delta}[1 - \Phi(-\delta)]$$

$$= \frac{2e^{\delta^2/2}}{\delta}\Phi(\delta)$$

Thus $G_0(x)$ is integrable. Now for validity of taking second order derivative, following similar techniques as earlier, we have

$$|A_2| = \left|\frac{f(x, \theta_0 + 2h) - 2f(x, \theta_0 + h) + f(x, \theta_0)}{h}\right|$$

$$\leq \left(\frac{e^{h(x-\theta_0)}-1}{h}\right)^2 f(x, \theta_0) \leq \frac{e^{2\delta|x-\theta_0|}}{\delta^2} f(x, \theta_0)$$

Therefore we take $G_1(x) = \dfrac{1}{\delta^2}\dfrac{1}{\sqrt{2\pi}} \exp\left\{\dfrac{-(x-\theta_0)^2}{2} + 2\delta\,|\,x - \theta_0\,|\right\}$ and

$\displaystyle\int_{R_1} G_1(x)\,dx = \frac{1}{\delta^2}$ (Mgf of $|\,W\,|$ at 2δ) $= \dfrac{2}{4\delta^2} e^{-4\delta^2/2}\,\Phi(2\delta)$ and here also $G_1(x)$ is integrable.

Exercise 2.2.1 (i) Consider $f(x, \theta) = \theta e^{-\theta x}$, $x > 0$, $\theta > 0$, which is the pdf of exponential distribution with mean $\dfrac{1}{\theta}$ (or failure rate θ). Show that differentiation under integral sign is valid twice and the Fisher information $I_X(\theta) = \dfrac{1}{\theta^2}$.

(ii) Consider $f(x, \theta) = \dfrac{1}{\pi}\dfrac{1}{1+(x-\theta)^2}$, $x \in R_1$. Show that the Fisher information $I(\theta) = 1/2$. Here validity of differentiation under integral sign is more difficult to prove than that in the above examples.

If the distribution of the r.v. X depends on an m-dimensional parameter $\theta = (\theta_1, \theta_2, \ldots, \theta_m)'$ then we define the Fisher information matrix under the conditions that

(a) The range of the r.v. X or the support of pdf $S_\theta = \{x \mid f(x, \theta) > 0\}$ does not depend on θ.

(b) Interchange of partial differentiation w.r.t. θ_r, θ_s and integration w.r.t. x is valid in the identity.

$$\int f(x, \theta)\,dx = 1 \;\forall\; \theta \in \Omega \subset R_m.$$

Anologous to (2.2.3) we can show that

$$E_\theta\left[\frac{\partial \log f(x, \theta)}{\partial \theta_r}\right] = 0,\; r = 1, 2, \ldots, m \qquad (2.2.9)$$

and $$E_\theta\left[\frac{-\partial^2 \log f(x, \theta)}{\partial\theta_r\,\partial\theta_s}\right] = E_\theta\left[\frac{\partial \log f}{\partial \theta_r}\cdot\frac{\partial \log f}{\partial \theta_s}\right] = J_{rs}(\theta)$$

We define then the Fisher information matrix $J_X(\theta)$ as the $(m \times m)$ matrix $((J_{rs}(\theta)))$. We observe that $J_X(\theta)$ is the variance covariance matrix of the m dimensional vector of score functions $\left(\dfrac{\partial \log f}{\partial \theta_1}, \ldots, \dfrac{\partial \log f}{\partial \theta_m}\right)'$ and as such $J_X(\theta)$ is a non-negative definite (nnd) matrix.

For a random sample of size n then analogous to (2.2.4) we have

$$J_{(X_1,\ldots,X_n)}(\theta) = nJ_X(\theta).$$

We will not go into details here but observe that if $T = (T_1, T_2, \ldots, T_k)'$ is a statistic with pdf $\{g(t, \theta), \theta \in \Omega \subset R_m\}$ then analogous to (2.2.6) we obtain the matrix equation

$$J_{(X_1,\ldots,X_n)}(\theta) = J_T(\theta) + E_\theta[J_{(X_1,\ldots,X_{n-k})|T=t}(\theta)]$$

There is loss of information due to reduction of data by constructing a statistic T in the sense that the difference between the Fisher information matrix of the sample and that of the statistic T is a nnd matrix and we say that there is no loss of information if $J_{(X_1,\ldots,X_n)}(\theta) = J_T(\theta)$ or $E_\theta[J_{(X_1,\ldots,X_{n_k})|T=t}(\theta)]$ is a null matrix. This is possible only when $J_{(X_1,\ldots,X_{n-k})|T=t}(\theta) = 0$ with probability one or $g_1(x_1, \ldots, x_{n-k} \mid T = t)$ does not depend on θ which implies that T is a sufficient statistic.

2.3 Defining Sufficient Statistic

The above motivation for sufficient statistic through information is conceptually very useful but checking sufficiency or otherwise of a statistic T in the above way requires verification that for any other statistic T_1

$$g(t_1, t, \theta) = g_0(t, \theta)\ g_1(t_1\ \theta \mid t) \tag{2.3.1}$$

where $g_1(t_1, \theta \mid t)$ the conditional distribution of T_1 given $T = t$ is such that it does not depend on θ for each $T = t$. On the other hand the result (2.1.3) namely the joint pdf of $(X_1, \ldots, Xn)'$ factorizes into

$$L(x, \theta) = g_0(T_1(x), \ldots, T_k(x), \theta)\ h(x) \tag{2.3.2}$$

where $x = (x_1, x_2, \ldots, x_n)'$ and $g_0(T_1(x), \ldots, T_k(x), \theta)$ denotes the pdf of $(T_1, \ldots, T_k)'$ under θ. This is a verifiable defination. In fact in (2.3.2), h is the conditional distribution of $(X_1, \ldots, X_n)'$ given $T = t$. Thus we propose the following working definition of a sufficient statistic.

Definition 2.3.1 Let $(X_1, X_2, \ldots, X_n)'$ be a random sample of size n on a r.v. X with pdf belonging to the class $\{f(x, \theta), \theta \in \Omega\}$. Then a statistic T with corresponding pdf belonging to $\{g_0(t, \theta), \theta \in \Omega\}$ is said to be sufficient for θ, or more precisely for the family of pdfs $\{f(x, \theta), \theta \in \Omega\}$ indexed by θ, if the following factorization holds

$$L(x, \theta) = g_0(T(x), \theta)\ h(x) \tag{2.3.3}$$

where $g_0\ (T(x), \theta)$ is the pdf of $T(x)$ under θ and $h(x)$ is a function of x only. The above factorization must hold for each $\theta \in \Omega$ and almost all x in $S_L = \bigcup_\theta \{x \mid L(x, \theta) > 0\}$ the support of $L(x, \theta)$.

EXAMPLE 2.3.1 Consider (X_1, X_2, X_3) i.i.d. $N(\theta, 1)$ then

$$L(x,\ \theta) = \left(\frac{1}{\sqrt{2\pi}}\right)^3 \exp\left\{-\sum_{i=1}^{3}\frac{(x_i-\theta)^2}{2}\right\},\ x \in R_3,\ \theta \in R_1$$

Here $S_\theta = R_3$ and therefore $S_L = R_3$.

Consider $T = X_1 + X_2 + X_3$ then $T \sim N(3\theta,\ 3)$ and

$$g_0(t,\ \theta) = \frac{1}{\sqrt{2\pi\cdot 3}} \exp\left\{-\frac{(t-3\theta)^2}{6}\right\} \text{ where } t = x_1 + x_2 + x_3$$

$$= \frac{1}{\sqrt{6\pi}} \exp\left\{-\frac{(\Sigma x_i - 3\theta)^2}{6}\right\},\ x \in R_3,\ \theta \in R_1$$

Taking logarithms consider the difference $\log L - \log g_0 = A$ then

$$A = c - \frac{1}{2}\sum_{i=1}^{3} x_i^2 + \theta \Sigma x_i - \frac{3\theta^2}{2} + \frac{(\Sigma x_i)^2}{6} - \theta(\Sigma x_i) + \frac{9\theta^2}{6}$$

$$= c - \frac{1}{2}\Sigma x_i^2 + (\Sigma x_i)^2/6$$

Thus A does not depend on θ and we have $L(x,\ \theta) = g_0(t,\ \theta)\ h(x)$ and therefore T is sufficient.

Next consider $T = (X_1,\ X_2 + X_3)'$ a two dimensional statistic then

$$g_0(t,\ \theta) = \frac{1}{\sqrt{2\pi}} \exp\left\{-\frac{(x_1-\theta)^2}{2}\right\}\left(\frac{1}{\sqrt{2\pi\cdot 2}}\right) \exp\left\{-\frac{(x_2+x_3-2\theta)^2}{4}\right\}$$

as $X_1 \sim N(\theta,\ 1)$ and $X_2 + X_3 \sim N(2\theta,\ 2)$ are independent. Again taking logs and considering $\log L - \log g_0 = A$ we have

$$A = c - \frac{1}{2}\sum_{i=2}^{3} x_i^2 + \theta \sum_{i=2}^{3} x_i - \theta^2 + \left(\sum_{2}^{3} x_i\right)^2/4 - \theta \sum_{2}^{3} x_i + \theta^2$$

$$= c - \frac{1}{2}\sum_{2}^{3} x_i^2 + \left(\sum_{2}^{3} x_i\right)^2/4$$

which is independent of θ and therefore $(X_1,\ X_2 + X_3)'$ is also sufficient. In a similar way it can be verified that $(X_2,\ X_1 + X_3)'$ and $(X_3,\ X_1 + X_2)'$ are sufficient for θ.

Next consider $\bar{X} = \frac{1}{3}\sum_{i=1}^{3} X_i$. Then $\bar{X} \sim N\left(\theta,\ \frac{1}{3}\right)$ and

$$g_0(\bar{x},\ \theta) = \frac{\sqrt{3}}{\sqrt{2\pi}} \exp\left\{-\frac{3(\bar{x}-\theta)^2}{2}\right\}$$

and $$L(x,\ \theta)/g_0(\bar{x},\ \theta) = \text{const.} \exp\left\{-\frac{1}{2}\Sigma x_i^2 + \frac{3}{2}\bar{x}^2\right\}$$

which does not depend on θ and therefore $\bar{X}$ is also sufficient. We note that ΣX_i and $\bar{X}$ are in one-one correspondence and as remarked earlier if one of them is sufficient so would be other. On the other hand if $T_1 = (X_1, X_2 + X_3)'$ and $T_2 = (X_2, X_1 + X_3)'$ then as seen above both are sufficient but T_1 and T_2 are not in one-one correspondence. Further T_1 and T_2 are not functions of each other. Note that $(X_1 + X_2 + X_3)$ is a function of T_1 as well as that of T_2 but T_1 or T_2 is not a function of $(X_1 + X_2 + X_3)$.

Now consider $T = (X_1 + 2X_2 + 3X_3) \sim N(6\theta, 14)$ and

$$g_0(t, \theta) = \frac{1}{\sqrt{2\pi \cdot 14}} \exp\left\{-\frac{(x_1 + 2x_2 + 3x_3 - 6\theta)^2}{2.14}\right\}, x \in R_3, \theta \in R_1$$

Again considering the difference $\log L(x, \theta) - \log g_0(t(x), \theta)$ we observe that coefficient of θ^2 is $-\frac{3}{2} + \frac{36}{28} \neq 0$ hence the above difference is not purely a function of x and therefore $(X_1 + 2X_2 + 3X_3)$ is not sufficient. Suppose $T = X_1 + 2X_2 - 3X_3$. Then $T \sim N(0, 14)$ and $g_0(t, \theta) = \frac{1}{\sqrt{2\pi \cdot 14}} \exp\left\{-\frac{(x_1 + 2x_2 - 3x_3)^2}{2.14}\right\}$, $x \in R_3$, $\theta \in R_1$ which does not depend on θ. Therefore obviously $L(x, \theta)/g_0(t(x), \theta)$ cannot be independent of θ and $(X_1 + 2X_2 - 3X_3)$ is not sufficient.

Exercise 2.3.1 (1) For X_1, X_2 i.i.d. $N(\theta, 1)$ using the above definition show that $T = lX_1 + mX_2$ is sufficient iff $l = m$.

(2) Generalize the above result for a random sample of size 3 from $N(\theta, 1)$.

(3) Let $f(x, \lambda) = \lambda e^{-\lambda x}$, $x > 0$, $\lambda > 0$. Let $(X_1, \ldots, X_n)$ be a random sample of size n then show that ΣX_i is sufficient.

(4) For a sample of size n from $N(\theta, \sigma^2)$ where σ^2 is known show that $\bar{X}$ is sufficient.

(5) For a sample of size n from $N(\theta, \sigma^2)$ where θ is known show that $T = \Sigma (X_i - \theta)^2 \sim \sigma^2\chi^2_n$ is sufficient for σ^2 but $S^2 = \Sigma (X_i - \bar{X})^2 \sim \sigma^2\chi^2_{n-1}$ is not sufficient.

(6) For a sample of size n from $N(\theta, \sigma^2)$ where $(\theta, \sigma^2)'$ is unknown show that $(\bar{X}, S^2)'$, is sufffficient for the vector valued parameter $(\theta, \sigma^2)'$, where S^2 is as defined in (5) above.

(7) Let the pmf of X be $P_\theta[X = 0] = \theta$, $P_\theta[X = 1] = 2\theta$ and $P_\theta[X = 2] = 1 - 3\theta$ where $0 < \theta < 1/3$. Let $(T_1, T_2, T_3)'$ denote the vector of observed frequencies of 0, 1 and 2 in a sample of size n. Show that $(T_1, T_2, T_3)'$ is a sufficient statistic. Let $T_4 = T_1 + T_2$ then using the result that $T_4 \sim B(n, 3\theta)$ show that T_4 is also sufficient.

EXAMPLE 2.3.2 Let $(X_1, X_2, \ldots, X_n)'$ be a random sample of size n on a continuous real r.v. X with pdf $\{f(x, \theta), \theta \in \Omega\}$. Then we can form order statistic $T = (X_{(1)}, \ldots, X_{(n)}))'$ where $X_{(1)} < X_{(2)} \ldots < X_{(n)}$ as $P_\theta[X_i = X_j] = 0$ for any $\theta \in \Omega$ i.e. there are no tied observations since the r.v. X is continuous. Now T is an n-dimensional statistic constructed from an n-dimensional r.v. $(X_1, \ldots, Xn)'$. However if the support of f is say an interval (a, b) then range of T is $a < X_{(1)} < X_{(2)} \ldots < X_{(n)} < b$ whereas that of $(X_1, X_2 \ldots X_n)'$ is $(a, b) \times (a, b) \ldots \times (a, b)$. The order statistic is a reduction of data in the sense that if $(x_{(1)} < \ldots < x_{(n)})$ is the observed order statistic i.e. if $T = t$ then $A_t = T^{-1}\{t\}$ is the set of all permutations of $(x_{(1)}, \ldots, x_{(n)})$ and consists of $n!$ points or equivalently any one of the $n!$ permutations of $(x_{(1)}, \ldots, x_{(n)})$ would give rise to the same order statistic $(x_{(1)}, \ldots, x_{(n)})'$. Note that the range space of the order statistic is a proper subset of the range space of the observations themselves. The joint pdf of the order statistic is

$$g_0(x_{(1)}, \ldots, x_{(n)}, \theta) = n! \prod_{i=1}^{n} f(x_{(i)}, \theta),\ a < x_{(1)} < x_{(2)} \ldots < x_{(n)} < b$$
$$= 0 \quad \text{o.w.}$$

Now $L(x, \theta) = \prod_{i=1}^{n} f(x_i, \theta)$, $a < x_i < b$, $i = 1, 2, \ldots, n$. Observe that $\prod_{i=1}^{n} f(x_{(i)}, \theta) = \prod_{i=1}^{n} f(x_i, \theta)$, since $(x_{(1)}, \ldots, x_{(n)})$ is a permutation of $(x_1, \ldots, x_n)'$. Thus we can write

$$L(x, \theta) = g_0(x_{(1)}, \ldots, x_{(n)}, \theta)\ h(x) \tag{2.3.2}$$

where $$h(x) = \frac{1}{n!} \quad \text{for } x \in A_t$$
$$= 0 \quad \text{o.w.}$$

where A_t denotes the set of $n!$ permutations of $(x_{(1)}, \ldots, x_{(n)})$.

Since $h(x)$ in (2.3.2) does not depend on θ it follows that the order statistic $(X_{(1)}, \ldots, X_{(n)})'$ is sufficient for θ.

Example 2.3.3 Let $(X_1, \ldots, X_n)$ be i.i.d. geometric r.v. with $f(x, \theta) = P_\theta(X = x) = (1 - \theta)\ \theta^x$, $x = 0, 1, 2, \ldots, 0 < \theta < 1$. Here, support S_θ is the set of all non-negative integers $\mathbb{N}$ and does not depend on θ. Since the r.v. X is discrete $P_\theta[X_i = X_j] > 0$ and we can not define the order statistic of the sample. Consider $T = \sum_{i=1}^{n} X_i$ then T is negative binomial with parameters n and θ with pmf $g_0(t, \theta) = P_\theta[T = t] = \binom{-n}{t} (1 - \theta)^n\ \theta^t$, $t = 0, 1, 2, \ldots$ Now

$L(x, \theta) = (1 - \theta)^n\ \theta^{\Sigma x_i}$ and $g_0(t, \theta) = \binom{-n}{t} (1 - \theta)^n \theta^t = P_\theta(A_t)$.

Therefore, we can write

$$L(x, \theta) = g_0(T(x), \theta)\ h(x)$$

where $$h(x) = \frac{1}{\binom{-n}{t}} \quad \text{if } \Sigma x_i = t \text{ or } x \in A_t$$
$$= 0 \quad \text{otherwise.}$$

Hence $T = \Sigma X_i$ is sufficient for θ. We observe that $P_\theta(A_t) = \binom{-n}{t}(1-\theta)^n \theta^t$ and if A is any subset of $S_L = \mathbb{N}^n$, then $P_\theta(A \mid A_t) = P_\theta(A \cap A_t)/g_0(t, \theta) = \sum_{x \in A \cap A_t} h(x)$ and thus the $h(x)$ in fact defines conditional distribution of $(X_1 \ldots X_n)$ given $T = t$.

Remark 2.3.1 Consider Example 2.3.2. Here $A_t = \{$all $n!$ permutations of $\{x_{(1)}, \ldots, x_{(n)}\}$ and as X is continuous r.v. $P_\theta(A_t) = 0$. Therefore the conditional probability of subset A of sample space $(a, b)^n$ given A_t, $P_\theta(A \cap A_t)/P_\theta(A_t)$ is an indeterminate form of the type $\frac{0}{0}$. However using methods of measure theoretic analysis, we can give a meaning to this indeterminate form in a consistent manner. Heuristically we can interprete the function $h(x) = \frac{1}{n!}$, $x \in A_t$ and zero otherwise as a discrete probability distribution on $(a, b)^n$ so that $P_\theta(A \mid A_t) = \frac{1}{n!}$ {number of permutations of $(x_{(1)}, \ldots, x_{(n)})$ belonging to A}. In the normal case discussed in Example 2.3.1 for $T = \sum_{i=1}^{3} X_i$ we have $P_\theta(A_t) = 0$ and therefore we have the same difficulty in defining the conditional distribution on R_3. Using techniques of multivariate normal distribution one can show that the conditional distribution of (X_1, X_2, X_3) given $T = t$ is a singular multivariate normal distribution on R_3 with mean vector $\left(\frac{t}{3}, \frac{t}{3}, \frac{t}{3}\right)$ and variance covariance matrix given by $\frac{1}{3}\begin{bmatrix} 2 & 1 & 1 \\ 1 & 2 & 1 \\ 1 & 1 & 2 \end{bmatrix}$ which has rank 2. i.e. the disribution is concentrated on a subspasce of dimension two given by $x_1 + x_2 + x_3 = t$. In both the cases above the conditional distributions are singular on the original sample space but can be interpreted as proper disributions on a lower dimensional space.

Because of the difficulties involved in defining the conditional distribution of X given T in the continuous case we have defined sufficiency of T using the factorization given by (2.3.1), which we will call as Neyman Factorizability Criterion. It is also possible to define T to be sufficient if the conditional distribution of X given T is independent of θ and then derive (2.3.1). On the other hand if (2.3.1) holds one can show that the conditional distribution of X given T does not depend on θ and thus the two definitions are equivalent. We will show this in the discrete case where conditional distribution of X given T can be defined very easily.

THEOREM 2.3.1 Under the set up given in Definition 2.3.1, the conditional distribution X given T is independent of θ if and only if the factorization (2.3.1) holds.

Consider $P_\theta[X = x \mid T = t] = P_\theta[X = x, T = t]/g_0(t(x), \theta)$ where $g_0(t(x), \theta) = \sum_{x \in A_t} L(x, \theta) > 0$. Then

$$\left.\begin{aligned} P_\theta[X = x \mid T = t] &= \frac{L(x, \theta)}{g_0(t(x), \theta)} \quad \text{if } x \in A_t \\ &= 0 \quad \text{o.w.} \end{aligned}\right\} \tag{2.3.4}$$

Now suppose (2.3.1) holds then it follows that

$$\begin{aligned} P_\theta[X = x \mid T = t] &= h(x) \text{ if } x \in A_t \\ &= 0 \quad \text{o.w.} \end{aligned}$$

which is independent of θ for each x and any t and therefore the conditional distribution of X given T does not depend on θ. On the other hand suppose that conditional distribution of X given T does not depend on θ. Consider $x \in S_L$ then as $\{A_t\} = T^{-1}(t)$ is a partition of S_L there exists a unique t depending on x but not on θ such that $x \in A_t$ and then for each $\theta \in \Omega$

$$\begin{aligned} P_\theta[X = x] &= L(x, \theta) = P_\theta[X = x \mid T = t]\; P_\theta[T = t] \\ &= W(x, t(x))\; g_0(t(x), \theta) \end{aligned}$$

since $P_\theta[X = x \mid T = t]$ does not depend on θ. Taking $W(x, t(x)) = h(x)$ we have the required factorization namely $L(x, \theta) = g_0(t(x), \theta)h(x)$.

This theorem can be used to show that a statistic T is not sufficient without explicitly obtaining the distribution of T as the following example shows.

EXAMPLE 2.3.4 Let (X_1, X_2) be i.i.d. Poisson with parameter λ and let $T = (X_1 + 2X_2)$. Consider $T = 0$ then $A_0 = (0, 0)$ and $P_\theta[X = x \mid T = 0] = 1$ if $x = (0, 0)$ and zero otherwise, which is independent of λ. However consider $T = 2$ then $A_2 = \{(2, 0), (1, 1)\}$

$$\begin{aligned} P[X_1 = 2, X_2 = 0) \mid T = 2] &= e^{-2\lambda}\frac{\lambda^2}{2!} \Big/ e^{-2\lambda}\frac{\lambda^2}{2!} + e^{-2\lambda}\lambda \\ &= \lambda/(2 + \lambda) \end{aligned}$$

which depends on λ. Therefore $(X_1 + 2X_2)$ can not be sufficient for λ.

EXAMPLE 2.3.5 Let (X_1, X_2) be i.i.d. $b(1, \theta)$ then it is easy to show that $(X_1 + X_2)$ is sufficient for θ by using Neyman factorizability criterion. Consider $T_1 = X_1 + 2X_2$. We can show that T is sufficient by arguing that (x_1, x_2) and $(x_1 + 2x_2)$ are in 1:1 correspondence since $(0, 0) \leftrightarrow 0$, $(0, 1) \leftrightarrow 2$, $(1, 0) \leftrightarrow 1$ and $(1, 1) \leftrightarrow 3$. Note that these points constitute the sample space of (X_1, X_2). There are many points at which (x_1, x_2) and $(x_1 + 2x_2)$ are not in one-one correspondence but the set of these points has probability zero. Further $T_2 = X_1 + 3X_2$ is also in 1:1 correspondence with

(X_1, X_2) and with T_1 on the sample space of (X_1, X_2). One can generalize this result for $T = aX_1 + bX_2$ where $a \neq b$.

Exercise 2.3.1 (1) Let (X_1, X_2, X_3) be i.i.d. $b(1, \theta)$ and let $T = X_1 + 2X_2 + 3X_3$. Show that (X_1, X_2, X_3) and T are not in 1:1 correspondence as $T^{-1}(3) = \{(1, 1, 0)\ (0, 0, 1)\}$. Obtain conditional distribution on the sample space given $T = 3$. Does this depend on θ? Show that except for the above two points T and (X_1, X_2, X_3) are in 1:1 correspondence.

(2) Let X_1, X_2 be i.i.d with pmf given by $f(0, \theta) = e^{-\theta}$, $f(1, \theta) = \theta e^{-\theta}$, $f(2, \theta) = 1 - e^{-\theta} - \theta e^{-\theta}$. This pmf can be obtained from the Poisson distribution with mean θ assuming that all observations greater than or equal to 2 are recorded as 2. Thus if X is Poisson we define $Z = \text{Min}\ (X, 2)$ then the pmf of Z is the pmf given above. Such an operation of recording observations is known as right censoring. We have seen that in case of non-censored obervations $(X_1 + X_2)$ is sufficient for θ. Show that in the above censored case, $P[X_1 = 0 \mid X_1 + X_2 = 2]$ depends on θ and $(X_1 + X_2)$ is not sufficient.

2.4 Obtaining Sufficient Statistic

The Neyman Factorizability criterion given in the previous section enables us to check whether a given statistic, real or vector valued one, is sufficient or not. However, it does not provide us any method to obtain the sufficient statistic which will allow maximum possible reduction of data. For example for a random sample of size n from $N(\theta, 1)$ we have $(X_1, \ldots, X_n)'$ observations themselves as well as the order statistic $(X_{(1)}, \ldots, X_{(n)})'$ and $\sum X_i$ are all sufficient for θ. Note that $\sum X_i$ is a further reduction of order statistic as it can be expressed as a function of $(X_{(1)}, \ldots, X_{(n)})'$ since $\sum x_i = \sum x_{(i)}$. A question naturally arises as to whether any further reduction of $\sum X_i$ is possible. We will answer both of these questions by providing a construction to obtain a sufficient statistic which represents maximum possible reduction of the data. As before our basic frame work remains the same namely $(X_1, \ldots, X_n)'$ are i.i.d. with pdf (pmf) belonging to the family $\{f(x, \theta), \theta \in \Omega\}$. We allow x and θ to be both vector valued of dimensions k and m although emphasis initially would be on the case $k = 1$ and $m = 1$.

Let $\mathscr{X}^{(n)}$ be the sample space of $(X_1, \ldots, X_n)'$ and let $T(x)$ be a statistic, a function which maps $\mathscr{X}^{(n)}$ into the range space $\mathscr{J}$. Recall that a function T induces a partition of $\mathscr{X}^{(n)}$ given by $\{A_t, t \in \mathscr{J}\}$ i.e. $A_{t_1} \cap A_{t_2} = \phi$ and $\bigcup_{t \in \mathscr{J}} A_t = \mathscr{X}^{(n)}$ where $A_t = \{x \mid T(x) = t\}$. Conversely any partition $\{B_u, u \in \mathscr{U}\}$ defines a function u which takes different values on the disjoint subsets B_l and B_k. Thus a function induces a partition of $\mathscr{X}^{(n)}$ and can also be defined by a partition of $\mathscr{X}^{(n)}$. On the other hand a partition of $\mathscr{X}^{(n)}$ defines and is in turn defined by an equivalence relation on $\mathscr{X}^{(n)}$. Thus a function T from $\mathscr{X}^{(n)}$ to τ can be defined by an equivalence relation. For a function T on $\mathscr{X}^{(n)}$ we say that $x \in \mathscr{X}^{(n)}$ is T equivalent to $y \in \mathscr{X}^{(n)}$, written as $x\ T\ y$ iff $T(x) = T(y)$ and this defines the partition $\{A_t, t \in \tau\}$. On the other hand if $\{B_u, u \in \mathscr{U}\}$ is a partition of $\mathscr{X}^{(n)}$ then we can define a function on $\mathscr{X}^{(n)}$ such that on B_u it takes the value u or $U(x) = u$ iff $x \in B_u$.

Now consider joint pdf of $(X_1, \ldots, X_n)'$ given by $L(x, \theta) = \prod_{i=1}^{n} f(x_i, \theta)$ and let S_L be the support of $L(x, \theta)$, as defined in the previous section. We now define a partition of $\mathscr{X}^{(n)}$ using $L(x, \theta)$ as follows. First notice that $\{S_L, S_L^c\}$ is a partition of $\mathscr{X}^{(n)}$. Now we define a further partition of S_L. Let x and $y \in S_L$. Then we say that x is likelihood equivalent to y, denoted by $x \overset{L}{\sim} y$ if there exists a positive constant k_1, independent of θ (but possibly depending on x, y) such that $L(x, \theta) = k_1 L(y, \theta)$ for all $\theta \in \Omega$. i.e. considered as functions of θ for fixed x and y, $L(x, \theta)$ and $L(y, \theta)$ are proportional to each other and we denote this by $x \overset{L}{\sim} y$. We observe that $\overset{L}{\sim}$ is an equivalence relation on S_L. For $x \overset{L}{\sim} x$ as we can take $k_1 = 1$. If $x \overset{L}{\sim} y$ then $y \overset{L}{\sim} x$ as we have $L(y, \theta) = \dfrac{1}{k_1} L(x, \theta)$. If $x \overset{L}{\sim} y$ and $y \overset{L}{\sim} z$ then $L(x, \theta) = k_1 k_2 L(z, \theta)$ and $x \overset{L}{\sim} z$. Thus $\overset{L}{\sim}$ defines an equivalence relation and a partition of S_L and together with S_L^c defines a partition of $\mathscr{X}^{(n)}$. We point out the fact that if T_1 and T_2 are such that they are in one-one correspondence then they define the same partition of $\mathscr{X}^{(n)}$.

EXAMPLE 2.4.1 Let $(X_1, X_2, \ldots, X_n)$ be i.i.d. with pdf given by

$$f(x, \theta) = \theta / x^{\theta+1},\ x \geq 1,\ \theta > 0$$

This is the well known Pareto distribution used as a model for income distribution. Here $L(x, \theta) = \dfrac{\theta^n}{(\pi x_i)^{\theta+1}}$, $x_i \geq 1$, $i = 1, 2, \ldots, n$, $\theta > 0$, $\mathscr{X}^n = [1, \infty)^n = S_L$ and $S_L^c = \phi$.

Now let $x, y \in S_L$ then $x \overset{L}{\sim} y$ iff $L(x, \theta)/L(y, \theta)$ is independent of θ. Consider $\dfrac{L(x, \theta)}{L(y, \theta)} = (\pi x_i)^{\theta+1}/(\pi y_i)^{\theta+1} = \left(\dfrac{\pi x_i}{\pi y_i}\right)^{\theta+1}$ and is independent of θ for every $\theta > 0$ iff $\pi x_i = \pi y_i$. Thus $\overset{L}{\sim}$ defines the funcion $T_1(x) = \pi x_i$ on $\mathscr{X}^{(n)}$. Consider $T_2(x) = \sum \log x_i = \log T_1$. Then T_1 and T_2 are in 1:1 correspondence as $T_1(x) = \exp(T_2(x))$. Note that the ranges of T_1 and T_2 are different but they induce the same portion of $\mathscr{X}^{(n)}$ in the sense that $T_1^{-1}(a) = T_2^{-1}(\log a)$ and $T_2^{-1}(b) = T_1^{-1}(e^b)$. Is T_1 or T_2 sufficient? To use Neyman Factorizability criteria we must obtain the pdf and it is easy to work with T_2. Noting that $\log X$ is exponential with mean $1/\theta$, we have $T_2 \sim G\left(n, \dfrac{1}{\theta}\right)$ or

$$g_0(t_2, \theta) = \theta^n \frac{1}{\Gamma(n)} e^{-t_2\theta}\, t_2^{n-1},\ t_2 \geq 0,\ \theta > 0$$

Hence $L(x, \theta)/g_0(t_2(x), \theta) = \dfrac{\Gamma(n)}{(\pi x_i)\,(\sum \log x_i)^{n-1}}$, $x_i \geq 1$ which is independent of θ and therefore T_2 is sufficient. Since T_1, T_2 are in one-one correspondence T_1 is also sufficient.

Recall that the pdf of T_1 can be obtained from that of T_2 by using the rule of transformation given by

$$g(t_1, \theta) = g_0(t_2(t_1), \theta)\left|\frac{dt_2}{dt_1}\right|$$

$$= g_0\,(\log t_1,\ \theta)\ \frac{1}{t_1},\ \text{as } t_2 = \log t_1$$

$$= \frac{\theta^n}{\Gamma(n)}\, e^{-\theta \log t_1}\,(\log t_1)^{n-1}\,\frac{1}{t_1},\ t_1 \geq 1$$

Hence $L(x, \theta)/g(t_1(x), \theta) = \dfrac{\Gamma(n)}{(\Sigma \log x_i)^{n-1}},\ x_i \geq 1.$

Showing that $T_1 = \pi X_i$ is also sufficient.

EXAMPLE 2.4.2 Let $(X_1, \ldots, X_n)$ be i.i.d. with pdf given by $f(x, \theta) = \dfrac{1}{\theta}$, $0 < x_i < \theta$. This is the well known uniform distribution over the interval $(0, \theta)$. Here the range of r.v. X depends on the parameter θ and $S_\theta = (0, \theta)$ and as $\Omega = (0, \infty)$ $S_L = (0, \infty)^n$ and $L(x, \theta) = \dfrac{1}{\theta^n}$, $0 < x_i < \theta$, $i = 1, 2, \ldots, n$. Let $x \in S_L$ and $\theta \in \Omega$. For each $x_i \leq x_{(n)} = \max\,(x_1, \ldots, x_n)$ we can write

$$L(x, \theta) = \frac{1}{\theta^n}\{\prod_{i=1}^{n} \psi(x_i, x_{(n)})\}\ \psi(x_{(n)}, \theta) \text{ where } \psi(a, b) = 1$$

if $a \leq b$ and zero otherwise.

Let $x \in S_L$ and $y \in S_L$ and let $y_{(n)} = \max\,(y_1, \ldots, y_n)$. Then

$$L(y, \theta) = \frac{1}{\theta^n}\{\prod_{i=1}^{n} \psi(y_i, y_{(n)})\}\ \psi(y_{(n)}, \theta)$$

Comparing $L(x, \theta)$ and $L(y, \theta)$ we observe that $x \overset{L}{\sim} y$ iff $\psi(x_{(n)}, \theta) = \psi(y_{(n), \theta})$ for all $\theta \in (0, \infty)$. Suppose that $x_{(n)} \neq y_{(n)}$ then either we must have $x_{(n)} < y(n)$ or $x_{(n)} > y_{(n)}$. Consider the case $x_{(n)} < y_{(n)}$ then for $\theta \in (x_{(n)}, y_{(n)})$, $\psi(x_{(n)}, \theta) = 1$ but $\psi(y_{(n)}, \theta) = 0$ and x is not likelihood equivalent to y. Similarly for the case $y_{(n)} < x_{(n)}$ we have for $\theta \in (y_{(n)}, x_{(n)})$, $\psi(y_{(n)}, \theta) = 1$ but $\psi(x_{(n)}, \theta) = 0$ and x is not likelihood equivalent to y. Thus $x \overset{L}{\sim} y$ iff $x_{(n)} = y_{(n)}$ or the likelihood equivalence defines the statistic $T = X_{(n)} = \text{Max}\,(X_1, \ldots, X_n)$.

To verify that $X_{(n)}$ is sufficient we note that $g_0(x_{(n)}, \theta) = \dfrac{n x_{(n)}^{n-1}}{\theta^n}\ \psi(x_{(n)}, \theta)$ and therefore we can write $L(x, \theta) = g_0(x_{(n)}, \theta)\ h(x)$ where $h(x) = \prod_{i=1}^{n} \psi(x_i, x_{(n)})/n x_{(n)}^{n-1}$ which is independent of θ. Hence $X_{(n)}$ is sufficient.

Exercise 2.4.1 (1) Let $(X_1, \ldots, X_n)$ be i.i.d. exponential with location θ having pdf $f(x, \theta) = \exp\{-(x - \theta)\}$, $x > \theta$. Show that the likelihood equivalence relation defines the statistic $T = X_{(1)} = \text{Min}(X_1 \ldots, X_n)$ and T is sufficient. Note that $g_0(x_{(1)}, \theta) = n \exp\{-n(x_{(1)} - \theta)\}$, $x_{(1)} > \theta$.

(2) Show that for a random sample of size n from the population with pdf $f(x, \theta) = \theta x^{\theta-1}$, $0 < x < 1$, $\theta > 0$ the likelihood equivalence defines the statistic $T = \sum \log X_i$ which is sufficient. What is the connection between this problem and Example 2.4.1?

(3) Show that for a random sample of size n from $b(1, \theta)$, $x \overset{L}{\sim} y$ iff $\sum x_i = \sum y_i$ and that $T = \sum X_i$ is sufficient.

(4) Show that the result of Exercise (3) above also holds for the Poisson distribution and the geometric distribution.

EXAMPLE 2.4.3 We now consider an example in which the $\overset{L}{\sim}$ leads to a two dimensional sufficient statistic for a real parameter θ. Let (X_1, X_2) be i.i.d. Cauchy $(\theta, 1)$ with pdf

$$f(x, \theta) = \frac{1}{\pi} \frac{1}{1 + (x - \theta)^2}, \; x \in R_1, \; \theta \in R_1$$

Here $S_L = R_2$ and $\Omega = R_1$. Now $x \overset{L}{\sim} y$ iff for any $\theta \in R_1$

$$[1 + (y_1 - \theta)^2]\,[1 + (y_2 - \theta)^2] = k[1 + (x_1 - \theta)^2]\,[1 + (x_2 - \theta)^2].$$

where k does not depend on θ. Expanding both sides and comparing coefficient of θ^4 we get $k = 1$, and using $k = 1$ and then comparing coefficients of θ^3 we get $y_1 + y_2 = x_1 + x_2$. Using these relations and equating coefficients of θ^2 we get $x_1 x_2 = y_1 y_2$. Comparing coefficients of θ and the constant term, we have the equations

$$y_1(1 + y_2^2) + y_2(1 + y_1^2) = x_1(1 + x_2^2) + x_1(1 + x_1^2)$$

$$(1 + y_1^2)\,(1 + y_2^2) = (1 + x_1^2)\,(1 + x_2^2)$$

The last two equations hold when $x_1 + x_2 = y_1 + y_2$ and $x_1 x_2 = y_1 y_2$ and do not lead to any further restriction on x, y. Thus $x \overset{L}{\sim} y$ implies that either $x_1 = y_1$, $x_2 = y_2$ or $x_1 = y_2$, $x_2 = y_1$ or y is a permutation of x. Thus $\overset{L}{\sim}$ relation defines in this case the order statistic $(x_{(1)}, x_{(2)})$, since the order statistic induces the same partition of R_2 as the permutation of (x_1, x_2). We have already seen that $(x_{(1)}, x_{(2)})$ is sufficient for θ.

EXAMPLE 2.4.3 In this example we show how $\overset{L}{\sim}$ defines a two dimensional sufficient statistics for a two dimensional parameter. Let $(X_1, \ldots, X_n)$ be i.i.d. $N(\theta_1, \theta_2)$ so that $S_L = R^n$ and $\Omega = (-\infty, \infty) \times (0, \infty)$. Now $L(x, \theta) =$ $\left(\frac{1}{\sqrt{2\pi\theta_2}}\right)^n \exp\left\{\frac{-\sum (x_i - \theta_1)^2}{2\theta_2}\right\}$. Consider $x, y \in R_n$ and

$$\log L(x, \theta) - \log L(y, \theta) = \frac{1}{2\theta_2}\{\Sigma (y_i - \theta_1)^2 - \Sigma (x_i - \theta_1)^2\}$$

$$= (\Sigma y_i^2 - \Sigma x_i^2)\frac{1}{2\theta_2} + \frac{\theta_1}{\theta_2}(\Sigma x_i - \Sigma y_i)$$

This does not depend on θ_1 or θ_2 iff $\Sigma x_i^2 = \Sigma y_i^2$ and $\Sigma x_i = \Sigma y_i$. Hence $\underset{\sim}{L}$ defines a two-dimensional statistic (T_1, T_2) where $T_1 = \Sigma X_i$ and $T_2 = \Sigma X_i^2$. Marginally $T_1 \sim N(\theta_1, n\theta_2)$ and ΣT_2 is noncentral χ^2 with n degrees of freedom and non-centrality parameter $\delta = n\theta_1^2/\theta_2$. The joint distribution of (T_1, T_2) is not easy to obtain. Hence we consider a one-one transformation of (T_1, T_2) given by $T_3 = \bar{X} = \frac{T_1}{n}$ and $T_4 = \Sigma (X_i - \bar{X})^2 = T_2 - T_1^2/n$

$$g_0(t_3, t_4, \theta_1, \theta_2) = \sqrt{\frac{n}{2\pi\theta_2}}\exp\left\{\frac{-n(t_3-\theta_1)^2}{2\theta_2}\right\}\frac{e^{-t_4/2\theta_2}\,t_4^{\frac{n-1}{2}-1}}{\Gamma\left(\frac{n-1}{2}\right)(2\theta)^{(n-1)/2}}$$

since $T_3 \sim N\left(\theta_1, \frac{\theta_2}{n}\right)$, $T_4 \sim \theta_2\chi^2_{n-1}$ and (T_3, T_4) are independent. Therefore $L(x, \theta_1, \theta_2) = g_0(t_3, t_4, \theta_1, \theta_2)\, h(x)$ where

$$h(x) = \frac{\Gamma\left(\frac{n-1}{2}\right)2^{n/2}}{\sqrt{(2\pi)^{(n-1)/2}}}\,\frac{1}{\{\Sigma (x_i - \bar{x})^2\}^{\frac{n-1}{2}-1}}$$

This shows that $(T_3; T_4)$ is sufficient for (θ_1, θ_2) and as (T_1, T_2) is in one-one correspondence with (T_3, T_4) we have $(T_1; T_2)$ is also sufficient.

Now we show that the likelihood equivalence always leads to sufficient statistic which represents the maximum possible reduction of data. Let T_1 and T_2 be two sufficient statistics then T_2 represents a further reduction of T_1 if T_2 can be expressed as a function of T_1 i.e. $T_2(x) = \varphi(T_1(x))$. In terms of partition of the sample space T_1 induces a finer partition than that induced by T_2. Recall that T_2 is a function of T_1 if $T_1(x) = T_1(y)$ implies that $T_2(x) = T_2(y)$. To illustrate, consider Example 2.3.1 where we had a random sample of size three from $N(\theta, 1)$. As seen there $T_1 = (X_1, X_2 + X_3)'$ is sufficient and $T_2 = (X_1 + X_2 + X_3)$ is a function of T_1. Observe that $T_3 = (X_2, X_1 + X_3)'$ is also sufficient and T_2 is a function of T_3 but neither T_1 is a function of T_3 nor T_3 is a function of T_1. Thus T_2 is a reduction of T_1 and T_3 both. Note that $T_2 = \frac{1}{2}(T_1 + T_3)$.

We now define the maximum possible reduction of the data or minimal sufficient statistic.

Definition 2.4.1 $M(x)$ is called a minimal sufficient statistic for θ provided

(i) $M(x)$ is sufficient for θ and
(ii) If $T(x)$ is any other sufficient statistic then

$$T(x) = T(y) \text{ implies that } M(x) = M(y).$$

Thus the minimal sufficient statistic $M(x)$ is a function of any other sufficient statistic and therefore represents the maximum possible reduction of data. We now show that the likelihood equivalence leads to minimal sufficient statistic.

THEOREM 2.4.1 Let T be sufficient for θ then for any $x, y \in S_L$ $T(x) = T(y)$ implies that $x \overset{L}{\sim} y$.

Since T is sufficient by Neyman factorizability criterion for any $x, y \in S_L$ for every $\theta \in \Omega$ we have

$$L(x, \theta) = g_0(T(x), \theta) \cdot h(x)$$

$$L(y, \theta) = g_0(T(y), \theta)\ h(y)$$

But $T(x) = T(y)$ implies that $g_0(T(x), \theta) = g_0(T(y), \theta)$ and therefore $L(x, \theta) = k\ L(y, \theta)$ where k does not depend on θ and thus $x \overset{L}{\sim} y$.

As seen earlier the likelihood equivalence defines a statistic $M(x)$ i.e. $M(x) = M(y)$ if and only if $x \overset{L}{\sim} y$ where $x, y \in S_L$. We now show that $M(x)$ is a sufficient statistic by proving that the conditional distribution on the sample space given $M(x)$ does not depend on θ and then use Theorem 2.3.1. We will consider the discrete case only.

THEOREM 2.4.2 Let the likelihood equivalence define a statistic $M(x)$ then $P_\theta[X = x \mid M(x) = m]$ does not depend on θ for any $\theta \in \Omega$ and every m in the range of M.

Let $A_m = \{x \mid M(x) = m\}$ then consider

$$P_\theta[X = x \mid M(x) = m] = \frac{P_\theta[X = x, M(x) = m]}{P_\theta(A_m)}$$

$$= \frac{L(x, \theta)}{\sum_{y \in A_m} L(y, \theta)} \quad \text{if } x \in A_m$$

$$= 0 \qquad x \notin A_m$$

However for x and $y \in A_m$ we have $x \overset{L}{\sim} y$ and $L(y, \theta) = kL(x, \theta)$ where k can depend on x, y say $k = k(x, y)$ then

$$P_\theta[X = x \mid M(x) = m] = \frac{1}{\sum_{y \in A_m} k(x, y)} = h(x)$$

which is independent of θ.

By Theorem 2.3.1 the statistics $M(x)$ induced by likelihood equivalence is sufficient for θ and it is minimal sufficient as by Theorem 2.4.1 it is a function of any other sufficient statistic. The likelihood equivalence thus gives us a constructive method to obtain minimal sufficient statistic.

We have already seen that if T_1 is sufficient and T_2 is in one-one correspondence with T_1 then T_2 is also sufficient. In the same way we point out that if the family $\{f(x, \theta), \theta \in \Omega\}$ is reparametrized by considering a one-one transformation of θ to φ then if T_1 is sufficient for θ then it continues to remain sufficient for φ and thus sufficiency is in fact a property of the family of distributions which can indeed be labelled or indexed by different parametrizations. To illustrate this point consider $(X_1, \ldots, X_n)$ i.i.d. $b(1, \theta)$ and consider $\phi = \log \dfrac{\theta}{1-\theta}$, the so called log-odds in favour of occurrence of an event E. Now $\dfrac{d\phi}{d\theta} = \dfrac{1}{\theta(1-\theta)} > 0$ and θ and ϕ are in one-one-correpondence. Further $\theta = \dfrac{e^{\phi}}{1+e^{\phi}}$ and as θ varies over $(0, 1)$, ϕ varies over $(-\infty, \infty)$. Now we have $L(x, \theta) = \theta^{\Sigma x_i}(1-\theta)^{n-\Sigma x_i}$ which gets transformed to $L\left(x, \dfrac{e^{\phi}}{1+e^{\phi}}\right) = L_1(x, \phi) = e^{\phi \Sigma x_i} \cdot \dfrac{1}{(1+e^{\phi})^n}$. Now consider likelihood equivalence in terms of ϕ. We then have $x \overset{L}{\sim} y$ iff $e^{\phi \Sigma x_i} = k e^{\phi \Sigma y_i}$ for all $\phi \in R_1$ which holds iff $\Sigma x_i = \Sigma y_i$. Looking at Neyman Factorizability criteria the relation $L(x, \theta) = g_0(T(x), \theta)\, h(x)$ gets transformed to $L_1(x, \phi) = g_0(T(x), \psi(\phi))\, h(x)$ where $\theta = \psi(\phi)$ is the inverse transformation.

In the next section we now consider some of the important class of pdfs which occur quite often in many practical situations and which admit minimal sufficient statistic of the same dimensionality as that of the parameter.

2.5 Exponential Family

First consider the case of a real parameter θ. Let X be a real or vector valued r.v. with pdf belonging to the class $\{f(x, \theta), \theta \in \Omega \subset R_1\}$ where the following regularity conditions hold.

C_1 : The support $S_\theta = \{x \mid f(x, \theta) > 0\}$ does not depend on θ i.e. $S_\theta = S$, $\forall\, \theta \in \Omega$.

C_2 : The parameter space Ω is an open interval of R_1, $\underline{\theta} < \theta < \overline{\theta}$.

C_3 : The pdf $f(x, \theta)$, is such that for $x \in S$ and $\theta \in \Omega$, $\log f(x, \theta) = u(\theta)\, k(x) + v(\theta) + w(x)$

where (a) $u(\theta)$ is twice differentiable with $\dfrac{du}{d\theta} \neq 0$.

(b) $\{1, k(x)\}$ are linearly independent over S i.e. $a_0 + a_1 k(x) = 0$ $\forall\, x \in S$ iff $a_0 = 0$ and $a_1 = 0$.

If the above conditions hold for $\{f(x, \theta), \theta \in \Omega\}$ then we say that the r.v. X or its pdf is of the exponential type or it belongs to a one-parameter exponential family. Many distributions used as models for practical situations belong to exponential family and the reader should verify that the families $\{N(\theta, 1), \theta \in R_1\}$, {Poisson (θ), $\theta > 0$} and {exponential distribution with mean θ, $\theta > 0$} are of the exponential type. As examples of family of pdfs which are not of exponential type we first consider $U(0, \theta)$. Here $S_\theta = (0, \theta)$ and therefore C_1 does not hold. On the other hand consider Laplace distribution with pdf $f(x, \theta) = \frac{1}{2} \exp\{-|x - \theta|\}$, $x \in R_1$, $\theta \in R_1$. Here C_1 and C_2 hold but $\log f(x, \theta) = -\log 2 - |x - \theta|$ can not be written in the form $u(\theta)\, k(x) + v(\theta) + w(x)$. Cauchy distribution with pdf $f(x, \theta) = \frac{1}{\pi} \frac{1}{1 + (x - \theta)^2}$, $x \in R_1$, $\theta \in R_1$ is also not of the exponential type since as in the case of Laplace distribution. C_1 and C_2 hold but C_3 does not hold.

Example 2.5.1 Consider the Pareto distribution with pdf $f(x, \theta) = \frac{\theta}{x^{\theta+1}}$, $x > 1$, $\theta > 0$. Here $S_\theta = (1, \infty)$ and $\Omega = (0, \infty)$ and C_1 and C_2 are satisfied. Consider $\log f(x, \theta) = \log \theta - (\theta + 1) \log x$ then $u(\theta) = -(\theta + 1)$, $k(x) = \log x$, $\frac{du}{d\theta} = -1$ and $\frac{d^2u}{d\theta^2} = 0$, so C_3 (a) holds. Consider $a_0 + a_1 \log x = \varphi(x)$ for $x \geq 1$. Taking $x = e$ and e^2 we have $a_0 + a_1 = 0$ and $a_0 + 2a_1 = 0$ which implies that $a_0 = 0$ and $a_1 = 0$ and C_3 (b) also holds. One can show the linear independence of $\{1, k(x)\}$ over S in a variety of ways. For example $\varphi(x) = 0$ for $x > 1$ then $\varphi'(x) = 0$, $\forall\, x > 1$ and as $\varphi'(x) = \frac{a_1}{x} = 0$, $\forall\, x > 1$ we have $a_1 = 0$ and therefore $a_0 = 0$. Thus $f(x, \theta) = \frac{\theta}{x^{\theta+1}}$, $x \geq 1$, $\theta > 0$ belongs to an exponential family.

Example 2.5.2 Let X be a geometric r.v. with pdf given by $f(x, \theta) = \theta(1 - \theta)^x$, $x = 0, 1, 2, \ldots, 0 < \theta < 1$. Then $S = \{0, 1, 2, \ldots\}$ which does not depend on θ and $\Omega = (0, 1)$, is an open interval of R_1. further $\log f(x, \theta) = \log \theta + x \log (1 - \theta)$ so that $u(\theta) = \log (1 - \theta)$ and $k(x) = x$. Then $\frac{du}{d\theta} = \frac{-1}{(1 - \theta)}$ and $\frac{d^2u}{d\theta^2} = \frac{-1}{(1 - \theta)^2}$ and $C_3(a)$ holds. Further $a_0 + a_1 x = 0$ for $x = 0, 1, 2, \ldots$ implies that $a_0 = 0$, $a_1 = 0$ as we can take $x = 0$ to conclude $a_0 = 0$ and then $x = 1$ to claim $a_1 = 0$.

Example 2.5.3 There are situations in biological assays when the probability of a favourable response at dosage level $d > 0$ is given by $\psi(\beta) = \frac{e^{\beta d}}{1 + e^{\beta d}}$ i.e. $P[X = 1] = \psi(\beta)$ and $P[X = 0] = 1 - \psi(\beta)$ Here $\beta \in R_1$ and $S = \{0, 1\}$ and C_1, C_2 are satisfied. Further

$$\log f(x, \beta) = x \log \psi(\beta) + (1 - x) \log [1 - \psi(\beta)]$$

$$= x \log \frac{\psi(\beta)}{1-\psi(\beta)} + \log [1 - \psi(\beta)]$$

$$= x\, d\, \beta - \log (1 + e^{\beta d})$$

Thus $k(x) = x$ and $u(\beta) = d\beta$. Again $\frac{du}{d\beta} = d \neq 0$ and $\frac{d^2u}{d\beta^2} = 0$ and $a_0 + a_1 x = 0$ for $x = 0$ and 1 implise that $a_0 = 0$ and $a_1 = 0$. Therefore $\{f(x, \beta), \beta \in R_1\}$ is an exponential family.

Let $(X_1, \ldots, X_n)$ be i.i.d. with pdf belonging to $\{f(x, \theta), \theta \in \Omega\}$ which forms an exponential family. Then the joint pdf of $(X_1, \ldots, X_n)$ is given by $L(x, \theta) = \prod_{i=1}^{n} f(x_i, \theta)$. Consider the family $\{L(x, \theta), \theta \in \Omega\}$ then $S_L = S^n$ and the parameter space for L is Ω, same as that for f. Further for $x \in S^n$ and $\theta \in \Omega$ we have

$$\log L(x, \theta) = u(\theta) \sum_{i=1}^{n} k(x_i) + nv(\theta) + \sum w(x_i)$$

Therefore defining $T(x) = \sum_{i=1}^{n} k(x_i)$ we have $\{L(x, \theta), \theta \in \Omega\}$ is an exponential family provided $a_0 + a_1 T(x) = 0$ for $x \in S^n$ implies that $a_0 = 0$ and $a_1 = 0$. Now consider $x_1 \in S$, $(x_2^0, \ldots, x_n^0)' \in S^{n-1}$ where $(x_2^0, \ldots, x_n^0)$ is a fixed point. Then $a_0 + a_1 T(x) = a_0 + a_1 \sum_{i=2}^{n} k(x_i^0) + a_1 k(x_1) = b_0 + a_1 k(x_1)$. But as $\{f(x_1, \theta), \theta \in \Omega\}$ is an exponential family we have $b_0 + a_1 k(x_1) = 0$ for $\forall\, x_1 \in S$ implies that $b_0 = 0$ and $a_1 = 0$. This in turn implies $a_1 = 0$ and $a_0 = 0$. Therefore $\{L(x, \theta), \theta \in \Omega\}$ is also an exponential family.

We now show that $T = \sum k(x_i)$ is a minimal sufficient statistic. Let $x, y \in S_L$ then if $x \overset{L}{\sim} y$, $\log L(x, \theta) - \log L(y, \theta) = u(\theta) [\sum k(x_i) - \sum k(y_i)] + [\sum w(x_i) - \sum w(y_i)]$ should not depend on θ. Therefore taking derivative w.r.t. θ and observing that $\frac{du}{d\theta} \neq 0$ we have $\sum k(x_i) - \sum k(y_i) = 0$.

Therefore $x \overset{L}{\sim} y$ iff $\sum k(x_i) = \sum k(y_i)$ and as the likelihood equivalence leads to a minimal sufficient statistic, $T = \sum k(x_i)$ is a minimal sufficient statistic.

We further show that the pdf of T is itself of the exponential type. Consider a transformation $t_1 = \sum_{i=1}^{n} k(x_i)$ and $t_2 = x_2, \ldots, t_n = x_n$. Then the joint pdf of $(t_1, \ldots, t_n)$ is given by

$$g(t_1, \ldots, t_n, \theta) = L(x, \theta) \left| \frac{\partial(t_1, \ldots, t_n)}{\partial(x_1, \ldots, x_n)} \right|^{-1}$$

$$= \exp \{u(\theta) t_1 + nv(\theta)\} \exp \{\sum w(h_i(t)\} \left| \frac{\partial(t_1, \ldots, t_n)}{\partial(x_1, \ldots, x_n)} \right|^{-1}$$

where $x_i = h_i(t)$ $i = 1, 2, \ldots, n$ is the inverse transformation. Thus $g(t, \theta) = \exp\{u(\theta)t_1 + nv(\theta)\}\, p(t_1, \ldots, t_n)$.

Therefore

$$g_0(t_1, \theta) = \exp\{\, u(\theta)t_1 + nv(\theta)\} \int p(t_1, \ldots, t_n)\, dt_2 \ldots dt_n$$

$$= \exp\{u(\theta)t_1 + nv(\theta) + w_1(t_1)\}$$

where $w_1(t_1) = \log [\int p(t_1, \ldots, t_n) dt_2 \ldots dt_n]$. Note that $\exp\{\Sigma\, w_i(h_i(t))\} > 0$, since we take the absolute value of the Jacobian $\left|\dfrac{\partial(t_1, \ldots, t_n)}{\partial(x_1, \ldots, x_n)}\right|^{-1}$ expressed in terms of $(t_1, \ldots, t_n)$, therefore $p(t_1, \ldots, t_n) > 0$ and its integral is positive. It then follows that $\{g_0(t_1, \theta), \theta \in \Omega\}$ is an exponential family.

These results can be generalized for the case of multiparameter exponential family which we define next. The condition C–1 remains the same. Since $\theta = (\theta_1, \ldots, \theta_m)'$ we now change C–2 as

C'–2 : $\theta \in \Omega$ is an open set of R_m containing an m-dimensional rectangle $\underline{\theta}_i < \theta_i < \bar{\theta}_i$, $i = 1, 2, \ldots, m$.

Similarly C–3 is changed to

$$C'\text{–3} : \log f(x, \theta) = \sum_{r=1}^{m} u_i(\theta)\, k_i(x) + v(\theta) + w(x)$$

where

(a) $u_i(\theta)$, $i = 1, 2, \ldots, m$ has partial derivatives of order two and the Jacobian $\left|\dfrac{\partial(u_1, \ldots, u_m)}{\partial(\theta_1, \ldots, \theta_m)}\right| \neq 0, \ \forall\ \theta \in \Omega.$

(b) $\{1, k_1(x), \ldots, k_m(x)\}$ are linearly independent over S. If the above conditions hold then we say that $\{f(x, \theta), \theta \in \Omega\}$ is an m-parameter exponential family. The reader can verify that the classical normal distribution $N(\theta_1, \theta_2)$ or two parameter Gamma distribution with pdf $f(x, \theta, \lambda) = \dfrac{1}{\Gamma(\lambda)} \cdot \dfrac{1}{\theta^{\lambda}}\, x^{\lambda-1} e^{-x/\theta}$, $x > 0$, $\theta > 0$, $\lambda > 0$ form two parameter exponential families.

We now show that if $(X_1, \ldots, X_n)'$ are i.i.d with the pdf belonging to m-parameter exponential family then $(T_1, \ldots, T_m)'$ where $T_r(x) = \sum_{l=1}^{n} k_r(x_l)$ is a minimal sufficient statistic. Consider $x, y \in S_L$ then $x \overset{L}{\sim} y$ iff

$$\log L(x, \theta) - \log L(y, \theta) = \sum_{r=1}^{m} u_r(\theta)\, T_r(x) + \sum_{l=1}^{n} (w(x_l) - w(y_l))$$

is independent of θ, this holds iff derivatives of RHS w.r.t. θ_i, $i = 1, 2, \ldots, m$ are all zero. Thus we must have

$$\sum_{r=1}^{m} \frac{\partial u_r(\theta)}{\partial \theta_i} [T_r(x) - T_r(y)] = 0, \; i = 1, 2, \ldots, m$$

However, as $\left| \frac{\partial(u_1, \ldots, u_m)}{\partial(\theta_1, \ldots, \theta_m)} \right| \neq 0$ we must have $T_r(x) = T_r(y)$ for $r = 1, 2, \ldots, m$. Thus $x \overset{L}{\sim} y$ iff $\sum_{l=1}^{n} k_r(x_l) = \sum k_r(y_l)$, $r = 1, 2, \ldots, m$ or $x \overset{L}{\sim} y$ defines an m-dimensional minimal sufficient statistic

$$\left(\sum_{l=1}^{n} k_1(x_l), \ldots, \sum_{l=1}^{n} k_m(x_l)\right)' = (T_1, \ldots, T_m)'.$$

As in the case of $m = 1$, we also have the property $\{g_0(t_1, \ldots, t_m, \theta), \theta \in \Omega\}$ is itself an m-parameter exponential family with pdf

$$g_0(t_1, \ldots, t_m, \theta) = \exp\left\{\sum_{r=1}^{m} u_r(\theta)\, t_r + nv(\theta) + w_2(t_1, \ldots, t_m)\right\}$$

EXAMPLE 2.5.4 Consider the classical regression problem in which we have the response $y_i = \alpha + \beta x_i + \varepsilon_i$, $i = 1, 2, \ldots, n$ where $(x_1, \ldots, x_n)$ are fixed levels and ε_i's represent random errors and are i.i.d. $N(0, \sigma^2)$. Then the joint pdf of $(Y_1, \ldots, Y_n)'$ is given by

$$L(y, \alpha, \beta, \sigma^2) = \left(\frac{1}{\sqrt{2\pi\sigma^2}}\right)^n \exp\left\{-\sum_{i=1}^{n} \frac{(y_i - \alpha - \beta x_i)^2}{2\sigma^2}\right\}$$

Here $S_L = R_n$, $\Omega = R_1 \times R_1 \times R_+$ and C_1 and C_2' hold. Now

$$\log L(y, \alpha, \beta, \sigma^2) = \frac{-n}{2} \log 2\pi\sigma^2 - \sum_{i=1}^{n} \frac{(\alpha + \beta x_i)^2}{2\sigma^2} - \frac{\sum y_i^2}{2\sigma^2} + \frac{\sum y_i(\alpha + \beta x_i)}{\sigma^2}$$

Therefore we can take $v(\alpha, \beta, \sigma^2) = -\frac{n}{2} \log 2\pi\sigma^2 - \sum_{i=1}^{n} \frac{(\alpha + \beta x_i)^2}{2\sigma^2}$ as $(x_1, \ldots, x_n)$ are fixed known constants. Further $w(y) = 0$ and

$$u_1(\alpha, \beta, \sigma^2) = \frac{-1}{2\sigma^2} \quad T_1(y) = \sum y_i^2$$

$$u_2(\alpha, \beta, \sigma^2) = \frac{\alpha}{\sigma^2} \quad T_2(y) = \sum y_i$$

$$u_3(\alpha, \beta, \sigma^2) = \frac{\beta}{\sigma^2} \quad T_3(y) = \sum x_i y_i$$

Now

$$\left| \frac{\partial(u_1, u_2, u_3)}{\partial(\alpha, \beta, \sigma^2)} \right| = \begin{vmatrix} 0 & 0 & 1/2\sigma^4 \\ 1/\sigma^2 & 0 & -\alpha/\sigma^4 \\ 0 & 1/\sigma^2 & -\beta/\sigma^4 \end{vmatrix} = \frac{1}{2\sigma^8} > 0$$

Consider $a_0 \sum y_i^2 + a_1 \sum y_i + a_2 \sum x_i y_i + a_3 = 0$, $\forall\ y \in R_n$. Take $y = 0$ then we have $a_3 = 0$. Take $(y, 0, \ldots, 0)$ then we have $a_0 y^2 + y(a_1 + a_2 x_1) = 0$, $\forall\ y \in R_1$ which implies that $a_0 = 0$, and $a_1 + a_2 x_1 = 0$. Next take $y = (0, y, \ldots, 0)'$ we have $a_1 + a_2 x_2 = 0$. If $x_1 \neq x_2$ this immediately implies that $a_1 = 0, a_2 = 0$. Thus under the additional condition (often implicitly assumed) that there is at least on pair (x_i, x_j) where $x_i \neq x_j$ we have $\{L(y, \alpha, \beta, \sigma^2), (\alpha, \beta) \in R_2, \sigma^2 \in R_+\}$ is a three parameter exponential family with $(\sum y_i^2, \sum y_i, \sum x_i y_i)'$ as minimal sufficient statistic for $(\alpha, \beta, \sigma^2)'$. The classical Lest Square estimators

$$\hat{\beta} = (\Sigma\, x_i y_i - \bar{x} \Sigma\, y_i) / \Sigma\, (x_i - \bar{x})^2$$

and $$\hat{\alpha} = \bar{y} - \hat{\beta}\bar{x} \text{ and } \hat{\sigma}^2 = \frac{1}{n-2}(\Sigma\, y_i^2 - \hat{\alpha}\Sigma\, y_i - \hat{\beta}\Sigma\, x_i y_i)$$

are functions of the minimal sufficient statistic. Observe that the condition $x_i \neq x_j$ for at least one pair (i, j) is equivalent to $\sum (x_i - \bar{x})^2 > 0$. Note that if all x_i's are equal to x then $(Y_1, \ldots, Y_n)$ are i.i.d. $N(\alpha + \beta x, \sigma^2)$ and $(\alpha, \beta, \sigma^2)$ is no longer a labelling parameter as we can determine (α_1, β_1) and (α_2, β_2) such that $\alpha_1 + \beta_1 x = \alpha_2 + \beta_2 x$ but $\alpha_1 \neq \alpha_2$ and $\beta_1 \neq \beta_2$.

EXAMPLE 2.5.5 Consider a bivariate discrete r.v. with pmf

$$f(x, y, \lambda, p) = \binom{x}{y} p^y (1-p)^{x-y} \frac{e^{-\lambda} \cdot \lambda^x}{x!},$$

$$y = 0, 1, 2, \ldots, x,\ x = 0, 1, 2, \ldots,\ \lambda > 0,\ 0 < p < 1.$$

This model arises when conditionally $(Y \mid X = x) \sim$ Binomial (x, p) and marginally $X \sim$ Poisson (λ) which is used in variety of situations. For example let X represent the number of telephone calls received in an exchange and Y denote the number of calls which are wrongly connected. Here, S the range of pmf does not depend on $\theta = (\lambda, p)$ and $\Omega = (0, \infty) \times (0, 1)$ and C–1 and C'–2 both hold. Further

$$\log f = \log \binom{x}{y} - \log x! - \log \lambda + x[\log \lambda + \log (1-p)] + y[\log p - \log (1-p)]$$

and we have

$$u_1(\lambda, p) = \log \lambda + \log (1-p) \quad k_1(x, y) = x$$

$$u_2(\lambda, p) = \log p - \log (1-p) \quad k_2(x, y) = y$$

Now,

$$\left| \frac{\partial (u_1, u_2)}{\partial (\lambda, p)} \right| = \begin{vmatrix} 1/\lambda & -1/(1-p) \\ 0 & 1/p(1-p) \end{vmatrix} = \frac{1}{\lambda p(1-p)} > 0$$

Next consider $a_0 + a_1x + a_2y = 0$ for all $(x, y) \in S$. Take $(x, y) = (0, 0)$ to obtain $a_0 = 0$ and then take $x = 1$, $y = 0$ to obtain $a_1 = 0$ which then implies $a_2 = 0$. Thus the pmf belongs to two parameter exponential family and $(\Sigma X_i, \Sigma Y_i)'$ is minimal sufficient statistic for $(\lambda, p)'$.

Exercise 2.5.1 (1) Let $f(x, \sigma) = \frac{1}{2\sigma} \exp \{- | x |/\sigma\}$, $x \in R_1$, $\sigma > 0$. Show that $\{f(x, \sigma), \sigma > 0\}$ is a one-parameter exponential family and $\Sigma | X_i | \sim G(n, 1/\sigma)$ is minimal sufficient statistic for σ.

(2) Let $f(x, \theta) = \theta x^{\theta - 1}$, $0 < x < 1$ and $\theta > 0$. Show that this pdf belongs to one parameter exponential family. Show that $\Sigma \log X_i$ is minimal sufficient and obtain its distribution

(3) Let $(X_1, \ldots, X_n)$ be a random sample of size n from $N(\mu, \sigma_1^2)$ and $(Y_1, Y_2, \ldots, Y_m)'$ be a random sample of size m from $N(\mu_2, \sigma_2^2)$ show that joint pdf of $(X, Y)'$ belongs to four parameter exponential family and $(\bar{X}, \bar{Y}, S_x^2, S_y^2)$ is minimal sufficient for $(\mu_1, \mu_2, \sigma_1^2, \sigma_2^2)$ where $S_x^2 = \Sigma (X_i - \bar{X})^2$, $S_y^2 = \Sigma (Y_i - \bar{Y})^2$.

(4) If in (3) above $\sigma_1^2 = \sigma_2^2 = \sigma^2$ show that we have a three parameter exponential family with $(\bar{X}, \bar{Y}, S_x^2 + S_y^2)$ as minimal sufficient statistic.

(5) If in (3) above $\mu_1 = \mu_2 = \mu$ then the joint pdf of $(X, Y)'$ is not an exponential family but $(\bar{X}, \bar{Y}, S_x^2, S_y^2)$ is minimal sufficient for $(\mu, \sigma_1^2, \sigma_2^2)$.

(6) Consider a power series distribution with pmf given by $f(x, \theta) = \frac{a(x)\theta^x}{b(\theta)}$ for $x = 0, 1, 2, \ldots$, $0 < \theta < \rho$ the radius of convergence, $a(x) \geq 0$ and $b(\theta) = \sum_{x=0}^{\infty} a(x)\theta^x$. Show that $\{f(x, \theta), 0 < \theta < \rho\}$ is an exponential family. Many distributions such as Binomial (n, θ), Poisson (θ), negative binomial with $f(x, \theta) = \binom{x + k - 1}{x - 1} \theta^k (1 - \theta)^x$ with k known, logarithmic series distribution with $f(x, \theta) = \frac{\theta^x}{x} \cdot \frac{1}{[- \log (1 - \theta)]}$, $x = 1, 2, \ldots$, $0 < \theta < 1$ are examples of power series distributions.

(7) Let $\{f(x, \theta), \theta \in \Omega\}$ be a m-parameter exponential family and let A be a subset of S where A does not depend on θ. Consider the distribution truncated to set A with pdf given by $f_A(x, \theta) = \frac{f(x, \theta)}{P_\theta(A)}$, $x \in A$. Show that $\{f_A(x, \theta), \theta \in \Omega\}$ is again an m-parameter exponential family.

2.6 Pitman Family

This section considers the case where $(X_1, \ldots, X_n)$ is a random sample of size n from $\{f(x, \theta), \theta \in \Omega \subset R_1\}$ where the range $S_\theta = \{x \mid f(x, \theta) > 0\}$ depends on θ. The prime examples of such a family are $U(0, \theta)$, uniform distribution over $(0, \theta)$ and exponential distribution with location θ with $f(x, \theta) = \exp \{- (x - \theta)\}$ for $x > \theta$. We have seen earlier in Example 2.4.2 that for $\{U(0, \theta), \theta > 0\}$, $X_{(n)}$ is minimal sufficient statistic for θ and the reader can verify that for the exponential distribution with location θ, $X_{(1)}$ is minimal sufficient for θ. The pdf of $U(0, \theta)$ has the form $f(x, \theta) = \frac{1}{\theta}$,

$0 < x < \theta$ and here the upper limit depends on θ where as for exponential distribution with location θ, $f(x, \theta) = e^{-x}/e^{-\theta}$, $\theta < x < \infty$ and the lower limit depends on θ. In both cases the pdf has the form $\dfrac{u(x)}{v(\theta)}$ over S_θ. We generalize this situation to define Pitman family.

Definition 2.6.1 Let $\{f(x, \theta), \theta \in \Omega \subset R_1\}$ be a family of pdfs such that $f(x, \theta) = \dfrac{u(x)}{v(\theta)}$, for $a(\theta) < x < b(\theta)$ where $a(\theta) < b(\theta)$ for every $\theta \in \Omega$ and Ω is an interval $(\underline{\theta}, \overline{\theta})$ then $\{f(x, \theta), \theta \in \Omega \subset R_1\}$ is a Pitman family.

(A) We now show that for $a(\theta) = a$, a constant, $X_{(n)}$ is minimal sufficient for θ and for the case $b(\theta) = b$, $X_{(1)}$ is minimal sufficient for θ.

Now $$L(x, \theta) = \frac{1}{[v(\theta)]^n} \prod_{i=1}^{n} u(x_i),\ a < x_i < b(\theta),\ i = 1, 2, \ldots, n$$

$$= \frac{1}{[v(\theta)]^n} \prod_{i=1}^{n} [u(x_i)\ \psi(x_i, a)\ \psi(b(\theta), x_i)],\ x \in R_n,\ \theta \in \Omega$$

where $\psi(c, d) = 1$ if $c > d$ and zero otherwise. Noting that $\prod_{i=1}^{n} \psi(b(\theta), x_i)$ $= \prod_{i=1}^{n} \psi(x_{(n)}, x_i)\ \psi(b(\theta), x_{(n)})$, $L(x, \theta) = \dfrac{1}{[v(\theta)]^n} [\pi u(x_i)\ \psi(x_i, a)\ \psi(x_{(n)}, x_i)]$ $\times\ \psi(b(\theta), x_{(n)})$. Let $x \in R_n$, $y \in R_n$, then $x \overset{L}{\sim} y$ iff

$$\psi(b(\theta), x_{(n)}) = \psi(b(\theta), y_{(n)}) \tag{2.6.1}$$

Now if $x_{(n)} \neq y_{(n)}$ then we can determine a value of θ such that $b(\theta)$ is between $x_{(n)}$ and $y_{(n)}$, and (2.6.1) can not hold. Hence $x \overset{L}{\sim} y$ iff $x_{(n)} = y_{(n)}$ or $\overset{L}{\sim}$ determines the statistic $T = X_{(n)}$. As already seen in Example 2.4.2, $X_{(n)}$ is minimal sufficient for θ.

(B) In a similar way if

$$f(x, \theta) = \frac{u(x)}{v(\theta)},\ a(\theta) < x < b \text{ then}$$

$$L(x, \theta) = \frac{1}{[v(\theta)]^n}\ \pi\ [u(x_i)\ \psi(b, x_i)\ \psi(x_i, a(\theta))],\ x \in R_n\ \theta \in \Omega$$

$$= \frac{1}{[v(\theta)]^n}\ \pi[u(x_i)\ \psi(b, x_i)\ \psi(x_i, x_{(1)})]\ \psi(x_{(1)}, a(\theta))$$

since $\pi\ \psi(x_i, a(\theta)) = \psi(x_{(1)}, a(\theta))$. Therefore

$$x \overset{L}{\sim} y \text{ iff } \psi(x_{(1)}, a(\theta)) = \psi(y_{(1)}, a(\theta)) \tag{2.6.2}$$

Again if $x_{(1)} \neq y_{(1)}$ we can determine a value of θ such that $a(\theta)$ is between $x_{(1)}$ and $y_{(1)}$ and (2.6.2) does not hold. Thus $\overset{L}{\sim}$ defines the minimal sufficient statistic $T = X_{(1)}$.

(C) When $a(\theta)$ and $b(\theta)$ both depend on θ then if $a(\theta) \downarrow$ and $b(\theta) \uparrow$, then $T(x) = \max \{a^{-1}(x_{(1)}), b^{-1}(x_{(n)})\}$ is minimal sufficient for θ. Here we have

$$L(x, \theta) = \frac{1}{[v(\theta)]^n} \pi[u(x_i)\ \psi(x_i, a(\theta))\ \psi(b(\theta), x_i)]$$

$$= \frac{1}{[v(\theta)]^n} \pi[u(x_i)\ \psi(x_i, x_{(1)})\ \psi(x_{(n)}, x_i)]\ \psi(x_{(1)}, a(\theta))\ \psi(b(\theta), x_{(n)})$$

Now observe that $a(\theta) < x_{(1)}$ iff $a^{-1}(x_{(1)}) < \theta$ since $a(\theta) \downarrow$

and $\qquad x_{(n)} < b(\theta)$ iff $b^{-1}(x_{(n)}) < \theta$ as $b(\theta) \uparrow$

Therefore, $a(\theta) < x_{(1)} < x_{(n)} < b(\theta)$ iff $\max \{a^{-1}(x_{(1)}), b^{-1}(x_{(n)})\} = T(x) < \theta$ and $\psi(x_{(1)}, a(\theta))\ \psi(b(\theta), x_{(n)}) = \psi(\theta, T(x))$. It then follows that $x \overset{L}{\sim} y$ iff $\psi(\theta, T(x)) = \psi(\theta, T(y))$. Again arguing as before we conclude that $T(x) = T(y)$ and $\overset{L}{\sim}$ defines the statistic $T = \max (a^{-1}(X_{(1)}), b^{-1}(X_{(n)}))$.

(D) When $a(\theta) \uparrow$ and $b(\theta) \downarrow$ then in a similar way we can show that $\overset{L}{\sim}$ defines the statistic $T = \min \{a^{-1}(x_{(1)}), b^{-1}(x_{(n)})\}$.

EXAMPLE 2.6.1 Consider $f(x, \theta) = \frac{1}{2\theta}$, $-\theta < x < \theta$. Here $S_\theta = (-\theta, \theta)$, $\Omega = (0, \infty)$ and $a(\theta) = -\theta \downarrow$ and $B(\theta) = \theta \uparrow$. Then $a^{-1}(x_{(1)}) = -x_{(1)}$ and $b^{-1}(x_{(n)}) = x_{(n)}$ and $T = \text{Max}\,(-X_{(1)}, X_{(n)})$ is minimal sufficient for θ. To obtain $g_0(t, \theta)$ consider the d.f. of T. Note that the range of T is $(0, \theta)$ and thus $G(t, \theta) = P[T \leq t] = 0$ if $t \leq 0$ and $= 1$ if $t > \theta$. For $0 < t \leq \theta$ we have

$$G(t, \theta) = \int_{-t}^{t} \int_{-t}^{x(n)} n(n-1)\,(x_{(n)} - x_{(1)})^{n-1} \frac{1}{(2\theta)^n}\, dx_{(1)}\, dx_{(n)}$$

Since $\max(-x_{(1)}, x_{(n)}) \leq t$ iff $-t < x_{(1)} < x_{(n)} < t$ and the joint pdf of $(X_{(1)}, X_{(n)})$ is the integrand above. Therefore

$$\begin{aligned} G(t, \theta) &= 0 && \text{if } t \leq 0 \\ &= t^n/\theta^n && \text{if } 0 < t \leq \theta \\ &= 1 && \text{if } t > \theta \end{aligned}$$

Observe that $G(t, \theta)$ is the d.f. of $Y_{(n)}$ where $(Y_1, \ldots, Y_n)$ are i.i.d. $U(0, \theta)$. Now we have $\max [-x_{(1)}, x_{(n)}] = \max (\,|x_1|, \ldots, |x_n|\,)$. Let $Y = |X|$ then as X varies over $(-\theta, \theta)$, Y varies over $(0, \theta)$ and $P[Y \leq y] = P[-y < X < y] = \int_{-y}^{y} \frac{1}{2\theta}\,dx = \frac{y}{\theta}$ for $0 < y < \theta$ and $(Y_1, \ldots, Y_n)$ are i.i.d. $U(0, \theta)$. The sufficient statistics for θ based on Y is

$$\max(Y_1, \ldots, Y_n) = \max(|X_1|, \ldots, |X_n|) = \max(-X_{(1)}, X_{(n)})$$

When $a(\theta)$ and $b(\theta)$ are not of the above four types then we show by an example that there is no one dimensional sufficient statistic for θ although $(X_{(1)}, X_{(n)})$ is minimal sufficient.

EXAMPLE 2.6.2 Let X_1, X_2 be i.i.d. $U(\theta, \theta + 1)$, $\theta \in R_1$ then $a(\theta) \uparrow$ and $b(\theta) \uparrow$ and $L(x, \theta) = 1$, $\theta < x_i < \theta + 1$, $i = 1, 2$ and if $X_{(1)}, X_{(2)}$ is the order statistic which is sufficient for θ, in view of the general result for samples on continuous r.v.

$$\begin{aligned} g_0(x_{(1)}, x_{(2)}, \theta) &= 2!\ \theta < x_{(1)} < x_{(2)} < \theta + 1 \\ &= 2!\ \psi(x_{(1)}, \theta)\ \psi(\theta, x_{(2)} - 1) \end{aligned}$$

Now

$$(x_{(1)}, x_{(2)}) \overset{L}{\sim} (y_{(1)}, y_{(2)}) \text{ iff } \psi(x_{(1)}, \theta)\ \psi(\theta, x_{(2)} - 1) = \psi(y_{(1)}, \theta), \psi(\theta, y_{(2)} - 1).$$

Now if either $x_{(1)} \neq y_{(1)}$ or $x_{(2)} \neq y_{(2)}$ we can find a value of θ in R_1 such that the above equality fails and therefore we have $(X_{(1)}, X_{(2)})$ is minimal sufficient.

It is an interesting exercise to obtain the distributions of the minimal sufficient statistic in the four cases discussed above and the reader is strongly recommended to derive these distributions using the distribution theory of order statistics. For further details we refer to Huzurbazar (1955).

3

Minimum Variance Unbiased Estimation

3.1 Unbiasedness

Consider a random sample of size n from $\{f(x, \theta), \theta \in \Omega \subset R_1\}$ and suppose that T is an estimator of the function $\psi(\theta)$ which is of interest. Let $\{g(t, \theta), \theta \in \Omega\}$ be the corresponding class of pdfs. Then we evaluate the performance of T as an estimator of $\psi(\theta)$ using the sampling distribution of T. The difference between $T(x)$ and $\psi(\theta)$ given by $e(x, \theta) = T(x) - \psi(\theta)$ for each fixed θ and given x is the residual or error. We regard errors of overestimation and underestimation on par and thus a natural quantity to look at is $| T(x) - \psi(\theta) |$ following Boscovitch and Laplace or $(T(x) - \psi(\theta))^2$ following Gauss and Legendre. Since $(T(x) - \psi(\theta))^2$ is a r.v. for each fixed θ, we consider $E_\theta[(T(x) - \psi(\theta))^2] = \text{MSE}(T|\theta)$ and prefer T_1 over T_2 if $\text{MSE}(T_1 | \theta) \leq \text{MSE}(T_2 | \theta)$ for $\forall\ \theta \in \Omega$. Recall that by Tchebychev's inequality for any estimator T, for any $\varepsilon > 0$

$$P_\theta[| T(x) - \psi(\theta) | < \varepsilon] \geq 1 - \frac{\text{MSE}(T | \theta)}{\varepsilon^2}$$

and $\text{MSE}(T | \theta)$ in a way indicates concentration of the distribution of T around $\psi(\theta)$ and we prefer an estimator with greater concentration around $\psi(\theta)$ which corresponds to smaller mean squared error. However just working with MSE criterion does not lead to the unique choice or recommendation of an estimator T as the following examples shows.

EXAMPLE 3.1.1 Consider $(X_1, \ldots, X_n)$ i.i.d. $N(\theta, 1)$, $T_1(x) = \bar{x}$ and $T_2(x) = \theta_0$. Then as $\bar{X} \sim N(\theta, 1/n)$ we have $\text{MSE}(T_1 | \theta) = 1/n$ and $\text{MSE}(T_2 | \theta) = (\theta - \theta_0)^2$. Now for values of $\theta \in \left(\theta_0 - \frac{1}{n}, \theta_0 + \frac{1}{n}\right)$, $\text{MSE}(T_2 | \theta) < \text{MSE}(T_1 | \theta)$ and for other values of θ we have $\text{MSE}(T_2 | \theta) > \text{MSE}(T_1 | \theta)$ and we cannot choose T_1 over T_2 or T_2 over T_1. Observe that T_1 is minimal sufficient where as T_2 does not depend on observations at all, and ignores the data completely.

Going back to Boscovitch's assumption that sums of residuals is zero and the consequent assumption on the error distributions discussed in Chapter 1, we could impose condition that the distribution of T should be symmetric around $\psi(\theta)$ and the centre of the distribution of T given by its first moment should coincide with $\psi(\theta)$, the parametric function that we want to estimate. Now if $E_\theta(T)$ exists and the distribution of T is symmetric around θ we have $E_\theta(T) = \psi(\theta)$ but $E_\theta(T) = \psi(\theta)$ does not necessarily imply that the distribution of T is symmetric around θ. We therefore impose the weaker condition namely

$$E_\theta(T) = \psi(\theta) \ \forall \ \theta \in \Omega \tag{3.1.1}$$

or $$\int g(t, \theta)\, dt = \psi(\theta) \ \forall \ \theta \in \Omega.$$

If T satisfies the above condition then we define T to be an unbiased estimator of $\psi(\theta)$. Usually there are several unbiased estimators of $\psi(\theta)$ and we denote by $U_{\psi(\theta)} = \{T \mid E_\theta(T) = \psi(\theta), \ \forall \ \theta \in \Omega\}$, the class of all unbiased estimators of $\psi(\theta)$.

EXAMPLE 3.1.2 Let $(X_1, \ldots, X_n)$ be i.i.d. $N(\theta, 1)$ then $\bar{X} \sim N(\theta, 1/n)$ and therefore $E_\theta(\bar{X}) = \theta \ \forall \ \theta \in R_1$ and $\bar{X} \in U_\theta$. Similarly if $T_a = \sum a_i X_i / \sum a_i$ then $T_a \sim N(\theta, \sum a_i^2/(\sum a_i)^2)$ and $T_a \in U_\theta$ for any $a \in R_n - \{0\}$. Since normal pdf is symmetric about mean, $T_a \in U_\theta$ has its pdf symmetric about θ.

On the other hand consider $(X_1, \ldots, X_n)$ i.i.d. exponential with mean θ having pdf $f(x, \theta) = \frac{1}{\theta} e^{-x/\theta}$, $x > 0$, $\theta > 0$. Let $T_1 = X_1$ then $T_1 \in U_\theta$ and the pdf of X_1 is same as $f(x, \theta)$ given above which is certainly not symmetric about θ. Further if $T_2 = \bar{X}$ then $T_2 \in U_\theta$ but its pdf given by $g(t_2, \theta) = \frac{n^n t_2^{n-1}}{\Gamma(n)\,\theta^n} e^{-nt_2/\theta}$, $t_2 > 0$, $\theta > 0$, is not symmetric about θ for moderate values of n. Note that due to central limit theorem we have T_2 is approximately normal with mean θ and variance θ^2/n and therefore for large n the pdf $g(t_2, \theta)$ is nearly symmetric about θ. Reader should verify that in case of $(X_1, \ldots, X_n)$ i.i.d. Poisson with mean θ, a similar phenomenon occurs.

We say that a parametric function $\psi(\theta)$ is estimable if $U_{\psi(\theta)}$ is not empty or there exists at least one unbiased estimator of $\psi(\theta)$. Note that if $U_{\psi(\theta)}$ contains two distinct elements T_1 and T_2 then it contains infinitely many as any convex combination of T_1, T_2 would also belong to $U_{\psi(\theta)}$. In general whether a function $\psi(\theta)$ is estimable or not is connected with the problem of solving the integral equation

$$\int T(x)\, L(x, \theta)\, dx = \psi(\theta), \ \forall \ \theta \in \Omega \tag{3.1.2}$$

where $L(x, \theta)$ and $\psi(\theta)$ are known and $T(x)$ is to be obtained. When $L(x, \theta)$ is pmf of a discrete r.v. (3.1.2) reduces to the equation

$$\sum_x \varphi(x)\, L(x, \theta) = \psi(\theta) \tag{3.1.3}$$

and particularly in case where $\{f(x, \theta), \theta \in \Omega\}$ is a power series distribution we can use the technique of equating coefficients of θ^r, $r = 0, 1, \ldots$ when $\psi(\theta)$ is an analytic function. We illustrate this by an example.

EXAMPLE 3.1.3 Let (X_1, X_2) be i.i.d. Bernoulli with $P(X_i = 1) = \theta$ and $P(X_i = 0) = 1 - \theta$. Then $L(x, \theta) = \theta^{x_1+x_2}(1-\theta)^{2-(x_1+x_2)}$. Consider $\psi(\theta) = \theta(1-\theta) = P(X_1 = 1 \text{ and } X_2 = 0)$. Then by using indicator function, define $T(x_1, x_2) = 1$ if $x_1 = 1$, $x_2 = 0$ and zero otherwise. Then $E(T) = \theta(1-\theta)$, $\forall\ \theta \in \Omega = (0, 1)$ and $\psi(\theta)$ is estimable. On the other hand let $\varphi(x_1, x_2)$ be an unbiased estimator of $\theta(1-\theta)$ then

$$\varphi(0, 0)(1-\theta)^2 + [\varphi(0, 1) +, \varphi(1, 0)]\,\theta(1-\theta) + \varphi(1, 1)\theta^2$$
$$= \theta(1-\theta) \qquad (3.1.4)$$

Since (3.1.4) holds for each $\theta \in (0, 1)$ coefficients of θ^r, $r = 0, 1, 2$ must be identical. Since θ^0 has zero coefficient on RHS we must have $\varphi(0, 0) = 0$. Equating coefficients of θ and θ^2 we have

$$\varphi(0, 1) + \varphi(1, 0) = 1$$

and
$$\varphi(1, 1) - [\varphi(0, 1) + \varphi(1, 0)] = -1$$

This gives $\varphi(1, 1) = 0$ and $\varphi(0, 1) + \varphi(1, 0) = 1$. The estimator $T(x_1, x_2)$ given above is a particular case of the general solution (corresponding to $\alpha = 0$) given by

$$\varphi(0, 0) = 0 = \varphi(1, 1)$$

and
$$\varphi(0, 1) = \alpha,\ \varphi(1, 0) = 1 - \alpha$$

where
$$\alpha \in R_1 \qquad (3.1.5)$$

Indeed (3.1.5) defines the entire class of unbiased estimators of $U_{\theta(1-\theta)}$.

Suppose we now take $\psi(\theta) = \theta^3$ then RHS of (3.1.4) is θ^3. Since the coefficient of θ^3 on RHS is unity and that on LHS is zero it follows that no matter how we define $\varphi(x_1, x_2)$, we can not have

$$\varphi(0, 0)(1-\theta)^2 + [\varphi(0, 1) + \varphi(1, 0)]\theta(1-\theta) + \varphi(1, 1)\theta^2$$
$$= \theta^3,\ \forall\ \theta \in (0, 1).$$

Therefore $\psi(\theta) = \theta^3$ is not estimable. However if instead of a sample of size two we have a sample of size three then we can show that θ^3 is estimable and U_{θ^3} is given by

$$\varphi(0, 0, 0) = 0,\ \varphi(1, 0, 0) + \varphi(0, 1, 0) + \varphi(0, 0, 1) = 0$$

$$\varphi(1, 1, 0) + \varphi(1, 0, 1) + \varphi(0, 1, 1) = 0 \text{ and } \varphi(1, 1, 1) = 1.$$

In the case of continuous r.v. X, the problem is more difficult and one has to use the technique of Laplace Transforms among others to obtain the solution and we will not go into further details.

Exercise 3.1.1 (1) Consider a Poisson distribution with zero class missing i.e. a discrete r.v. X with pmf $f(x, \lambda) = \frac{\lambda^x}{x!}\frac{1}{e^{\lambda} - 1}$, $x = 1, 2, \ldots$. This model arises in a Poisson situation in which the event $X = 0$ is not observable. Show that the parameter λ is estimable by solving the equation $\sum_{x=1}^{\infty} \frac{\lambda^x}{x!} T(x) = \lambda(e^{\lambda} - 1)$. Expand $\lambda(e^{\lambda} - 1)$ in a power series and equate coefficients to obtain $T(1) = 0$, $T(x) = x$, $x = 2, 3. \ldots,$.

(2) In the above case show that $\frac{1}{\lambda}$ is not estimable.

(3) Let $(X_1, \ldots, X_n)$ be i.i.d. exponential with mean θ and let $T = \sum_{i=1}^{n} X_i$ be minimal sufficient statistic for θ with pdf given by $g(t, \theta) = \frac{1}{\Gamma(n)\theta^n} t^{n-1}e^{-t/\theta}$, $t > 0$, $\theta > 0$. Show that for any $r > 0$, θ^r is estimable with $u_r(t) = \frac{\Gamma(n)}{\Gamma(n+r)} t^r$ as an unbiased estimator of θ^r. Let $P_k(\theta) = \sum_{r=0}^{k} a_r \theta^r$ be a polynomial degree k in θ. Then $\sum_{r=0}^{k} a_r u_r(t) \in U_{P_k(\theta)}$.

(4) In the above example show that $\frac{n-1}{n}\frac{1}{t} \in U_{1/\theta}$ if $n > 1$ (For $n = 1$, using Laplace transform theory one can show that $\frac{1}{\theta}$ is not estimable i.e. $U_{1/\theta}$ is empty).

We point out that Gauss had used a different definition of unbiasedness in the problem of direct or indirect measurements on the "magnitudes of interest". Thus in the simplest linear model $X_i = \theta + \varepsilon_i$, $i = 1, 2, \ldots, n$, Gauss defined an estimator $T(x_1, \ldots, x_n)$ to be unbiased for θ, if $T(\theta, \ldots, \theta) = \theta$. Thus if we happen to observe θ without any error i.e. $\varepsilon_i = 0$, $i = 1, 2, \ldots, n$ then the estimator T must coincide with θ. When ε_i's are continuous $P(\varepsilon_i = 0) = 0$, $i = 1, 2, \ldots, n$. However if ε_i are discrete then we can have an estimator T which is Gauss unbiased but not unbiased as defined by the condition $E(T) = \theta$, $\forall\ \theta \in \Omega$. For example if error distribution is $P(\varepsilon_i = -1) = \frac{1}{6}$, $P(\varepsilon_i = +1) = \frac{2}{6}$ and $P(\varepsilon_i = 0) = \frac{1}{2}$, then $\bar{X}$ is Gauss unbiased but $\bar{X} \notin U_{\theta}$.

3.2 Best Linear Unbiased Estimator (BLUE)

In the context of repeated direct measurement model, where $X_i = \theta + \varepsilon_i$, $i = 1, 2, \ldots, n$, Gauss proved an interesting and important result that within the class of linear unbiased estimators, $\bar{X}$ is BLUE. The result holds under the assumption that ε_i, $i = 1, 2, \ldots, n$ are independent with $E(\varepsilon_i) = 0$ and Var $(\varepsilon_i) = \sigma^2$ which implies that $E(X_i) = \theta$ and Var $(X_i) = \sigma^2$. Consider the sub-class of U_θ, namely linear unbiased estimators, $U_\theta(a) = \{\sum a_i X_i \mid \sum a_i = 1\}$. Then $\bar{X}$ is "Best" within $U_\theta(a)$ in the sense that for any $T(a) \in U_\theta(a)$, we have Var $(\bar{X}) = \frac{\sigma^2}{n} \leq$ Var $(T(a)) = \sum a_i^2 \sigma^2$. The result is easy to prove. The restriction $\sum a_i = 1$ follows from the requirement that $E(\sum a_i X_i) = \theta \sum a_i = \theta$, $\forall\ \theta \in \Omega$ which implies that $\sum a_i = 1$. Now Var $(T(a)) = \sigma^2 \sum a_i^2$ and is to be minimized subject to a linear restriction that $\sum a_i = 1$. One may use Cauchy-Schwartz inequality $(\Sigma\, a_i b_i)^2 \le \Sigma\, a_i^2 \cdot \Sigma\, b_i^2$ or use Lagrange's method of undetermined multipliers to show that the minimum of $\Sigma\, a_i^2$ subject to constraint $\Sigma\, a_i = 1$ occurs at $a_1 = a_2 = \ldots = a_n = \frac{1}{n}$.

We point out the fact that the above result of Gauss, also known as Gauss-Markov theorem, does not assume a fully parametric model for the error distribution or equivalently no specific distributional form for the distribution of X_i is assumed. The only assumptions made are $E(X_i) = \theta$ and Var $(X_i) = \sigma^2$ and X_i's are uncorrelated but not necessarily mutually independent and even not necessarily i.i.d. as they can differ in higher order moments. Observe that if we use the definition of unbiasedness as originally proposed by Gauss then $T(a) = \Sigma\, a_i X_i$ is unbiased if $\Sigma\, a_i\ \theta = \theta$, $\forall\ \theta \in \Omega$ or $\Sigma\, a_i = 1$. In this case also we minimize $\Sigma\, a_i^2$ subject to $\Sigma\, a_i = 1$ and $\bar{X}$ is BLUE of θ.

EXAMPLE 3.2.1 Let $X_i = \theta + \varepsilon_i$ where ε_i are i.i.d. with zero expectation and finite second order moments. These assumptions allow various distributions such as Normal, Laplace, Uniform, and Euler.

(1) $f(\varepsilon) = \frac{1}{\sqrt{2\pi\sigma^2}} \exp\{-\varepsilon^2/2\sigma^2\}\ \varepsilon \in R_1\ E(\varepsilon) = 0$ Var $(\varepsilon) = \sigma^2$

(2) $f(\varepsilon) = \frac{1}{2\sigma} \exp\{-|\varepsilon|/\sigma\},\ \varepsilon \in R_1\ E(\varepsilon) = 0$ Var $(\varepsilon) = 2\sigma^2$

(3) $f(\varepsilon) = \frac{1}{2\sigma},\ -\sigma < \varepsilon < \sigma \qquad E(\varepsilon) = 0$ Var $(\varepsilon) = \sigma^2/3$

(4) $f(\varepsilon) = \frac{3}{4\sigma^3}(\sigma^2 - \varepsilon^2),\ -\varepsilon < \sigma < \varepsilon,\ E(\varepsilon) = 0$ Var $(\varepsilon) = \sigma^2/5$

For all the above distributions we have $\bar{X}$ is BLUE of θ. On the other hand the result is not applicable if errors are Cauchy distributed with pdf $f(\varepsilon) = \frac{\sigma}{\pi}\, \frac{1}{\sigma^2 + \varepsilon^2},\ \varepsilon \in R_1$.

EXAMPLE 3.2.2 The result that $\bar{X}$ is BLUE of θ holds when $(X_1, \ldots, X_n)$ are i.i.d. Poisson with mean θ, or exponential with mean θ. As seen earlier the distribution of X is not symmetric around θ and the observations are not arising out of repeated measurements model $x_i = \theta + \varepsilon_i$.

The restriction to the class of linear unbiased estimators may create some problems in that such a sub-class of U_θ may be empty but U_θ itself may not be empty. We consider two examples to illustrate this point.

EXAMPLE 3.2.3 Let $(X_i, \ldots, X_n)$ be i.i.d. with pdf $f(x, \theta) = \exp\{-(x - \theta)\}$, $x \ge \theta$ and $\theta \in R_1$. Then $E(X - \theta) = 1$ and $E(X) = \theta + 1$ or $E(X - 1) = \theta$. Hence U_θ is not empty. However if we restrict to class of linear estimators $T(a) = \sum_{i=1}^{n} a_i X_i$ then $E(\Sigma\, a_i X_i) = \Sigma\, a_i(\theta - 1) = \Sigma\, a_i \theta - \Sigma\, a_i$. Then $T(a) \in$

U_θ iff $\theta \Sigma a_i - \Sigma a_i = \theta$, $\forall\ \theta \in R_1$ or iff $\Sigma a_i = 1$ and $\Sigma a_i = 0$ which is impossible. This proves the result that the class of linear unbiased estimators is empty. Of course we can generalize the class of linear estimators by defining it to consist of $T_1 = a_0 + \sum_{i=1}^{n} a_i X_i$, $(a_0, a_1, \ldots, a_n), \in R_{n+1}$ i.e. allowing non-homogeneous linear estimators. In this problem one can show that $(\bar{X} - 1)$ is BLUE of θ in the larger class of non-homogeneous linear unbiased estimators of θ.

EXAMPLE 3.2.4 Let $(X_1, \ldots, X_n)$ be i.i.d. with pdf $f(x, \theta) = \theta x^{\theta-1}$, $0 < x < 1$, and $0 < \theta < 1$. Then $E(X) = \frac{\theta}{\theta+1}$ and $E(T(a_0, \ldots, a_n)) = a_0 + \frac{\theta}{\theta+1} \Sigma a_i$ and $T(a_0, \ldots, a_n) \in U_\theta$ iff $a_0 (\theta + 1) + \theta \Sigma a_i = \theta(\theta + 1)$ for every $\theta \in (0, 1)$. On comparing coefficients of θ^2 this leads to contradiction $1 = 0$ and thus the class of non-homogeneous (and therefore homogeneous) linear unbiased estimators is empty. On the other hand U_θ is not empty if $n \geq 2$. This follows from the fact that $Y_i = -\log X_i$ are i.i.d. exponential with mean $\frac{1}{\theta}$ and therefore $T = \Sigma Y_i$ is Gamma $\left(n, \frac{1}{\theta}\right)$ and $E\left(\frac{n-1}{T}\right) = \theta$. Note that T is also minimal sufficient as $\{f(x, \theta), \theta \in (0, 1)\}$ is a one parameter exponential family. Note that $\frac{n-1}{T}$ is not a linear estimator even in $(Y_1, \ldots, Y_n) = (\log X_1, \ldots, \log X_n)$.

Another difficulty with the BLUE is the fact that where as $\bar{X}$ may be BLUE of θ there may exist non-linear estimators in U_θ whose variance could be smaller than that of $\bar{X}$. Again we illustrate this situation by an example.

EXAMPLE 3.2.5 Let (X_1, X_2, X_3) be i.i.d. $U(\theta - 1, \theta + 1)$ then $f(x, \theta) = \frac{1}{2}$, for $\theta - 1 < x < \theta + 1$ and $E(X) = \theta$ and Var $(X) = \frac{1}{3}$. Therefore $\bar{X}$ is BLUE of θ with Var $(\bar{X}) = \frac{1}{9}$. Consider the order statistic $X_{(1)}, X_{(2)}, X_{(3)}$. We consider $T_1 = (X_{(1)} + X_{(3)})/2$ and $T_2 = X_{(2)}$ as estimators of θ. Let $Y = \frac{X - \theta + 1}{2}$ then $Y \sim U(0, 1)$ and let $(Y_{(1)}, Y_{(2)}, Y_{(3)})$ correspond to $(X_{(1)}, X_{(2)}, X_{(3)})$. Then it is easy to show that $E(Y_{(1)}) = \frac{1}{4}$, $E(Y_{(2)}) = \frac{1}{2}$, $E(Y_{(3)}) = \frac{3}{4}$. Therefore both $T_1 = \frac{1}{2}(X_{(1)} + X_{(3)})$ and $T_2 = X_{(2)}$ are unbiased for θ. Now we can show that Var $(Y_{(1)}) = \frac{3}{80}$ = Var $(Y_{(3)})$ and Cov $(Y_{(1)}, Y_{(3)}) = \frac{1}{80}$. Using the relation $X_{(i)} = 2Y_{(i)} + \theta - 1$, we obtain Var $\frac{(X_{(1)} + X_{(3)})}{2} = \frac{1}{10}$ which is smaller than that of BLUE $\bar{X}$ which is $\frac{1}{9}$. Note that Var $(X_{(2)})$ =

4 Var $(Y_{(2)}) = \frac{1}{5}$ and although $X_{(2)} \in U_\theta$, Var $(X_{(2)}) >$ Var $(\bar{X}) >$ Var $\left(\frac{(X_{(1)} + X_{(3)})}{2}\right)$.

The approach based on BLUE has its roots in linear models where it is natural to restrict attention to linear unbiased estimators of parameters of interest. Particularly when the errors are assumed to be normally distributed, the method of least squares leads to the BLUE which also happens to be Minimum Variance Unbiased (MVU) estimator. If the errors are not normal then of course this result is not true as the above examples show. In the next section we consider the problem of MVU estimation of θ (or of $\psi(\theta)$) using the approach based on Cramer-Rao Lower Bound (CRLB) to the variance of an unbiased estimator.

3.3 Cramer-Rao Inequality and Its Applications

Let X be a random vector $(X_1, \ldots, X_n)'$ with joint pdf of X belonging to the class $\{f(x, \theta), \theta \in \Omega \subset R_1\}$. Assume that this class satisfies the regularity conditions given in Sec. 2.2 so that Fisher Information $I_X(\theta)$ is well defined. Recall that these conditions are

(1) $S_\theta = \{x \mid f(x, \theta) > 0\}$ does not depend on θ or $S_\theta = S$.

(2) The identity $\int_S f(x, \theta)\, dx = 1 \;\forall\; \theta \in \Omega$ can be differentiated under integral sign twice.

Consider $U_\psi = \{T(x) \mid E(T(x)) = \psi(\theta), \forall\; \theta \in \Omega\}$ which we assume to be non-empty. We will now establish the Cramer-Rao Lower Bound (CRLB) for Var (T) under the following additional regularity condition:

(3) $T \in U_\psi$ is such that differentiation under integral sign is valid at least once or for every $\theta \in \Omega$ we have

$$\frac{\partial}{\partial\theta}\int_S T(x) f(x, \theta)\, dx = \int_S T(x) \cdot \frac{\partial \log f(\theta)}{\partial\theta},\; f(x, \theta)\, dx = \frac{\partial\psi}{\partial\theta} \tag{3.3.1}$$

THEOREM 3.3.1 Under the above regularity conditions, we have

$$\text{Var}\,(T) \geq \left(\frac{d\psi}{d\theta}\right)^2 \Big/ I_X(\theta)$$

Noting that $E\left(\frac{\partial \log f(x, \theta)}{\partial\theta}\right) = 0$ the condition (3) implies that

$$\text{Cov}\left(\frac{\partial \log f}{\partial\theta}, T\right) = \frac{d\psi}{d\theta}.$$

Now using the fact that $(\text{Cov})^2 \leq$ product of variances, we have

$$\left(\frac{d\psi}{d\theta}\right)^2 \leq \text{Var}\,(T)\, I_X(\theta),$$

since $\text{Var}\left(\frac{\partial \log f}{\partial \theta}\right) = I_X(\theta)$. We have therefore

$$\text{Var}(T) \geq \left(\frac{d\psi}{d\theta}\right)^2 \Big/ I_X(\theta) \tag{3.3.2}$$

The RHS of (3.3.2) is called as the CRLB for Var (T).

We have already seen in the Section 2.2 on Fisher Information, that verifying the condition (2) above is not always easy. In a similar way verifying the condition (3) of differentiation under integral sign in $E_\theta(T) = \psi(\theta)$ is not always easy. Further the CRLB of variance of an unbiased estimator $T \in U_\psi$ would be a valid lower bound for the class U_ψ, as opposed to an individual element $T \in U_\psi$, provided the condition (3) holds for every $T \in U_\psi$. Thus we assume that the condition (3) holds for every $T \in U_\psi$ as condition (4).

THEOREM 3.3.2 If conditions (1) through (4) hold then for any $T \in U_\psi$, $\text{Var}(T) \geq \left(\frac{d\psi}{d\theta}\right)^2 \Big/ I_X(\theta)$ and if there exists a $T^* \in U_\psi$ for which the equality holds, then T^* is MVUE of $\psi(\theta)$.

The above theorem presents a complete solution to the problem of MVUE in case the model is such that the regularity conditions (1) through (4) are satisfied and existence of T^* is assured.

EXAMPLE 3.3.1 As an example we consider the case of a random sample of size n from $b(1, \theta)$ or $(X_1, \ldots, X_n)$ are i.i.d. with $P(X_i = 1) = \theta$, $P(X_i = 0) = 1 - \theta$, $0 < \theta < 1$. Then the joint pmf of X is $f(x, \theta) = \theta^{\Sigma x_i}(1 - \theta)^{n - \Sigma x_i}$, $x_i = 0, 1$, $i = 1, 2, \ldots, n$, $0 < \theta < 1$. Here S_θ is the set of all 2^n sequences of zeroes and ones and the condition (1) holds. Further the identity $\sum_{x \in s} T(x)\theta^{\Sigma x_i}(1 - \theta)^{n - \Sigma x_i} = \theta$ has 2^n terms on LHS. Therefore, differentiation under the summation sign is valid and the conditions (1) through (4) hold. Therefore for any $T \in U_\theta$, $\text{Var}(T) \geq \frac{\theta(1 - \theta)}{n}$ since $\frac{d\psi}{d\theta} = 1$ and $I_X(\theta) = \frac{n}{\theta(1 - \theta)}$. Now in view of the fact that $(X_1, \ldots, X_n)$ are i.i.d. with $E(X_i) = \theta$ and $\text{Var}(X_i) = \theta(1 - \theta)$, $\bar{X}$ is BLUE of θ with $\text{Var}(\bar{X}) = \frac{\theta(1 - \theta)}{n}$ and thus $\bar{X}$ is MVUE. In this case S consisted of a finite number of elements and the conditions (3) and (4) were easy to verify as the derivative of a sum of finitely many terms is the sum of their derivatives. However in the continuous case such as $N(\theta, 1)$ or the Poisson case wherein an infinite series is involved, the verification of conditions (3) and (4) is not easy. We will however show in the following example that for a sample of size n in $N(\theta, 1)$ the conditions (1) through (4) hold and $\bar{X}$ is in fact MVUE of θ.

EXAMPLE 3.3.2 Let $(X_1, \ldots, X_n)$ be i.i.d. $N(\theta, 1)$ then using the results of Section 2.2 we claim that conditions (1) and (2) hold and $I_X(\theta) = n$. Now consider any $T(x) \in U_\theta$. Then the fact that $\Sigma\,(x_i - \theta)^2 = \Sigma\,(x_i - \bar{x})^2 + n(\bar{x} - \theta)^2$ and $|A_1| = \left| T(x) \dfrac{L(x, \theta_0 + h) - L(x, \theta_0)}{h} \right|$, using arguments similar to Example 2.2.2 we have for $0 < |h| \le \delta$

$$|A_1| \le |T(x)| \exp\left\{\frac{-\Sigma\,(x_i - \bar{x})^2}{2}\right\} \exp\left\{\frac{-n(\bar{x} - \theta_0)^2}{2} + n|\bar{x} - \theta_0|\right\}$$

$$= G_0(x)$$

$$\int_{R_n} G_0(x)\,dx \le \frac{1}{\delta}\left\{\int_{R_n} |T(x)|\,L(x, \theta_0 + \delta)\,dx + \int_{R_n} |T(x)|\,L(x, \theta_0 - \delta)\,dx\right\} \tag{3.3.3}$$

Since $E(T(x)) = \theta \;\forall\; \theta \in R_1$ the integrals on RHS of (3.3.3) exist and therefore $G_0(x)$ is integrable and the conditions (3) and (4) hold. Therefore for any $T \in U_\theta$, Var $(T) \ge \dfrac{1}{n}$. The CRLB of $\dfrac{1}{n}$ is attained by Var $(\bar{X})$ and thus $\bar{X}$ is not only BLUE but is also MVUE.

We recommend to an enterprising reader to work out similar results for a sample of size n from Poisson with mean θ and also for exponential distribution with mean θ to show that CRLB for estimating θ is attained by $\bar{X}$ and $\bar{X}$ is not only BLUE but is also MVUE of θ. Indeed one can show that for random sample of size n from a distribution belonging to one parameter exponential family, the CRLB result holds but we will not pursue the problem.

Suppose that the model $\{L(x, \theta), \theta \in \Omega\}$ and $T \in U_\psi$ are such that CRLB result holds. Then we call T^* to be a Minimum Variance Bound Unbiased (MVBU) estimator of $\psi(\theta)$ provided

$$\text{Var}\,(T^*) = \left(\frac{d\psi}{d\theta}\right)^2 \Big/ I_X(\theta) \tag{3.3.4}$$

Now Var (T^*) attains the CRLB iff $\left[\text{cov}\left(T^*, \dfrac{\partial \log L}{\partial \theta}\right)\right]^2 = \text{Var}\,(T^*)\,I_X(\theta)$ or the correlation coefficient between T^* and $\dfrac{\partial \log L}{\partial \theta}$ is $\pm$ 1. Thus (3.3.4) holds iff $\dfrac{T^* - \psi(\theta)}{\sqrt{\text{Var}\,(T^*)}} = \pm \dfrac{\partial \log L}{\partial \theta} \Big/ \sqrt{I_X(\theta)}$. Noting that $\sqrt{\text{Var}\,(T^*)} = \left|\dfrac{d\psi}{d\theta}\right| \Big/ \sqrt{I_X(\theta)}$ we must have

$$T^* = \psi(\theta) \pm \frac{\left|\frac{\partial \psi}{d\theta}\right|}{I_X(\theta)} \frac{\partial \log L}{\partial \theta}. \tag{3.3.5}$$

Observe that the RHS of (3.3.5) can be computed once the model $\{L(x, \theta), \theta \in \Omega\}$ and the function $\psi(\theta)$ is specified and we can immediately check whether or not there exists a T^* which is MVBUE for $\psi(\theta)$.

EXAMPLE 3.5.3 Let $(X_1, \ldots, X_n)$ be i.i.d. Poisson (θ) then

$$L(x, \theta) = e^{-n\theta} \frac{\theta^{\Sigma x_i}}{\pi x_i!} \text{ and } \frac{\partial \log L}{\partial \theta} = -n + \frac{\Sigma x_i}{\theta} = \frac{n(\bar{x} - \theta)}{\theta}.$$

Further $I_X(\theta) = \frac{n}{\theta}$. Suppose we want to estimate $\psi(\theta) = \theta$ then $\frac{\partial \psi}{d\theta} = 1$ and equation (3.3.5) becomes $T^* = \theta \pm \frac{\theta}{n} \frac{n(\bar{x} - \theta)}{\theta}$. If we take positive sign on RHS, $T^* = \bar{x}$ and therefore $\bar{X}$ attains CRLB and is BLUE, MVBUE as well as MVUE of θ. On the other hand suppose we want to estimate $\psi(\theta) = e^{-\theta} = P[X = 0]$ then estimator defined by $T(x) = 1$ if $X_1 = 0$ and zero otherwise is unbiased for $\psi(\theta)$ and U_ψ is not empty. Further $\left|\frac{d\psi}{d\theta}\right| = e^{-\theta}$. Therefore equation (3.3.5) becomes $T^* = e^{-\theta} \pm \frac{e^{-\theta}\,\theta}{n} \cdot \frac{n(\bar{x} - \theta)}{\theta}$ and whether we choose positive or negative sign on RHS, in both cases T^* is a function of $\bar{x}$ and θ and it is not a statistic. Hence we conclude that there does not exist an MVBU estimator of $\psi(\theta) = e^{-\theta}$. The CRLB for $\psi(\theta) = e^{-\theta}$ however is well defined and is given by $\theta e^{-2\theta}/n$. The question whether there exists MVUE in U_ψ remains however open. Another way to write (3.3.5) is to observe that

$$\frac{\partial \log L}{\partial \theta} = \Lambda(\theta)\,[T^* - \psi(\theta)] \tag{3.3.6}$$

where
$$\Lambda(\theta) = \left\{\left|\frac{d\psi}{d\theta}\right| \Big/ I_X(\theta)\right\}^{-1}.$$

Assuming that $\Lambda(\theta)$, $\psi(\theta)$ and $\Lambda(\theta)\,\psi(\theta)$ are integrable w.r.t. θ, integrating (3.3.6) w.r.t. θ, we get

$$\log L = u(\theta)\, T^*(x) + v(\theta) + w(x) \text{ where}$$

$$u(\theta) = \int \Lambda(\theta)\, d\theta \text{ and } v(\theta) = -\int \Lambda(\theta)\, \psi(\theta)\, d\theta.$$

Thus $\{L(x, \theta), \theta \in \Omega \subset R_1\}$ is a one-parameter exponential family with $T^*(x)$ as minimal sufficient statistic such that $E(T^*(x)) = -\frac{dv}{d\theta} \Big/ \frac{du}{d\theta} = \psi(\theta)$.

Conversely if $\{L(x(\theta)), \theta \in \Omega\}$ is a one-parameter exponential family with pdf given by

$$\log L\ (x,\ \theta) = u(\theta)\ T(x) + v(\theta) + w(x) \tag{3.3.7}$$

then one can show that $\psi(\theta) = E(T(x)) = -\dfrac{dv}{d\theta}\Big/\dfrac{du}{d\theta}$, $T(x)$ is MVBUE as well as MVUE of $\psi(\theta)$. We can show that conditions (1) and (2) hold and we have

$$\frac{\partial \log L}{\partial \theta} = u'(\theta)[T(x) - \psi(\theta)] \text{ where } u' \text{ denotes the derivative of } u \text{ w.r.t. } \theta.$$

$$\frac{\partial^2 \log L}{\partial \theta^2} = u''(\theta)\ [T(x) - \psi(\theta)] - u'(\theta)\frac{d\psi}{d\theta}.$$

Hence $I_X(\theta) = E\left(\dfrac{-\partial^2 \log L}{\partial \theta^2}\right) = u'(\theta)\dfrac{d\psi}{d\theta}$ since $E(T(x)) = \psi(\theta)$. But $E\left(\dfrac{\partial \log L}{\partial \theta}\right)^2 = I_X(\theta) = [u'(\theta)]^2 \text{ Var } (T(x))$ therefore

$$\text{Var } (T(x)) = \frac{I_X(\theta)}{[u'(\theta)]^2} = \frac{I_X(\theta)}{[I_X(\theta)]^2\Big/\left(\dfrac{d\psi}{d\theta}\right)^2} = \left(\frac{d\psi}{d\theta}\right)^2\Big/ I_X(\theta).$$

A detailed proof of the above result particularly the proof that for $T_1 \in U_\psi$ conditions (3) and (4) hold will not be attempted here. An enterprising reader can attempt this by following techniques of Examples 2.2.2 and 3.3.2 which show that for $N(\theta, 1)$ family conditions (1) through (4) hold. The techniques illustrated in the above examples are to be applied after making a one-one parametric transformation $\phi = u(\theta)$ and writing (3.3.7) in the form

$$\log L_1(x,\ \phi) = \phi\ T(x) + v_1(\phi) + w(x).$$

It now follows that even in one-parameter exponential family the only parametric function which admits an MVU estimator whose variance attains the CRLB, is the function $\psi(\theta) = E(T(x))$, where T is minimal sufficient statistic. As mentioned in Example 3.3.3, the question still remains open as to whether a function $\eta(\phi)$ admits an MVU estimator which is not MVBUE i.e. whose variance does not attain CRLB. In the next two sections we will present a partial solution to the problem of MVU estimation using Rao-Blackwell-Lehmann Scheffè approach.

3.4 Rao-Blackwell Theorem

This theorem was a notable advance towards the solution of the problem of MVU estimation of an estimable parametric function $\psi(\theta)$. Suppose that the family $\{L(x,\ \theta),\ \theta \in \Omega\}$ is such that it admits a sufficient statistic T with corresponding pdf $\{g(t,\ \theta),\ \theta \in \Omega\}$. Let $T_1 \in U_\psi$, then first we observe that if we take conditional expectation of T_1 given $T = t$ then $E(T_1 \mid T = t) = \varphi_1(t)$

does not depend on θ for each t and therefore $E(T_1 \mid T) = \varphi_1(T)$ is a statistic. We then show that $\varphi_1(T) \in U_\psi$ and further Var $(T_1) \geq$ Var $(\varphi_1(T))$ for every $\theta \in \Omega$. Thus $\varphi_1(T)$ is an improved estimator obtained from T_1 by taking conditional expectation for fixed T, a process called as Rao-Blackwellization of T_1 w.r.t. a sufficient statistic T. Now if T is not minimal sufficient and M is minimal sufficient statistic then we can apply the process of Rao-Blackwellization to $T_2 = \varphi_1(T)$ and obtain $E(T_2 \mid M) = \varphi_2(M)$ such that $\varphi_2(M) \in U_\psi$ and Var $(\varphi_1(T)) \geq$ Var $(\varphi_2(M))$ for each $\theta \in \Omega$. Thus the theorem implies that the search for the MVU estimator of $\psi(\theta)$ be restricted to those elements of U_ψ which depend on X only through the minimal sufficient statistic M, or MVU estimator of $\psi(\theta)$ must be a function of M. Now we state and prove Rao-Blackwell Theorem.

THEOREM 3.4.1 Let $\{L(x, \theta), \theta \in \Omega\}$ be such that a statistic M is minimal sufficient for θ and U_ψ is nonempty. Let $T_1 \in U_\psi$ and $\varphi_1(M) = E(T_1 \mid M)$. Then

(a) $\varphi_1(M) \in U_\psi$ and (b) Var $(\varphi_1(M)) \leq$ Var (T_1), $\forall\ \theta \in \Omega$.

Let $g(t, m, \theta)$ be the joint pdf of T_1 and M. Then as M is sufficient we have $g(t_1, m, \theta) = g_0(m, \theta)\, g_1(t_1 \mid m)$ for each $\theta \in \Omega$. Now $\varphi_1(m) = E(T_1 \mid M = m)$ $= \int t_1 g_1(t_1 \mid m)\, dt_1$ and therefore

$$E(\varphi_1(M)) = \int [\int t_1 g_1(t_1 \mid m)\, dt_1]\, g_0(m, \theta)\, dm$$

$$= \iint t_1 g(t_1, m, \theta)\, dt_1 dm$$

$$= E(T_1) = \psi(\theta) \text{ and } \varphi_1(M) \in U_\psi.$$

Now

$$\text{Var}\ (T_1) = \iint (t_1 - \psi(\theta))^2\, g(t_1, m, \theta)\, dt_1\, dm$$

$$= \iint [t_1 - \varphi_1(m) + \varphi_1(m) - \psi(\theta)]^2\, g_1(t_1 \mid m)\, g_0(m, \theta)\, dt_1\, dm.$$

Expanding the square in the integrand we have RHS $= I_1 + I_2 + I_3$. Consider each I_r, $r = 1, 2, 3$ separately.

$$I_1 = \iint (t_1 - \varphi_1(m))^2\, g_1(t_1 \mid m)\, g_0(m, \theta)\, dt_1 dm$$

$$= \int [\int (t_1 - \varphi_1(m))^2\, g_1(t_1 \mid m)\, dt_1]\, g_0(m, \theta)\, dm$$

$$= E[\text{Var}\ (T_1 \mid M = m)].$$

Note that $\int (t_1 - \varphi_1(m))^2\, g_1(t_1 \mid m)\, dt_1 = \text{Var}\,(T_1 \mid M = m)$ is the conditional variance of T_1 given $M = m$ since $E(T_1 \mid M = m) = \varphi_1(m)$.

Similarly, let

$$I_2 = \iint (\varphi_1(m) - \psi(\theta))^2\, g_1(t_1 \mid m)\, g_0(m, \theta)\, dt_1 dm$$

$$= \int [\varphi_1(m) \;-\; \psi(\theta)]^2\, g_0(m, \theta)\, dm$$

$$= \text{Var}\,[\varphi_1(M)].$$

Now consider

$$I_3 = 2\iint [t_1 - \varphi_1(m)]\,[\varphi_1(m) - \psi(\theta)]\, g_1(t_1 \mid m)\, g_0(m, \theta)\, dt_1 dm$$

$$= 2\int [\varphi_1(m) - \psi(\theta)][\int (t_1 - \varphi_1(m))\, g_1(t_1 \mid m)\, dt_1]\, g_0(m, \theta)\, dm$$

$$= 0$$

Now $\int (t_1 - \varphi_1(m))\, g_1(t_1 \mid m)\, dt_1 = 0$ for each fixed m follows from the fact that $E(T_1 \mid M = m) = \varphi_1(m)$ for each fixed m. Thus $\text{Var}\,(T_1) = \text{Var}\,(\varphi_1(M)) + E[\text{Var}\,(T_1 \mid M = m)]$. As $\text{Var}\,(T_1 \mid M = m) \geq 0$ we have $E[\text{Var}\,(T_1 \mid M = m)] \geq 0$ and therefore

$$\text{Var}\,(T_1) \geq \text{Var}\,(\varphi_1(M)) \tag{3.4.1}$$

Observe that equality holds in (3.4.1) iff $\text{Var}\,(T_1 \mid M = m) = 0$ or the conditional distribution of T_1 given $M = m$ is concentrated at its conditional expectation $\varphi_1(m)$, i.e. T_1 is already a function of the minimal sufficient statistic M.

EXAMPLE 3.4.1 Let $(X_1, \ldots, X_n)$ be i.i.d. $N(\theta, 1)$ then we know that $\bar{X}$ is minimal sufficient for $\{L(x, \theta), \theta \in R_1\}$. Let $\psi_1(\theta) = \theta$ then observing that $\bar{X} \sim N(\theta, 1/n)$ we have $\bar{X} \in U_\theta$ and is itself function of minimal sufficient statistic. On the other hand suppose we take $T_1 = X_1$ then $T_1 \in U_\theta$. As the conditional distribution of T_1 given $\bar{X} = \bar{x}$ is $N(\bar{x}, 1 - 1/n)$ it follows that $E(T_1 \mid \bar{X} = \bar{x}) = \varphi(\bar{x}) = \bar{x}$ for each $\bar{x}$. Hence Rao-Blackwellization of $T_1 = X_1$ leads to $\bar{X}$. Note that if we had started with any $T_i = X_i$, $i = 1, 2, \ldots, n$, its Rao-Blackwellization would also lead to same statistic $\varphi_i(\bar{x}) = \bar{x}$, $i = 1, 2, \ldots, n$. Now let $T = \sum l_i X_i$ where $\sum l_i = 1$ so that $T \in U_\theta$. Then $(T, \bar{X})'$ is BVN with mean $(\theta, \theta)'$ and covariance matrix $\begin{pmatrix} \sum l_i^2 & 1/n \\ 1/n & 1/n \end{pmatrix}$ and the conditional distribution of T given $\bar{X} = \bar{x}$ is $N(\bar{x}, \sum l_i^2 - 1/n)$. Therefore Rao-Blackwellization of T, any linear unbiased estimator of θ leads to $\bar{X}$

only. Note that we have already seen that $\bar{X}$ is BLUE as well as MVUE since its variance attains the CRLB. Thus one should expect that for any $T \in U_\theta$ Rao-Blackwellization with respect to the minimal sufficient statistic $\bar{X}$ should lead to $\bar{X}$ itself.

Next consider the situation where we want to estimate $\psi_2(\theta) = \theta^2$. Then again observing that $E(\bar{X}^2) = \theta^2 + \frac{1}{n}$ we have $\varphi_2(\bar{X}) = \bar{X}^2 - \frac{1}{n} \in U_{\theta^2}$ which is a function of minimal sufficient statistic $\bar{X}$. Let now $T_1 = X_1^2 - 1$ then as $E(X_1^2) = \theta^2 + 1$ we have $T_1 \in U_{\theta^2}$ and $E(X_1^2 - 1 \mid \bar{x}) = \bar{x}^2 - \frac{1}{n}$ as the conditional distribution of X_1 given $\bar{X} = \bar{x}$ is $N\left(\bar{x}, 1 - \frac{1}{n}\right)$ and therefore $E(X_1^2 - 1 \mid \bar{x}) = \bar{x}^2 - \frac{1}{n}$, for each $\bar{x}$ fixed. Indeed Rao-Blackwellization of $X_i^2 - 1$, $i = 1, 2, \ldots, n$ will lead to $\varphi_i(\bar{X}) = \bar{X}^2 - \frac{1}{n}$. If we take $T' = \frac{1}{n} \Sigma X_i^2 - 1$ which belongs to U_{θ^2}, its Rao-Blackwellization would again lead to $\bar{X}^2 - 1/n$. The question whether $\bar{X}^2 - 1/n$ is MVUE of θ^2 can not be decided using the CRLB result. One can show that there does not exist an unbiased estimator of θ^2, whose variance attains the CRLB. This follows from the fact that $\frac{\partial \log L}{\partial \theta} = n(\bar{x} - \theta)$ and $T = \theta^2 \pm \frac{|2\theta|}{n} n(\bar{x} - \theta)$, and for any choice of sign, positive or negative, T does not become purely a function of $\bar{X}$. Moreover CRLB for $\psi(\theta) = \theta^2$ is $4\theta^2/n$ where as

$$\operatorname{Var}\left(\bar{X}^2 - \frac{1}{n}\right) = \operatorname{Var}(\bar{X}^2) = E(\bar{X}^4) - [E(\bar{X}^2)]^2$$

$$= \frac{3}{n^2} + \frac{6\theta^2}{n} + \theta^4 - \left(\theta^2 + \frac{1}{n}\right)^2 = \frac{4\theta^2}{n} + \frac{2}{n^2} > \frac{4\theta^2}{n}.$$

EXAMPLE 3.4.2 Let $(X_1, \ldots, X_n)$ be i.i.d. Poisson (λ) where $M = \Sigma X_i \sim$ Poisson $(n\lambda)$ is minimal sufficient statistic for λ. Suppose we want to estimate $\psi(\lambda) = e^{-\lambda} = P[X = 0]$. Define $T_1(X_1) = 1$ if $X_1 = 0$ and $T_1(X_1) = 0$ if $X_1 \neq 0$. Then $E(T_1) = P[X_1 = 0] = e^{-\lambda}$, and $T_1 \in U_\psi$. Now consider

$$\begin{aligned}\varphi_1(m) &= E(T_1 \mid M = m) \\ &= P[X_1 = 0 \mid M = m] \\ &= \frac{P[X_1 = 0 \text{ and } \Sigma X_i = m]}{P[\Sigma X_i = m]} \\ &= P[X_1 = 0 \text{ and } \sum_{i=2}^{n} X_i = m] / P(\sum_{1}^{n} X_i = m).\end{aligned}$$

But X_1 and $\sum_2^n X_i \sim$ Poisson $[(n-1)\lambda]$ are independent. Therefore

$$\varphi_1(m) = \frac{e^{-\lambda} \cdot e^{-(n-1)\lambda}[(n-1)\lambda]^m/m!}{e^{-m\lambda}(n\lambda)^m/m!} = \left(\frac{n-1}{n}\right)^m.$$

Thus $\quad \varphi_1(M) = \left(1 - \frac{1}{n}\right)^M, \; M = 0, 1, 2, \ldots$

On the other hand following methods of Section 3.1, we have if $\varphi(M) \in U_\psi$ then

$$\sum_{m=0}^{\infty} \varphi(m)\, e^{-n\lambda} \frac{(n\lambda)^m}{m!} = e^{-\lambda}, \; \lambda > 0$$

or $\quad \sum_{m=0}^{\infty} \frac{\varphi(m)n^m}{m!} \lambda^m = e^{(n-1)\lambda} = \sum_{m=0}^{\infty} \frac{\lambda^m}{m!}(n-1)^m, \; \forall \lambda > 0$

Comparing coefficients of λ^m on both sides we have

$$\frac{\varphi(m)n^m}{m!} = \frac{(n-1)^m}{m!}, \; m = 0, 1, 2, \ldots$$

or $\quad \varphi(m) = \frac{(n-1)^m}{n^m}, \; m = 0, 1, 2, \ldots$

What we have shown here is that within U_ψ there exists only one unbiased estimator of ψ which depends on $(X_1, \ldots, X_n)$ through minimal sufficient statistic $M = \sum X_i$. This implies that in this case, for any $T \in U_\psi$ the Rao-Blackwellization must lead to one and the same estimator $\varphi(M)$ defined above since for every $T \in U_\psi$, $E(T \mid M = m) = \varphi(m)$ for each m. Therefore $\varphi(M)$ obtained above must be MVUE. If such a situation were to occur in Ex. 3.4.1 we could claim that $\bar{X}$ is MVUE of θ and so is $\bar{X}^2 - \frac{1}{n}$ of θ^2. We assure the reader that this indeed is true and is a consequence of the property of completeness of the minimal sufficient statistic which holds in special models such as one-parameter exponential family. Observe that $\{N(\theta, 1), \theta \in R_1\}$ and $\{\text{Poisson}(\lambda), \lambda > 0\}$ are both one parameter exponential families. In the next section we study the completeness property and show how the problem of MVUE of $\psi(\theta)$ can be resolved in case the model $\{L(x, \theta), \theta \in \Omega\}$ admits a minimal sufficient statistic which has completeness property.

Exercise 3.4.1 (1) Let $(X_1, \ldots, X_n)$, $n \geq 2$ be i.i.d. Bernoulli with $P(X_i = 1) = \theta$. Let

$$T_1(X_1, X_2) = 1 \text{ if } X_1 = 1 \text{ and } X_2 = 0$$
$$= 0 \text{ otherwise}$$

Show that $T_1(X_1, X_2) \in U_{\theta(1-\theta)}$ and obtain Rao-Blackwellized version of T_1 say $\varphi_1(t)$ w.r.t. the minimal sufficient statistic $T = \Sigma X_i \sim$ Bin (n, θ). By solving the equation

$$\sum_{t=0}^{n} \varphi(t) \binom{n}{t} \theta^t (1-\theta)^{n-t} = \theta(1-\theta), \; 0 < \theta < 1,$$

show that the Rao-Blackwellized version of T_1, $\varphi_1(t)$ is the unique unbiased estimator of $\theta(1-\theta)$ i.e. $\varphi(t) = \varphi_1(t)$.

(2) In Example 3.4.2 let $\psi(\lambda) = e^{-\lambda} \dfrac{\lambda^r}{r!}$. Define $T_1(X_1) = 1$ if $X_1 = r$ and zero otherwise. Obtain $\varphi_1(m)$, the Rao-Blackwellized version of T_1 and show that it is the unique solution of

$$\sum_{m=0}^{\infty} \varphi(m)\, e^{-n\lambda} \frac{(n\lambda)^m}{m!} = e^{-\lambda} \cdot \frac{\lambda^r}{r!}, \; \lambda > 0.$$

(3) In Example 3.4.2 define $T_1(X_1, X_2) = 1$ if $X_1 = 0, X_2 = 0$. Show that $T_1 \in Ue^{-2\lambda}$ and Rao-Blackwellize T_1. Show that $\varphi_1(m)$ is the unique solution of $E[\varphi(m)] = e^{-2\lambda}$, $\forall\, \lambda > 0$.

(4) In Example 3.4.1 define $T_1(X_1) = 1$ if $a < X_1 < b$ and zero otherwise. Then $E(T_1) = \Phi(b-\theta) - \Phi(a-\theta)$. Rao-Blackwellize T_1 and obtain $\varphi_1(\bar{X})$. In quality control problem if the product is declared as satisfactory if $a < X_1 < b$ where a and b are specification limits of X_1 say diameter of the head of a screw, then $\Phi(b-\theta) - \Phi(a-\theta)$ represents the fraction of satisfactory items out of total production produced by the normal process with mean θ and variance one.

(5) Let $(X_1, \ldots, X_n)$ be i.i.d. $U(0, \theta)$ and let $T = X_{(n)}$ be the minimal sufficient statistic. Let $X_{(1)}$ be the first component of the order statistic. Obtain $T_1(X_{(1)})$ such that $E(T_1(X_{(1)})) = \theta$ and Rao-Blackwellize it to obtain $\varphi_1(X_{(n)}) = \dfrac{n+1}{n} X_{(n)}$.

(6) Let $(X_1, \ldots, X_n)$ be i.i.d. exponential with mean θ, a model often used to describe the failure time of items of various types e.g. a T.V. tube or an electric bulb. Let $\psi(\theta) = e^{-x_0/\theta} = P[X > x_0] = R[x_0]$ the reliability of the item at X_0, i.e. probability of failure free operation of the item until x_0. Define $T_1(X_1) = 1$ if $X_1 > x_0$ and zero otherwise then $T_1 \in U_\psi$. Rao-Blackwellize T_1 w.r.t. the minimal sufficient statistic $T = \; X_i$.

3.5 Completeness and Lehmann-Scheffe Theorem

Let M be minimal sufficient statistic for $\{L(x, \theta), \theta \in \Omega \subset R_1\}$ and let $\psi(\theta)$ be an estimable function i.e. U_ψ is not empty. Now if $T_1 \in U_\psi$ then by Rao-Blackwellization we obtain $E(T_1 \mid M) = \varphi_1(M) \in U_\psi$. As seen in special cases (Ex. 3.4.1 or 3.4.2) $\varphi_1(M)$ is the unique element in U_ψ which depends on X through the minimal sufficient statistic M. Let $\{g_0(m, \theta), \theta \in \Omega\}$ be the corresponding class of pdfs of M under θ then we wish to characterize the situation where the integral equation $\int \varphi(m)\, g_0(m, \theta)\, dm = \psi(\theta)$, $\forall\, \theta \in \Omega$ for given $g_0(m, \theta)$ and $\psi(\theta)$ has a unique solution $\varphi_1(m)$. Now observe that if $\varphi_1(M) \in U_\psi$ and $\varphi_2(M) \in U_\psi$ then $E[\varphi_1(M) - \varphi_2(M)] = 0$, $\forall\, \theta \in \Omega$. If this implies that $\varphi_1(M) = \varphi_2(M)$ with probability one for each $\theta \in \Omega$, then $\varphi_1(M)$ is essentially unique. Consider estimating a function

$\psi(\theta) = 0$, $\forall\ \theta \in \Omega$ and let U_0^M denote the class of all unbiased estimators of zero which are functions of M. Then we require that the integral equation

$$\int \varphi_0(M)\ g_0(m,\ \theta)\ dm = 0,\ \forall\ \theta \in \Omega$$

holds iff $\varphi_0(M) = 0$ w.p. 1 under each $\theta \in \Omega$, that is only unbiased estimator of zero in U_0^M is the estimator identically equal to zero. The reader will notice the obvious analogy with the solution of system of linear equations where we have to solve $Ax = b$ and we initially consider the homogeneous linear equations $Ax = 0$. When the matrix A is of full rank then the equation $Ax = 0$ has the unique solution $x = 0$ and consequently we have also the unique solution for the equation $Ax = b$ given by $x = A^{-1}b$. Here of course we assume that A is a square matrix $m \times m$ and x and b are $mx1$ vectors. We therefore define the completeness of a minimal sufficient statistic M as follows.

Definition 3.5.1 A minimal sufficient statistic M for the class of pdfs $\{L(x,\ \theta),\ \theta \in \Omega\}$ is said to be complete if $E_\theta[\varphi(M)] = 0$, $\forall\ \theta \in \Omega$ implies that $\varphi(M) = 0$ with probability one under each $\theta \in \Omega$.

EXAMPLE 3.5.1 Let $(X_1, \ldots, X_n)$ be i.i.d. $B(1,\ \theta)$ then as seen earlier $M = \Sigma X_i \sim B(n,\ \theta)$ is minimal sufficient for θ. We now show that M is complete. Let $\varphi(M)$ be an unbiased estimator of zero then

$$\sum_{m=0}^{n} \varphi(m) \binom{n}{m} \theta^m (1-\theta)^{n-m} = 0 \quad 0 < \theta < 1$$

Now L.H.S. of above equation is a polynomial of degree n which vanishes identically for $\theta \in (0, 1)$. Therefore coefficient of any power of θ must be zero. Consider the constant term (coefficient of θ^0). This is included in $\varphi(0)\binom{n}{0}\theta^0(1-\theta)^n$ only and is given by $\varphi(0)$ as $(1-\theta)^n = 1 - \binom{n}{1}\theta + \binom{n}{2}\theta^2 + \ldots + (-\theta)^n$. Therefore $\varphi(0) = 0$ and $\sum_{m=1}^{n} \varphi(m)\binom{n}{m}\theta^m(1-\theta)^{n-m} = 0$. Cancelling one power of θ and again looking at the constant term on LHS we have $\varphi(1) = 0$. Repeating the process we arrive at the result $\varphi(m) = 0$ for $m = 0, 1, 2, \ldots, n$ or $\varphi(M) = 0$ w.p. 1 under each $\theta \in (0, 1)$.

EXAMPLE 3.5.2 Let $(X_1, \ldots, X_n)$ be i.i.d. $\mathbb{P}(\lambda)$ then $M = \Sigma X_i \sim \mathbb{P}(n\lambda)$ and if we consider the identity $E[\varphi(M)] = 0$, $\forall\ \lambda > 0$ we have

$$\sum_{m=0}^{\infty} \varphi(m)\ e^{-n\lambda}\ \frac{(n\lambda)^m}{m!} = 0, \quad \forall\ \lambda > 0$$

$$\sum_{m=0}^{\infty} \frac{\varphi(m)\, n^m}{m!}\ \lambda^m = 0, \quad \forall\ \lambda > 0.$$

Now LHS is a power series in λ (analytic function of λ) which vanishes over a non-degenerate interval $(0, \infty)$. Hence coefficients of each power λ^m, $m = 0, 1, 2, \ldots$ should be equal to zero and therefore $\dfrac{\varphi(m)n^m}{m!} = 0$ for each $m = 0, 1, 2, \ldots$. As $\dfrac{n^m}{m!} \neq 0$ for any m we have $\varphi(m) = 0$ for each m, and thus the minimal sufficient statistic $M = \Sigma X_i$ is complete in this case also.

EXAMPLE 3.5.3 Let $(X_1, \ldots, X_n)$ be a random sample of size n from a power series distribution with $f(x, \theta) = \dfrac{a(x)\theta^x}{b(\theta)}$, $x = 0, 1, 2, \ldots$ and $\Omega = (0, \rho)$. As seen earlier the distribution belongs to one parameter exponential family with minimal sufficient statistic $M = \Sigma X_i$ having a power series distribution given by

$$g_0(m, \theta) = \frac{c(m)\theta^m}{[b(\theta)]^n}, \; m = 0, 1, 2, \ldots$$

Now
$$E[\varphi(m)] = \sum_{m=0}^{\infty} \varphi(m) \frac{c(m)\theta^m}{[b(\theta)]^n} = 0 \text{ or}$$

$$\sum_{m=0}^{\infty} \varphi(m)\, c(m)\theta^m = 0, \quad \forall\ \theta \in (0, \rho) \text{ which implies that}$$

$$\varphi(m)\, c(m) = 0, \; m = 0, 1, 2, \ldots$$

i.e.
$$\varphi(m) = 0 \text{ when } c(m) > 0 \text{ or } \varphi(M) \text{ is zero w.p. } 1$$

Note that $P(M = m) = 0$, $\forall\ \theta \in (0, \rho)$ if $c(m) = 0$.

It is not very easy to prove the completeness property of the minimal sufficient statistic M which is a continuous r.v. with pdf $\{g_0(m, \theta), \theta \in \Omega\}$. We will indicate below a heuristic proof of completeness of $\bar{X}$, the sample mean in case of a random sample of size n from $N(\theta, 1)$ which can be generalized to the case of random sample of size n from $\{f(x, \theta), \theta \in \Omega\}$ which forms a one parameter exponential family.

EXAMPLE 3.5.4 Let $(X_1, \ldots, X_n)$ be a random sample from $\{N(\theta, 1), \theta \in R_1\}$, then $\bar{X}$ is minimal sufficient with $g_0(\bar{x}, \theta) = \sqrt{\dfrac{n}{2\pi}} \exp\left\{-\dfrac{n(\bar{x}-\theta)^2}{2}\right\}$, $\bar{x} \in R_1$, $\theta \in R_1$. Let $\varphi(\bar{x})$ be an unbiased estimator of zero so that $E[\varphi(\bar{x})] = 0$, $\forall\ \theta \in R_1$ which implies that

$$\int_{R_1} \sqrt{\frac{n}{2\pi}}\, \varphi(\bar{x}) \exp\left\{-\frac{n(\bar{x}-\theta)^2}{2}\right\} d\bar{x} = 0, \forall\ \theta \in R_1.$$

or
$$\int_{R_1} \varphi(\bar{x}) \exp\left\{-\frac{n\bar{x}^2}{2}\right\} e^{\theta\bar{x}}\, d\bar{x} = 0, \qquad (3.5.1)$$

Define $\varphi_+(\bar{x}) = \text{Max}\ (\varphi(\bar{x}), 0)$ and $\varphi_-(\bar{x}) = \text{Max}\ (0, -\varphi(\bar{x}))$ then $\varphi(\bar{x}) = \varphi_+(\bar{x}) - \varphi_-(\bar{x})$ and $|\varphi(\bar{x})| = \varphi_+(\bar{x}) + \varphi_-(\bar{x})$. Recall that $E[\varphi(\bar{x})]$ exists iff $E[|\varphi(\bar{x})|]$ exists. Thus

$$\int \varphi_+(\bar{x}) \exp\left\{-\frac{n\bar{x}^2}{2}\right\} e^{\theta\bar{x}}\, d\bar{x} = \int \varphi_-(\bar{x}) \exp\left\{-\frac{n\bar{x}^2}{2}\right\} e^{\theta\bar{x}}\, d\bar{x}.$$

In particular at $\theta = 0$ we have

$$\int \varphi_+(\bar{x}) \exp\left\{-\frac{n\bar{x}^2}{2}\right\} d\bar{x} = \int \varphi_-(\bar{x}) \exp\left\{-\frac{n\bar{x}^2}{2}\right\} d\bar{x} = C_0.$$

Observe that $h_+(\bar{x}) = \dfrac{\varphi_+(\bar{x}) \exp\left\{-\frac{n\bar{x}^2}{2}\right\}}{C_0} \geq 0$ and $\int h_+(\bar{x})\, d\bar{x} = 1$ and $h_+(\bar{x})$ is a pdf over R_1. Similarly we have $h_-(\bar{x})$ is also a pdf over R_1. Then we have $\forall\ \theta \in R_1$.

$$Q_+(\theta) = \int h_+(\bar{x})\, e^{\theta\bar{x}}\, d\bar{x} = Q_-(\theta) = \int h_-(\bar{x})\, e^{\theta\bar{x}}\, d\bar{x} \tag{3.5.2}$$

However $Q_+(\theta)$ is the mgf of the pdf given by $h_+(u)$ and $Q_-(\theta)$ is the mgf of the pdf given by $h_-(u)$ and these two mgfs agree over an interval of R_1 containing origin. We now have a theorem which asserts that if two mgfs agree over an interval containing origin then their pdfs must be same or $h_+(\bar{x}) = h_-(\bar{x}),\ \forall\ \bar{x} \in R_1$. This however implies that $\varphi_+(\bar{x}) = \varphi_-(\bar{x})$ for $\bar{x} \in R_1$ from which it follows that $\varphi(\bar{x}) = 0$ for $\forall\ \bar{x} \in R_1$ or $\varphi(\bar{X}) = 0$ w.p.1 under each $\theta \in \Omega$.

Following the similar arguments one could show that for one parameter exponential family the minimal sufficient statistic $M = \sum_{i=1}^{n} k(x_i)$ is complete. Observe that

$$g_0(m, \theta) = \exp\{u(\theta)\, m + nv(\theta) + w_1(m)\}.$$

Using now the fact that $\dfrac{du}{d\theta} \neq 0$ we make 1 : 1 transformation of the parameter space $\theta \to \eta = u(\theta)$ then we have

$$g_0'(m, \eta) = \exp\{\eta m + v_1(\eta) + w_1(m)\}.$$

and we have for all $\eta \in \Omega_1$

$$\int \varphi(m)\, g_0'(m, \eta)\, dm = \int \varphi(m) \exp\{\eta m + v_1(\eta) + w_1(m)\}\, dm = 0$$

Thus $$\int h_+(m) e^{\eta m}\, dm = \int h_-(m)\, e^{\eta m}\, dm,\ \forall\ \eta \in \Omega_1 \tag{3.5.3}$$

where $$h_+(m) = \frac{\varphi_+(m) \exp\{w_1(m)\}}{c_0}$$

and $$h_-(m) = \frac{\varphi_-(m) \exp\{w_1(m)\}}{c_0}$$

and c_0 is the norming constant which makes $h_+(m)$ and $h_-(m)$ as pdfs. Equation (3.5.3) implies that mgf of $h_+(m)$ and $h_-(m)$ agree over an interval containing zero and thus $h_+(m) = h_-(m)$, $\forall$ m which implies that $\varphi_+(m) = \varphi_-(m)$, $\forall$ m and hence $\varphi(m) = 0$ w.p. 1.

In the heuristic proof given above we have slurred over many fine points. Indeed proving completeness of $M = \sum k(X_i)$ in one parameter exponential family is essentially same as proving uniqueness property of Laplace transform of an integrable function which plays a crucial role in applications of Mathematics to Engineering and Physics among others. It also indicates that property of completeness is a deep property yielding the following powerful theorem due to Lehmann and Scheffe.

THEOREM 3.5.2 Let $\{L(x, \theta), \theta \in \Omega\}$ admit a minimal sufficient statistic M which is complete. Suppose U_ψ is not empty then MVUE of $\psi(\theta)$ is the unique element $\varphi(m) \in U_\psi$.

To prove the theorem we observe that by Rao-Blackwell theorem the search for MVUE can be restricted to U_ψ^M, unbiased estimators of $\psi(\theta)$ which are functions of M. In view of completeness of M, U_ψ^M consists of a single element $\varphi(M)$ only and it follows that $\varphi(M)$ is MVUE.

EXAMPLE 3.5.5 As $\bar{X} \sim N(\theta, 1/n)$, $\theta \in R_1$ is an exponential family it follows that $\bar{X}$ is minimal sufficient and complete. Therefore, by Rao-Blackwell Lehemann-Scheffe theorems (RBLS) using Example 3.4.1 it follows that $\bar{X}$ is MVUE for θ and $\bar{X}^2 - \frac{1}{n}$ is MVUE for θ^2. Similarly, as $\sum X_i \sim \mathbb{P}(n\lambda)$ is minimal complete sufficient statistic, for a random sample of size n from $\mathbb{P}(\lambda)$, from Example 3.4.2, it follows that $\frac{M}{n} = \bar{X}$ is MVUE for λ and $\varphi(M) = \left(\frac{n-1}{n}\right)^M$ is MVUE of $e^{-\lambda}$. The minimal sufficient statistic $\sum X_i = M$ in sampling from $\{b(1, \theta), 0 < \theta < 1\}$, $\{\mathbb{P}(\lambda), \lambda > 0\} \cdot$ {exponential distribution with mean θ} can be shown to be complete as the pdfs belong to exponential family. Therefore the unbiased estimators derived by Rao-Blackwellization in the Exercise 3.4.1 except (5) will yield MVUE of the corresponding function ψ.

EXAMPLE 3.5.6 Let $(X_1, \ldots, X_n)$ be random sample of size n from the Pareto distribution $f(x, \theta) = \frac{\theta}{x^{\theta+1}}$, $x > 1$, $\theta > 0$. Then as Pareto distribution belongs to one parameter exponential family $M = \sum \log X_i$ is minimal complete sufficient statistic with pdf $g_0(m, \theta) = \frac{1}{\Gamma(n)} \theta^n e^{-m\theta} m^{n-1}$, $m > 0$,

$\theta > 0$. Suppose $\psi(\theta) = \theta$. Then we observe that $E[m^r] = \dfrac{\Gamma(n+r)}{\Gamma(n)} \dfrac{1}{\theta^r}$ for any $r > 0$, as well as for any r such that $n + r > 0$ or for $r > -n$. Therefore so long as $n \geq 2$, $E(m^{-1}) = \dfrac{\Gamma(n-1)}{\Gamma(n)} \theta$ or $\dfrac{n-1}{M} \in U_\theta$ and $\varphi(M) = \dfrac{n-1}{\Sigma \log X_i}$ is MVU of θ. Indeed $\varphi_r(M) = M^r$ would be MVUE of $\dfrac{\Gamma(n+r)}{\Gamma(n)} \dfrac{1}{\theta^r}$ for any $r > -n$ or $\dfrac{\Gamma(n)}{\Gamma(n+r)} M^r$ would be MVUE of $\dfrac{1}{\theta^r}$.

EXAMPLE 3.5.7 Let $(X_1, \ldots, X_n)$ be a random sample of size n from a geometric distribution with pmf $f(x, \theta) = \theta(1-\theta)^x$, $x = 0, 1, 2, \ldots, 0 < \theta < 1$. As $\{f(x, \theta), \theta \in \Omega\}$ is a one-parameter exponential family (in fact a power series distribution) we have $M = \Sigma X_i$ is minimal sufficient statistic with pmf

$$g_0(m, \theta) = \binom{m+n-1}{n-1} \theta^n (1-\theta)^m, \; m = 0, 1, 2, \ldots, 0 < \theta < 1$$

which itself belongs to one parameter exponential family. Suppose we want to estimate θ then by RBLS theorems we must solve the equation

$$\sum_{m=0}^{\infty} \varphi(m) \binom{m+n-1}{n-1} \theta^n (1-\theta)^m = \theta$$

or

$$\sum_{m=0}^{\infty} \varphi(m) \binom{m+n-1}{n-1} \phi^m = (1-\phi)^{-(n-1)} \text{ where } \phi = 1 - \theta$$

$$= \sum_{m=0}^{\infty} \binom{-n+1}{m} (-1)^m \phi^m.$$

Equating coefficients of ϕ^m for each m we have

$$\varphi(m) = \binom{-n+1}{m} (-1)^m \Big/ \binom{m+n-1}{n-1}$$

$$= \binom{m+n-2}{m} \Big/ \binom{m+n-1}{n-1}$$

$$= \frac{n-1}{m+n-1}, \; m = 0, 1, 2, \ldots$$

Therefore we have $\varphi(M) = \dfrac{n-1}{M+n-1}$, $M = 0, 1, 2, \ldots$ is MVUE of θ.

We note that sometimes in SQC, inverse binomial sampling method is used i.e. rather than taking a fixed lot for inspection and then determining number of defectives in the lot, we sample until n satisfactory/defective items are found. The sample size in excess of n, say M has a negative binomial distribution given by $g_0(m, \theta)$ where θ is the probability of item being satisfactory/defective.

We recommend to readers to Rao-Blackwellize $T_1(X_i) = 1$ if $X_1 = 0$ and zero otherwise to obtain the above MVUE. The basic property we have to use is the fact that if $X_1, \ldots, X_k$ are i.i.d. geometric then $(X_1 + X_2 + \ldots + X_k)$ is negative binomial with pdf

$$g(k, r, \theta) = \binom{k+r-1}{k-r} \theta^k (1-\theta)^r, \; r = 0, 1, 2, \ldots$$

We also point out that RB and LS theorems are sometimes paraphrased as a single theorem stating that.

THEOREM 3.5.3 (RBLS Theorem): If M is minimal complete sufficient statistic for $\{L(x, \theta), \theta \in \Omega\}$ then any statistic $\varphi(M)$ is MVUE of its expectation $E_\theta[\varphi(M)]$.

As in the case of one parameter exponential family we can show that in case of one parameter Pitman family where a one dimensional minimal sufficient statistic T exists then T is complete. We again give a heuristic proof in case of $X_{(n)} = \text{Max}\,(X_1, \ldots, X_n)$ where $(X_1, \ldots, X_n)$ are i.i.d. $U(0, \theta)$, $\theta > 0$. Let $\varphi(X_{(n)}) \in U_0$ then

$$\int_0^\theta \varphi(x_{(n)})\, n x_{(n)}^{n-1}\, dx_{(n)} = 0 \;\forall\; \theta \in (0, \infty)$$

or equivalently for each $\theta > 0$

$$\int_0^\theta \varphi_+(x_{(n)})\, n x_{(n)}^{n-1}\, dx_{(n)} = \int_0^\theta \varphi_-(x_{(n)})\, n x_{(n)}^{n-1}\, dx_{(n)}$$

or $H_+(0, \theta) = H_-(0, \theta)$ where $H_+(0, \theta)$ and $H_-(0, \theta)$ is the integral of $\varphi_+(x_{(n)})$ and $\varphi_-(x_{(n)})$. If φ_+ and φ_- are continuous, by fundamental theorem of integral calculus $H_+(0, \theta) = H_-(0, \theta)$ implies that $\varphi_+(\theta) = \varphi_-(\theta)$ for each $\theta > 0$ or $\varphi_+(x_{(n)}) = \varphi_-(x_{(n)})$, $x_{(n)} > 0$. Therefore $\varphi(x_{(n)}) = 0$ for each $x_{(n)} > 0$. If φ_+ and φ_- are not necessarily continuous one can use measure theoretic arguments based on the fact that if two non-negative set functions agree over all intervals of $(0, \infty)$ then they must agree on each Lebesgue set of $(0, \infty)$ and in turn the derivatives of such set functions w.r.t. Lebesgue measure must agree on $(0, \infty)$ or $\varphi_+ = \varphi_-$. Therefore, we can claim that $\varphi(x_{(n)}) = 0$ on $(0, \infty)$.

Again here we observe that completeness of minimal sufficient statistic is a deep property and its proof requires basic results in measure theory and integration and theory of analytic functions.

EXAMPLE 3.5.8 Using the fact that $X_{(n)}$ is minimal complete sufficient statistic for θ in $U(0, \theta)$, it follows from RBLS theorem that $\frac{n+1}{n} X_{(n)}$ is MVUE of θ. Suppose we want to obtain MVUE of $\frac{\theta^2}{12}$ the variance of

$U(0, \theta)$ distribution. Then observing that $E(X_{(n)}^2) = \dfrac{n\theta^2}{n+2}$ it follows that $\dfrac{n+2}{12n} X_{(n)}^2$ is MVUE of $\dfrac{\theta^2}{12}$.

Similarly, in the case of sampling from exponential distribution with location θ, i.e. $f(x, \theta) = \exp\{-(x - \theta)\}$, $x \geq \theta$ we have $X_{(1)}$ is minimal sufficient statistic and $X_{(1)} - \dfrac{1}{n}$ would be MVUE of θ.

Consider a random sample of size n from $U(-\theta, \theta)$ where $\theta > 0$. Then as seen in Example 2.6.1 we have $T = \text{Max}(-X_{(1)}, X_{(n)})$ is minimal sufficient with pdf $g_0(t, \theta) = \dfrac{nt^{n-1}}{\theta^n}$, $0 < t < \theta$. T is thus complete. Then as $E(T) = \dfrac{n}{n+1}\theta$ we have MVUE of the upper limit θ is $\dfrac{n+1}{n} T$ and that of lower limit $-\theta$ is $-\dfrac{n+1}{n} T$.

EXAMPLE 3.5.9 As an example of a minimal sufficient statistic which is not complete consider a sample of size 2 from the Cauchy distribution with pdf $f(x, \theta) = \dfrac{1}{\pi} \dfrac{1}{1 + (x - \theta)^2}$, $x \in R_1$, $\theta \in R_1$. As seen in Example 2.4.3, the order statistic $(X_{(1)}, X_{(2)})$ is minimal sufficient for θ. We now show that this two dimensional minimal sufficient statistic for the one dimensional parameter θ is not complete. Consider $T = X_{(2)} - X_{(1)}$ then $T = (X_{(2)} - \theta) - (X_{(1)} - \theta) = Y_{(2)} - Y_{(1)}$ where $(Y_{(1)}, Y_{(2)})$ is the order statistic of a sample of size two from Cauchy distribution with $\theta = 0$. As the distribution of T does not depend on θ and $P_\theta[a < T < b] = p$ does not depend on θ, for any given interval (a, b). Now define $\varphi(X_{(1)}, X_{(2)}) = 1 - p$ if $a < T < b$ and $\varphi(X_{(1)}, X_{(2)}) = -p$ if either $T < a$ or $T > b$ then $E[\varphi X_{(1)}, X_{(2)}] = 0$ for each $\theta \in R_1$. This shows that the minimal sufficient statistic $(X_{(1)}, X_{(2)})$ is not complete. We remark here that for a sample of size n from Cauchy distribution with location θ, we can show that the order statistic $(X_{(1)}, \ldots, X_{(n)})$ is minimal sufficient. Consider $T_1 = X_{(n)} - X_{(1)}$ or in general any linear function of the components of the order statistic $T(l) = \sum l_i X_{(i)}$ with $\sum l_i = 0$. We can write $T(l) = \sum l_i Y_{(i)}$ where $Y_{(i)} = X_{(i)} - \theta$ and therefore $(Y_{(1)}, \ldots, Y_{(n)})$ has the distribution of order statistic of a sample of size n from $\theta = 0$ and as such $T(l)$ has distribution independent of θ. We can now select a function $\varphi(X_{(1)}, \ldots, X_{(n)}) = \varphi(\sum l_i X_{(i)})$ with zero expectation to claim that the order statistic of a sample of size n is minimal sufficient but not complete in the Cauchy case.

The problem of MVU estimation is satisfactorily solved when the model is such that it admits a minimal sufficient statistic which is complete. This concept was introduced by Lehmann and Scheffe (1950) and they also showed that for most of the commonly occurring situation a search for a sufficient statistic which is complete would be enough i.e. minimality in fact follows. We will not go deeper in the issue but instead in the next section obtain a necessary and sufficient condition for T to be MVUE for $\psi(\theta)$ without any reference to sufficiency and completeness. The condition can be used in some situations where minimal complete sufficient statistic does not exist.

3.6 A Necessary and Sufficient Condition for MVUE

Let $(X_1, X_2, \ldots, X_n)$ have joint pdf given by $\{L(x, \theta), \theta \in \Omega\}$ and $U_\psi = \{T(x) \mid E(T(x)) = \psi(\theta), \forall\ \theta \in \Omega\}$ and $U_0 = \{u_0(x) \mid E(u_0(x)) = 0, \forall\ \theta \in \Omega\}$. We assume that Var $(T(x)) < \infty$ and Var $(u_0(x)) < \infty$.

THEOREM 3.6.1 A necessary and sufficient condition that T^* in U_ψ is MVUE of $\psi(\theta)$, is given by

$$\text{Cov}\ (T^*, u_0) = 0,\ \forall\ \theta \in \Omega,\ \forall\ u_0 \in U_0 \tag{3.6.1}$$

Assume that (3.6.1) holds and consider any $T \in U_\psi$ then as $T^* \in U_\psi$, we have $E(T^* - T) = 0$, $\forall\ \theta \in \Omega$ and therefore Cov $(T^*, T^* - T) = 0$, $\forall\ \theta \in \Omega$. Hence it follows that Cov (T^*, T) = Var (T^*). Now

$$\begin{aligned} \text{Var}\ (T^* - T) &= \text{Var}\ (T^*) + \text{Var}\ (T) - 2\ \text{Cov}\ (T^*, T) \\ &= \text{Var}\ (T) - \text{Var}\ (T^*) \\ &\geq 0,\ \forall\ \theta \in \Omega. \end{aligned}$$

Thus for any $T \in U_\psi$, Var $(T) \geq$ Var (T^*) $\forall\ \theta \in \Omega$ and therefore $T^* \in U_\psi$ is MVUE of $\psi(\theta)$. This completes the proof of sufficiency of the condition (3.6.1).

To prove necessity we follow the method of reductio ad absurdum. Suppose that T^* is MVUE but (3.6.1) does not hold. Then there exists a $u_0 \in U_0$ and a $\theta_0 \in \Omega$ such that Cov $(T^*, u_0) \neq 0$ at θ_0. Now for any λ real, $T^* + \lambda u_0 \in U_\psi$ and

$$\text{Var}\ (T^* + \lambda u_0) = \text{Var}\ (T^*) + \lambda^2\ \text{Var}\ (u_0) + 2\lambda\ \text{Cov}\ (T^*, u_0).$$

Now let $\lambda_{\theta_0} = -\ \text{Cov}_{\theta_0}(T^*, u_0)/\text{Var}_{\theta_0}(u_0)$, then at $\theta = \theta_0$

$$\text{Var}\ (T^* + \lambda u_0) = \text{Var}\ (T^*) - [\text{Cov}\ (T^*, u_0)]^2/\text{Var}\ (u_0)$$

which shows that Var $(T^* + \lambda_{\theta_0} u_0) <$ Var (T^*) at $\theta = \theta_0$ contradicting the datum that T^* is MVUE. This completes the proof of necessity of the condition (3.6.1).

In general, determination of U_0 class of unbiased estimators of zero is most difficult. Note that if we are able to specify U_0 then $U_\psi = \{T_1 + \alpha u_0 \mid u_0 \in U_0, \alpha \in R_1\}$ or U_ψ is completely specified if we know at least one unbiased estimator T_1 of $\psi(\theta)$ and U_0. This situation is similar to that where the set of solutions of non-homogeneous linear equations $Y = AX + b$ consists of $y_1 + \lambda y_0$, where y_1 is a particular solution and y_0 is a solution of the homogeneous linear equations $Y = AX$. We now consider an example to illustrate the application of the Theorem 3.6.1.

EXAMPLE 3.6.1 Let X be a discrete r.v. with pmf given by $P[X = -1] = \alpha$, $P[X = x] = (1-\alpha)^2\alpha^x$, $x = 0, 1, 2, \ldots$, where $0 < \alpha < 1$. Consider a sample of size one only. Then $x \overset{L}{\sim} y$ iff $L(x, \alpha)/L(y, \alpha)$ is independent of α. We claim that $x \overset{L}{\sim} y$ iff $x = y$. Let $x = -1$ and $x \neq y$. Then $L(-1, \alpha)/L(y, \alpha) = \dfrac{\alpha^{1-y}}{(1-\alpha)^2}$ for any $y = 0, 1, 2, \ldots$ and depends on α. Further let $x \neq y$ be two non-negative integers then $x \overset{L}{\sim} y$ iff α^{x-y} does not depend on α or $x = y$. Thus X itself is the minimal sufficient statistic for α.

We now determine U_0. Let $u(x) \in U_0$, then $u(-1)\,\alpha + \sum_{x=0}^{\infty} u(x)\,(1-\alpha)^2\alpha^x = 0$ for every $\alpha \in (0, 1)$. We now use the technique of equating coefficients of powers α^γ, $\gamma = 0, 1, 2, \ldots$. The constant term is $u(0)$ and hence $u(0) = 0$. The coefficient of α is $u(-1) + u(1) = 0$ or $u(-1) = -u(1)$. The coefficient of α^γ, $\gamma \geq 2$ is given by

$$u(\gamma) - 2u(\gamma - 1) + u(\gamma - 2) = 0 \text{ or } \Delta^2(u(\gamma)) = 0.$$

This is a finite difference equation of order 2 and as $\dfrac{d^2y}{dx^2} = 0$ implies that y is linear or $y = a + bx$, $\Delta^2 u(\gamma) = 0$ implies that for $\gamma \geq 2$, $u(\gamma) = a + b\gamma$, where constants a, b are determined by the initial conditions $u(0) = 0$ and $u(-1) + u(1) = 0$ i.e. $a = 0$ and b can be any real number. Thus $u(\gamma) = b\gamma$, $\gamma = -1, 0, 1, \ldots$ $b \in R_1$ is the general solution and $U_0 = \{cX \mid c \in R_1\}$.

Otherwise we can determine $u(\gamma)$ recursively. Now $u(0) = 0$ and let $u(1) = -u(-1) = c$. Then as $\Delta^2 u_2 = 0$ we have $u(2) = 2c$ and $u(3) = 2u(2) - u(1) = 3c$, and so on. Thus we are led to $U_0 = \{cX \mid c \in R_1\}$ and the minimal sufficient statistic X is not complete.

When minimal sufficient statistic is not complete strange situations occur as we now illustrate. Consider estimation of the indexing parameter α. Then $T_1(X) = 1$ if $X = -1$ and zero otherwise is unbiased for α and $U_\alpha = \{T_1(x) + \lambda u_0 \mid \lambda \in R_1,\, u_0 \in U_0\}$. But as $U_0 = \{cX \mid c \in R_1\}$ we have $U_\alpha = \{T_1(x) + cX \mid c \in R_1\} \cdot T^* \in U_\alpha$ is MVUE of α iff Cov $(T^*, cX) = 0$, $\forall\, c \in R_1$. But we have $T^* = T_1(x) + c^*X$, for some constant c^* and Cov $(T^*, cX) = c^*c$ Var $(X) + c$ Cov $(X, T_1(X))$. We can show that Cov $(X, T_1(X)) = -\alpha$ and Var $(X) = \dfrac{2\alpha}{1-\alpha}$. Therefore c^*c Var $(X) + c$ Cov $(X, T_1(X)) = 0$, $\forall\, c \in R_1$, holds iff $c^* = \dfrac{1}{2(1-\alpha)}$. This choice of c^* depends on α the parameter and is not same for all values of α. Thus although U_α is not empty MVUE of α does not exist.

On the other hand consider $\psi(\alpha) = (1 - \alpha)^2 = p[X = 0]$. Then $T_2(x) = 1$ if $X = 0$ and zero otherwise belongs to U_ψ. Further for any $c \in R_1$ we have for any α, Cov $(T_2(X), cX) = 0$ and $T_2(X)$ is MVUE of $(1 - \alpha)^2$.

Next we consider some important consequences of the Theorem 3.6.1, namely uniqueness and additivity of MVUE.

THEOREM 3.6.2 Let $T_1 \in U_\psi$ be MVUE of $\psi(\theta)$ then T_1 is essentially unique.

Let $T_2 \in U_\psi$ be also MVUE for $\psi(\theta)$. Then $T_1 - T_2 \in U_0$ and Cov $(T_1, T_1 - T_2) = 0 =$ Cov $(T_2, T_1 - T_2)$. i.e. Cov $(T_1, T_2) =$ Var $(T_1) =$ Var (T_2) $\forall\ \theta \in \Omega$. This implies that

Var $(T_1 - T_2) =$ Var $(T_1) - 2$ Cov $(T_1, T_2) +$ Var $(T_2) = 0$, $\forall\ \theta \in \Omega$ or $T_1 = T_2$ w.p. 1 for each $\theta \in \Omega$.

THEOREM 3.6.3 Let T_i be MVUE for $\psi_i(\theta)$, $i = 1, 2, \ldots, k$. Then $\sum_{i=1}^{k} a_i T_i$ is MVUE for $\psi(\theta) = \sum_{i=1}^{k} a_i \psi_i(\theta)$.

Let $u_0 \in U_0$ then Cov $(T_i, u_0) = 0$, $i = 1, 2, \ldots, k\ \forall\ \theta \in \Omega$. Therefore Cov $(\sum a_i T_i, u_0) = \sum a_i$ Cov $(T_i, u_0) = 0$, $\forall\ \theta \in \Omega$, $\forall\ u_0 \in U_0$ and $\sum a_i T_i$ is MVUE of $\sum a_i \psi_i(\theta)$.

THEOREM 3.6.4 Let $\{L(x, \theta), \theta \in \Omega\}$ be such that M is minimal sufficient and complete and $T(x)$ is any statistic such that $E(T(x))$ does not depend on θ and Var $(T) = \sigma_T^2(\theta) < \infty$. Suppose that $\varphi(M)$ is a statistic with finite variance then Cov $(\varphi(M), T(x)) = 0$.

By RBLS theorem $\varphi(M)$ is MVUE of $E_\theta[\varphi(M)]$ and $[T(x) - k_0] \in U_0$ where $k_0 = E_\theta(T(x))$, $\forall\ \theta \in \Omega$. Hence by Theorem 3.6.1 we have Cov $[\varphi(M), T(x)] = 0$, $\forall\ \theta \in \Omega$.

We now consider some applications of the above theorems by way of examples.

EXAMPLE 3.6.2 Let $(X_1, \ldots, X_n)$ be a random sample of size n from the double exponential distribution (Laplace) with zero mean and scale θ^r. Then $f(x, \theta) = \dfrac{1}{2\theta} \exp\{-|x|/\theta\}$, $x \in R_1$, $\theta > 0$ belongs to one parameter exponential family with $M = \sum_{i=1}^{n} |X_i|$ as minimal complete sufficient statistic for θ with $g_0(m, \theta) = \dfrac{1}{\Gamma(n)\theta^n} e^{-m/\theta} m^{n-1}$, $m > 0$, $\theta > 0$. Suppose we want to estimate θ^r then observing that $E(M^r) = \dfrac{\Gamma(n+r)}{\Gamma(n)}\theta^r$ we have $\dfrac{\Gamma(n) M^r}{\Gamma(n+r)}$ is MVUE for θ^r, $r = 1, 2, 3, \ldots$. Now by Theorem 3.6.3

$\sum_{r=1}^{k} a_r \dfrac{\Gamma(n) M^r}{\Gamma(n+r)}$ would be MVUE for $\sum_{r=1}^{k} a_r \theta^r$. We can thus obtain MVUE of any polynomial of degree k in θ using Theorem 3.6.3.

Observing that $E(X) = 0$ we have $T(l) = \sum_{i=1}^{n} l_i X_i \in U_0$ and by Theorem 3.6.4, $\text{Cov}\left(M, \sum_{i=1}^{n} l_i X_i\right) = 0$ or $\Sigma \mid X_i \mid$ and $\Sigma\, l_i X_i$ would be uncorrelated. Note that Var $(T(l)) = 2\, \Sigma\, l_1^2 \theta^2$ depends on θ and the distribution of T depends on θ.

Since distribution of X is symmetric we can take any odd function T_1 such that $T_1(-x) = -T_1(x)$, $x \in R_1$ (e.g. Sin x) with finite variance. Then Theorem 3.6.4 asserts that Cov $(M, T_1(X_i)) = 0$ for each $i = 1, 2, \ldots, n$ or Cov $(M, \Sigma\, l_i T_1(X_i)) = 0$. For example we can take $T_1(X_i) = X_i^{2r-1}$, $r \geq 1$ any odd power of x to claim that $\text{Cov}\left(M, \sum_{i=1}^{n} l_i X_i^{2r-1}\right) = 0$, $\forall\ \theta \in \Omega$. Taking $\varphi(M)$ to be any power of M, we have Cov $(M^s, \sum_{i=1}^{n} l_i X_i^{2r-1}) = 0$ as M^s would be MVUE for $E(M^s) = \dfrac{\Gamma(n+s)}{\Gamma(n)} \theta^s$. Using the first result of this example we have $\text{Cov}\left(\sum_{s=1}^{k} a_s M^s, \sum_{i=1}^{n} l_i X_i^{2r-1}\right) = 0$.

EXAMPLE 3.6.3 Let $(X_1, \ldots, X_n)$ be i.i.d. $N(\theta, 1)$ then $\bar{X}$ is minimal complete sufficient statistic. Let $e_i = X_i - \bar{X}$ be the i-th residual. Then $E(e_i) = 0$ and $e_i \in U_0$. Therefore Cov $(\bar{X}, e_i) = 0$ for any $i = 1, 2, \ldots, n$. We note that not only $E(e_i) = 0$ but $e_i \sim N\left(0, 1 - \dfrac{1}{n}\right)$ and its distribution is independent of θ. Again joint distribution of $\bar{X}$ and $X_i - \bar{X}$ is bivariate normal with mean vector $(\theta, 0)'$ and covariance matrix $\begin{pmatrix} \frac{1}{n} & 0 \\ 0 & \frac{n-1}{n} \end{pmatrix}$ and $\bar{X}$ and $X_i - \bar{X}$ are stochastically independent a much stronger result than Cov $(X_i - \bar{X}, \bar{X}) = 0$. Similarly let $S^2 = \Sigma (X_i - \bar{X})^2$ then $S^2 \sim \chi^2_{n-1}$ and its distribution is independent of θ and therefore Cov $(\bar{X}, S^2) = 0$ or using the fact that $\varphi(\bar{X})$ is MVUE of its expectation and $\eta(S^2)$ has expectation independent of θ, it follows that Cov $(\varphi(\bar{X}), \eta(S^2)) = 0$, Of course we already know that $\bar{X}$ and S^2 are stochastically independent and therefore $\varphi(\bar{X})$ and $\eta(S^2)$ would also be stochastically independent implying thereby Cov $(\varphi(\bar{X}), \eta(S^2)) = 0$.

The above example is an illustration of a theorem due to Basu (1955) which states that

THEOREM 3.6.5 (Basu's Theorem). Let M be minimal complete sufficient statistic for $\{L(x, \theta), \theta \in \Omega\}$ and let T be a statistic such that the distribution of T does not depend on θ. Then M and T are stochastically independent.

Consider $P(T \leq t) = F(t)$. Define an indicator function $I(x) = 1$ if $T(x) \leq t$ and zero otherwise. Then $E_\theta(I(x)) = F(t)$ for all $\theta \in \Omega$. Now consider $\eta(m) = E_\theta[I(x) \mid M = m] = P(T(x) \leq t \mid M))$. Then $E_\theta(\eta(M)) = E_\theta(I(x)) = F(t)$ or $E_\theta(\eta(M) - F(t)) = 0 \ \forall \ \theta \in \Omega$. Since M is complete this implies that $\eta(M) = F(t)$ w.p.1 under each $\theta \in \Omega$. Therefore $P[T \leq t \mid M = m] = P[T \leq t]$ for each pair t, m and for every $\theta \in \Omega$ therefore T and M are independent.

A statistic T whose distribution does not depend on θ is called as ancillary statistic and Basu's theorem can be paraphrased as saying that if M is minimal sufficient and complete statistic and T is ancillary then T and M are stochastically independent.

4

Simultaneous Estimation of Several Parameters

4.1 Optimality Criteria

In this chapter we consider the problem of simultaneous estimation of say a vector of k functions $\psi(\theta) = (\psi_1(\theta), \ldots, \psi_k(\theta))'$, $k > 1$ where functions $\{1, \psi_1, \ldots, \psi_k\}$ are linearly independent over Ω the parameter space and where $\theta = (\theta_1, \ldots, \theta_m)'$, $m \geq 1$ is itself a real or vector valued parameter. The vector parametric function ψ is estimated by a vector valued statistic $T = (T_1, \ldots, T_k)'$. The criterion of unbiasedness for real valued function $\psi(\theta)$ considered in Chapter 3, i.e. the case $k = 1$, can be easily extended by defining vector valued statistic T to be unbiased for vector valued ψ if

$$E(T_i) = \psi_i(\theta),\ i = 1, 2, \ldots, k,\ \forall\ \theta \in \Omega \tag{4.1.1}$$

Thus T is unbiased for ψ if it is componentwise unbiased. For example if we have a random sample of size n from $N(\theta, \sigma^2)$ then as $E(\bar{X}) = \theta$ and $E(S^2/n - 1) = \sigma^2$ where $S^2 = \Sigma\,(X_i - \bar{X})^2$ we have $\left(\bar{X}, \dfrac{1}{n-1}\Sigma\,(X_i - \bar{X})^2\right)'$ is unbiased for $(\theta, \sigma^2)'$. From results of Chapter 3 we know that $\bar{X}$ is MVUE for θ and $S^2/(n-1)$ is MVUE for σ^2. However we must now define the optimality criterion of the "Minimum Variance" for a vector valued statistic T unbiased for vector valued function ψ. Such a criterion must be related to the Variance Covariance matrix of T. Thus we assume that the estimator T is such that its Variance Covariance matrix M_T exists and is positive definite (pd). This is analogous to the requirement in case $k = 1$, where we assumed that $T \in U_\psi$ is such that Var $(T) = \sigma_T^2(\theta)$ exists and is positive. In case of $k = 1$ for $T_1, T_2 \in U_\psi$, T_1 is preferred to T_2 if Var $(T_1) \leq$ Var (T_2), $\forall\ \theta \in \Omega$. In a similar way if $T_1 = (T_{11}, \ldots, T_{1k})'$ and $T_2 = (T_{21}, \ldots, T_{2k})'$ are both unbiased for $\psi(\theta) = (\psi_1(\theta), \ldots, \psi_k(\theta))'$, T_1 is preferred to T_2 if $M_{T_2} - M_{T_1}$ is non-negative definite (nnd) for every $\theta \in \Omega$.

Definition 4.1.1 $T^* \in U_\psi$ is MVUE for ψ if $M_T - M_{T^*}$ is nnd $\forall\ \theta \in \Omega$, $\forall\ T \in U_\psi$.

We note the following important consequences of the above definition. Since $M_T - M_{T^*}$ is nnd it follows that for any i, Var $(T_i^*) \leq$ Var (T_i), $\forall\ \theta \in \Omega$ where $T_i, T_i^* \in U_{\psi_i}$. Thus each T_i^* is MVUE of $\psi_i(\theta)$ and the MVUness property of T^* for ψ as defined above implies componentwise MVUness of T^* for ψ. We have already seen in Chapter 3 that if T_i^*, $i = 1, 2, \ldots, k$ are MVUE for ψ_i, $i = 1, 2, \ldots, k$, then any linear combination $T_a^* = \sum_{i=1}^{k} a_i T_i^*$ is MVUE of $\psi_a = \sum_{i=1}^{k} a_i \psi_i$. Therefore for any $T \in U_\psi$

$$\text{Var}(T_a^*) = a' M_{T^*} a \le \text{Var}(T_a) = a' M_T a$$

One can conversely define T^* to be MVUE for ψ if for any $a \in R_k$ and any $T \in U_\psi$, $a' M_T a \ge a' M_{T^*} a \ \forall \ \theta \in \Omega$, which implies that $(M_T - M_{T^*})$ is nnd. Therefore an equivalent definition of MVUness of T^* can be given as

Definition 4.1.2 $T^* \in U_\psi$ is MVUE for ψ iff for any $a \in R_k$ and any $T \in U_\psi \cdot a' M_{T^*} a \le a' M_T a, \ \forall \ \theta \in \Omega$.

We note that the definition of MVUness of T^* for ψ immediately implies that for any $T \in U_\psi$ and $\forall \ \theta \in \Omega$

(i) $| M_{T^*} | \le | M_T |$ (4.1.2)

(ii) $Tr(M_{T^*}) \le Tr(M_T)$ (4.1.3)

(iii) $\lambda_1(M_{T^*}) \le \lambda_1(M_T)$ (4.1.4)

where $\lambda_1(A)$ denotes the largest eigen-value or the largest characteristic root of a positive definite matrix A and $| A |$ denotes its determinant and $Tr(A)$ denotes the trace of A. Historically speaking the optimality of vector unbiased estimator T was defined by Cramèr (1946) using minimization of the determinant of the Variance-Covariance matrix $| M_T |$, which was later called as generalized variance. Cramèr arrived at this criterion through the concept of an ellipsoid of concentration which we shall consider next. The geometrical interpretation will also throw more light on the optimality criteria defined by (4.1.2), (4.1.3) and (4.1.4) which are termed as *D*-optimality, *T*-optimality and *E*-optimality respectively. The definition (4.1.1) or equivalently (4.1.2) of MVUness will be called as *M*-optimality.

4.2 Ellipsoid of Concentration

Let T be a real statistic unbiased for a real parametric function $\psi(\theta)$ with variance $\sigma_T^2(\theta)$. Then as observed in Chapter 3, the MVUE T^* of ψ has highest concentration of probability around ψ, as a consequence of Tchebychev's inequality. More formally for each $T \in U_\psi$ consider a r.v. W which is uniformly distributed over the interval $(\psi(\theta) - \sqrt{3}\sigma_T(\theta), \psi(\theta) + \sqrt{3}\,\sigma_T(\theta))$ so that $E(W) = \psi(\theta)$ and $\text{Var}(W) = \sigma_T^2(\theta)$. Then W^* the r.v. corresponding to the MVUE T^* has the property that the length of the interval of the range of W^*, $2\sqrt{3}\,\sigma_{T^*}(\theta) \le 2\sqrt{3}\,\sigma_T(\theta)$ which is the length of the interval of the range of W, corresponding to any other $T \in U_\psi$.

We now generalize this property for a k-dimensional vector valued statistic T unbiased for a k-dimensional vector parametric function ψ. For each such $T \in U_\psi$ we consider a k-dimensional random vector W, which is uniformally distributed over a k-dimensional bounded connected set $S \subset R_k$ so that $E(W) = \psi$ and $M_w = M_T$ i.e. W is centered at ψ and has same Variance Covariance matrix as that of T. However in R_k there are many choices for such a region S whereas in R_1 such a region must only be an interval of the type $(\psi(\theta) - a, \psi(\theta) + a)$. In R_2 we could take rectangle, a circle or an ellipse. The choice of a rectangle or a circle is not as suitable for manipulation as an ellipse, and therefore we take S to be an ellipse with centre $\psi(\theta)$. Below we illustrate the algebra/calculus for the case $k = 2$ which can be generalized for $k \geq 3$ in a straight-forward manner.

Let $A = \begin{pmatrix} a_{11} & a_{12} \\ a_{12} & a_{22} \end{pmatrix}$ be a symmetric positive definite matrix and let $Y_1 = (W_1 - \psi_1(\theta))$ and $Y_2 = (W_2 - \psi_2(\theta))$. Then the r.v. $(Y_1, Y_2)'$ is assumed to be uniformly distributed over the region bounded by the ellipse defined by

$$a_{11}y_1^2 + 2a_{12}y_1y_2 + a_{22}y_2^2 = c^2 \quad \text{or} \quad y'Ay = c^2 \tag{4.2.1}$$

The elements of A and the value of c^2 are to be chosen such that $M_Y = M_T$.

Since A is symmetric pd there exists an orthogonal matrix P such that $P'AP = (\Lambda) = \begin{pmatrix} \lambda_1 & 0 \\ 0 & \lambda_2 \end{pmatrix}$ and the corresponding linear transformation is $Z = Py$ and the ellipse given by (4.2.1) now becomes ellipse in the standard form

$$\frac{\lambda_1 z_1^2}{c^2} + \frac{\lambda_2 z_2^2}{c^2} = 1 \tag{4.2.2}$$

Let $u_1 = \dfrac{\sqrt{\lambda_1}}{c} z_1$, $u_2 = \dfrac{\sqrt{\lambda_2}}{c} z_2$ then (4.2.2) becomes the circle

$$u_1^2 + u_2^2 = 1 \tag{4.2.3}$$

Now as the area of the ellipse defined by (4.2.1) is $\dfrac{c^2\pi}{|A|^{1/2}}$ the joint pdf of $(Y_1, Y_2)'$ is given by

$$f(y_1, y_2) = \frac{|A|^{1/2}}{c^2\pi}, \; y'Ay \leq c^2.$$

By rule of transformation and using the fact that P is an orthogonal matrix, the joint pdf of $(Z_1, Z_2)'$ is given by

$$f(z_1, z_2) = \frac{|A|^{1/2}}{c^2\pi}, \quad \frac{\lambda_1 z_1^2}{c^2} + \frac{\lambda_2 z_2^2}{c^2} \leq 1$$

and that of $(U_1, U_2)'$ is given by

$$f(u, u_2) = 1/\pi \text{ over the circle } u_1^2 + u_2^2 \leq 1$$

Note that $$dz_1 dz_2 = \frac{c^2}{\sqrt{\lambda_1 \lambda_2}} du_1 du_2 \text{ and } \lambda_1 \lambda_2 = |A|.$$

Now we have $E(U) = 0$ which implies that $E(Z) = 0$ and $E(Y) = 0$. Further, $M_U = \frac{1}{4} I$, where I is the identity matrix. Therefore, $M_Z = \frac{c^2}{4} \Lambda^{-1} = \frac{c^2}{4} PA^{-1}P'$ since $P^{-1} = P'$. Now $M_Z = E(ZZ') = E(PYY'P') = PM_TP'$. Hence the matrix A must be selected such that

$$\frac{c^2}{4} PA^{-1}P' = PM_TP' \tag{4.2.4}$$

In particular if we take $c^2 = 4$ we have $A^{-1} = M_T$ or $A = M_T^{-1}$ and the ellipse of concentration corresponding to T given by (4.2.1) is

$$y' M_T^{-1} y = 4 \tag{4.2.5}$$

For a k-dimensional case we proceed similarly and take k dimensional ellipsoid $y'Ay = c^k$, with $c > 0$. The elements of A and c are to be chosen such that $M_Y = M_T$. Define $Z = PY$ where $P'AP = \Lambda = \text{diag}(\lambda_1, \ldots, \lambda_k)$ and take $u_i = \frac{\sqrt{\lambda_i}}{c} Z_i$, $i = 1, 2, \ldots, k$ to get k-dimensional sphere $\sum_{i=1}^{k} u_i^2 = 1$. The distributions of U, Z and Y are obtained by using the formula for the volume of a sphere and an ellipsoid in k-dimensions, to obtain pdfs as:

$$f(u_1, \ldots, u_k) = \Gamma\left(\frac{k}{2} + 1\right) / \pi^{k/2}, \sum u_i^2 \leq 1$$

$$f(z_1, \ldots, z_k) = \frac{|A|^{1/2} \Gamma\left(\frac{k}{2} + 1\right)}{\pi^{k/2} c^k}, \sum \lambda_i z_i^2 \leq c^k$$

$$f(y_1, \ldots, y_k) = \frac{|A|^{1/2} \Gamma\left(\frac{k}{2} + 1\right)}{\pi^{k/2} c^k}, \ y'Ay \leq c^k$$

Now by straightforward calculations we can show that $E(U) = 0$ which implies that $E(Z) = 0$ and $E(Y) = 0$ and $M_U = \frac{1}{k+2} I$. Further $M_Z = \frac{c^k}{k+2} \Lambda^{-1} = \frac{c^k}{k+2} PA^{-1}P'$ and therefore we should select $c > 0$ and A such that

$$\frac{c^k}{k+2} PA^{-1}P' = PM_TP'$$

Taking $c^k = k + 2$, we have $A^{-1} = M_T$ or $A = M_T^{-1}$ and the ellipsoid of concentration corresponding to T is

$$y' M_T^{-1} y = k + 2 \tag{4.2.6}$$

The volume of the ellipsoid of concentration of T is $V_T = \dfrac{c^k \pi^{k/2}}{\Gamma\left(\frac{k}{2}+1\right) | M_T^{-1} |}$

which is proportional to $| M_T |$ and Cramèr recommended the criterion of minimizing V_T over $T \in U_\psi$. This is equivalent to minimizing $| M_T |$ the generalized variance of T for $T \in U_\psi$ which leads to the D-optimal estimator $T \in U_\psi$.

Following the above approach of Cramèr, Lehmann and Scheffè (1950) proposed a preference relation between T_1 and $T_2 \in U_\psi$. They recommended preferring T_1 to T_2 if for any $a \in R_m$ and $\forall\ \theta \in \Omega$

$$a' M_{T_1}^{-1}\ a \geq a'\ M_{T_2}^{-1}\ a \tag{4.2.7}$$

or $(M_{T_1}^{-1} - M_{T_2}^{-1})$ is nnd which is equivalent to the criterion where we prefer T_1 to T_2 if $(M_{T_2} - M_{T_1})$ is nnd. This leads to the M-optimality criterion which we have adopted as the definition 4.1.1.

In a similar way Trace-optimality can be described geometrically as minimizing the sum of half axes of the ellipsoid of concentration and E-optimality can be interpreted as minimizing the largest half major axis of the ellipsoid of concentration. This is a consequence of the fact that the characteristic roots of A which are $(\lambda_1, \ldots, \lambda_k)$ are reciprocals of characteristic roots of M_T^{-1} which are $(1/\lambda_1, \ldots, 1/\lambda_k)$.

4.3 Klebanov-Linnik-Rukhin Theorem

Section 3.6 showed that a necessary and sufficient condition for $T^* \in U_\psi$ a real statistic to be MVUE of a real parametric function ψ, is that $\text{Cov}(T^*, u) = 0$, $\forall\ u \in U_0$, $\forall\ \theta \in \Omega$. We now consider a generalization of this condition for a vector valued statistic $T^* \in U_\psi$, to be M-optimal for ψ, or equivalently MVUE of ψ.

Let $T = (T_1, \ldots, T_k)' \in U_\psi$ where $\psi = (\psi_1, \ldots, \psi_k)'$ and $\theta = (\theta_1, \ldots, \theta_m)'$ where $m \geq 1$. Let $U = (u_1, \ldots, u_k)' \in U_0^{(k)}$ defined by $E(U) = 0$, $\forall\ \theta \in \Omega$. Let M_{TU} be the covariance matrix of T and U i.e. $M_{TU} = E[(T - \psi)\ (U')]$. Then Klebenov-Linnik-Rukhin theorem states that

THEOREM 4.3.1 (K-L-R THEOREM). A necessary and sufficient condition for $T^* \in U_\psi$ to be M-optimal for ψ is $M_{T^*U} = M_{UT^*} = 0$, $\forall\ U \in U_0^{(k)}$, and $\forall \theta \in \Omega$.

We have already seen that M-optimality of T^* implies that for each $i = 1, \ldots, k$, T_i^* is MVUE of ψ_i and therefore for any u_j, such that $E(u_j) = 0$, $\forall\ \theta \in \Omega$ we have

$$\text{Cov}\ (T_i^*, u_j) = 0\ \forall\ \theta \in \Omega,\ i = 1, 2, \ldots, k, j = 1, 2, \ldots, k \quad (4.3.1)$$

Therefore $M_{T^*U} = M_{UT^*} = 0$, $\forall\ u \in U_0^{(k)}$, $\forall\ \theta \in \Omega$.

On the other hand if $M_{T^*U} = 0 = M_{UT^*}$ consider $U = T - T^*$ where $T \in U_\psi$ then

$$M_{T^*(T-T^*)} = 0 \text{ or } M_{T^*T} = M_{T^*T^*} = M_{TT^*} = M_{T^*} \quad (4.3.2)$$

Let M_{T^*-T} be the Variance Covariance matrix of $T^* - T$. Then

$$M_{T^*-T} = M_{T^*} - M_{T^*T} - M_{TT^*} + M_T$$

$$= M_T - M_{T^*} \text{ in view of (4.3.2)}$$

But as M_{T^*-T} is nnd, we have $M_T - M_{T^*}$ is nnd for any $T \in U_\psi$ and $\forall\ \theta \in \Omega$ or T^* is M-optimal for ψ.

Corollary 4.3.1 T^* is M-optimal for ψ if and only if T_i^* is MVUE of ψ_i for each $i = 1, 2, \ldots, k$.

Corollary 4.3.2 If T_1 and T_2 are M-optimal for ψ_1 and ψ_2, respectively, then $T^* = AT_1 + BT_2$ is M-optimal for $A\psi_1 + B\psi_2$ where A, B are real matrices of dimension $l \times k$.

Corollary 4.3.3 If T_1 and T_2 are both M-optimal for ψ then $T_1 = T_2$ with probability one, $\forall\ \theta \in \Omega$.

As a consequence of Corollary 4.3.1 we can construct an M-optimal estimator for $\psi = (\psi_1, \ldots, \psi_k)'$ by first obtaining MVUE, T_i^* of each ψ_i using the methods given in Chapter 3 and then form $T^* = (T_1^*, \ldots, T_k^*)'$ which will be M-optimal for ψ. Thus, we use Rao-Blackwell Lehmann-Scheffe′ Theorem, particularly in case of multiparameter exponential families to obtain M-optimal estimators of a vector of parametric functions of interest. We now consider a few examples to illustrate this procedure.

EXAMPLE 4.3.1 Consider one way Analysis of Variance or where we have k samples of size n_i each from $N(\mu_i, \sigma^2)$ so that

$$X_{ij} = \mu_i + \varepsilon_{ij},\ j = 1, 2, \ldots, n_i,\ i = 1, 2, \ldots, k$$

where ε_{ij} are i.i.d $N(0, \sigma^2)$. By straightforward calculations (which the reader should carry out) we can show that the joint pdf of $N = \sum n_i$ r.v.s X_{ij} is a $(k + 1)$ dimensional exponential family with $(T_1, T_2, \ldots, T_k, S^2)'$ as minimal sufficient and complete statistic where

$$T_i = \bar{X}_i,\ i = 1, 2, \ldots, k \text{ and } S^2 = \sum S_i^2,$$

$S_i^2 = \sum_{j=1}^{n_i} (X_{ij} - \bar{X}_i)^2$. Note that $\bar{X}_i$, $i = 1, 2, \ldots, k$ and S_i^2, $i = 1, 2, \ldots, k$ are mutually independent. Further $\bar{X}_i \sim N(\mu_i, \sigma^2/n)$ and $S_i^2 \sim \sigma^2\chi^2_{n_i-1}$. Therefore

$\bar{X}_i$ is MVUE for μ_i for $i = 1, 2, \ldots, k$ and it follows that $(\bar{X}_i, \ldots, \bar{X}_k)'$ the vector of sample means, is M–optimal or MVUE estimator of $(\mu_1, \ldots, \mu_k)'$ the vector of k-population means. Now $S^2 \sim \sigma^2\chi^2_{N-k}$ where $N = \sum n_i$, as $S_i^2 \sim \sigma^2\chi^2_{n_i-1}$ are mutually independent. Therefore, $\hat{\sigma}^2 = \sum_{i=1}^{k} S_i^2/N - k = S^2/(N-K)$ is MVUE of σ^2. Suppose we want to estimate $\psi_{ij} = (\mu_i - \mu_j)/\sigma$, i.e. the difference between the i-th and the j-th population means measured in units of population standard deviation σ. Then $c(\bar{X}_i - \bar{X}_j)/S$, where c is chosen such that $E(c/S) = 1/\sigma$ would be MVUE of ψ_{ij}. Note that $E[(\bar{X}_i - \bar{X}_j)c/S] = E(\bar{X}_i - \bar{X}_j) \cdot E(c/S)$ because of mutual independence. Using the fact that $S^2 \sim \sigma^2\chi^2_{N-k}$ the value of c can be determined and we can show that $c = \Gamma\left(\frac{N-k}{2}\right)\sqrt{2}\Big/\Gamma\left(\frac{N-k-1}{2}\right)$. Vector $(\hat{\psi}_{12}, \hat{\psi}_{13}, \ldots, \hat{\psi}_{1k})'$ is then M-optimal estimator of $(\psi_{12}, \ldots, \psi_{1k})'$.

EXAMPLE 4.3.2 (Chandrasekar, 1983). We consider now an application of K-L-R Theorem to a case in which class U_0 can be determined. Consider a single observation from $U(\alpha, \beta)$ where Ω is defined by $-\infty < \alpha \le \alpha_0 < \beta_0 \le \beta < \infty$. Then X has pdf $f(x, \alpha, \beta) = 1/(\beta - \alpha)$ over Ω. Let $u(X)$ be an unbiased estimator of zero then

$$\int_\alpha^\beta u(x)\,dx = 0 \;\forall\; (\alpha, \beta)' \in \Omega.$$

In particular taking $\alpha = \alpha_0$ and $\beta = \beta_0$, $u(x)$ must be such that

$$\int_{\alpha_0}^{\beta_0} u(x)\,dx = 0 \tag{4.3.3}$$

As $\int_\alpha^\beta u(x)\,dx = \int_\alpha^{\alpha_0} u(x)\,dx + \int_{\alpha_0}^{\beta_0} u(x)\,dx + \int_{\beta_0}^{\beta} u(x)\,dx = 0$ in view of (4.3.3) and taking $\beta = \beta_0$ we must have $\int_\alpha^{\alpha_0} u(x)\,dx = 0,\ \forall\ \alpha \le \alpha_0$ which implies that $u(x) = 0$ for $x \le \alpha_0$. Therefore $\int_{\beta_0}^{\beta} u(x)\,dx = 0,\ \forall\ \beta \ge \beta_0$ which implies that $u(x) = 0$ for $x \ge \beta_0$. The arguments here are same as those indicated in the proof of completeness of $X_{(n)}$ for $U(0, \theta)$, $\theta > 0$. Thus $\int_\alpha^{\alpha_0} u(x)\,dx = 0$ implies that $\int_\alpha^{\alpha_0} u^+(x)\,dx = \int_\alpha^{\alpha_0} u^-(x)\,dx$ for all $\alpha \le \alpha_0$ i.e. the measures assigned by $u^+(x)$ and $u^-(x)$ for any interval (α, α_0) are same which implies that $u^+(x) = u^-(x)$ for $x \le \alpha_0$ or $u(x) = 0$ for $x \le \alpha_0$. Similarly we can show that $u(x) = 0$ for $x \ge \beta_0$. Thus

$$U_0 = \{u(x) \mid u(x) = 0 \text{ if } x \notin (\alpha_0, \beta_0), \int_{\alpha_0}^{\beta_0} u(x)\, dx = 0\} \quad (4.3.4)$$

Now consider any estimator $T(x)$ such that it remains constant on (α_0, β_0). Then Cov $(T(x), u) = 0$ for $\forall\, u \in U_0$ and $\forall\, (\alpha, \beta) \in \Omega$ and therefore $T(x)$ is MVUE of $E(T(x))$. For example let

$$T_1(x) = \frac{\alpha_0 + \beta_0}{2}, \quad x \in (\alpha_0, \beta_0)$$

$$= x, \quad x \notin (\alpha_0, \beta_0)$$

Then

$$E(T_1(x)) = \frac{1}{\beta - \alpha}\left[\int_{\alpha}^{\alpha_0} x\, dx + \int_{\beta_0}^{\beta} x\, dx + \frac{(\alpha_0 + \beta_0)}{2}(\beta_0 - \alpha_0)\right]$$

$$= \frac{1}{(\beta - \alpha)}\left[\frac{\alpha_0^2 - \alpha^2}{2} + \frac{\beta^2 - \beta_0^2}{2} + \frac{\beta_0^2 - \alpha_0^2}{2}\right]$$

$$= \frac{\alpha + \beta}{2}$$

Thus $T(x)$ is MVUE of $\frac{\alpha + \beta}{2}$ which is the mean of the distribution. Note that X itself is unbiased for $\frac{\alpha + \beta}{2}$ but X is not MVUE for $\frac{\alpha + \beta}{2}$. Similarly consider k-th raw moment $\mu_k'(\alpha, \beta) = \frac{(\beta^{k+1} - \alpha^{k+1})}{(\beta - \alpha)(k + 1)}$. Define $T_k(x) = c$ if $x \in (\alpha_0, \beta_0)$ and x^k if $x \notin (\alpha, \beta)$. Then straightforward calculations show that

$$E(T_k(x)) = \frac{c(\beta_0 - \alpha_0)}{(\beta - \alpha)} + \frac{\beta^{k+1} - \beta_0^{k+1}}{(k + 1)(\beta - \alpha)} + \frac{\alpha_0^{k+1} - \alpha^{k+1}}{(k + 1)(\beta - \alpha)}$$

$$= \frac{\beta^{k+1} - \alpha^{k+1}}{(k + 1)(\beta - \alpha)} + \frac{1}{(\beta - \alpha)}\left[c(\beta_0 - \alpha_0) + \frac{\alpha_0^{k+1} - \beta_0^{k+1}}{(k + 1)}\right]$$

If we take $c = \frac{\beta_0^{k+1} - \alpha_0^{k+1}}{(k + 1)(\beta_0 - \alpha_0)}$ then

$$E(T_k(x)) = \frac{\beta^{k+1} - \alpha^{k+1}}{(k + 1)(\beta - \alpha)} = \mu_k'(\alpha, \beta)$$

As Cov $(T_k(x), u) = 0\ \forall\, u \in U_0$ and $(\alpha, \beta) \in \Omega$ each $T_k(x)$ is MVUE of

$\mu_k'(\alpha, \beta)$. It now follows that $(T_1(x), \ldots, T_k(x))'$ would be M-optimal for the first k-moments $(\mu_1'(\alpha, \beta), \ldots, \mu_k'(\alpha, \beta))'$.

EXAMPLE 4.3.3 A model that commonly occurs in modelling failure time distribution is described by the pdf

$$f(x, \mu, \sigma) = \frac{1}{\sigma} \exp \left\{ \frac{(x - \mu)}{\sigma} \right\}, \; x \geq \mu, \mu > 0, \sigma > 0$$

Here μ denotes the minimum life before which the item would not fail. The pdf also describes prognosis of a disease where X would denote the serum level and μ is the threshold point. The model is a combination of Pitman family and Exponential family since for μ known it becomes one parameter exponential family and for σ known it belongs to Pitman family where the lower end depends on the parameter μ.

As the r.v. X is continuous, a sufficient statistic is given by the order statistic $(X_{(1)}, \ldots, X_{(n)})$ the failure times of the first failure, second failure etc. Using the one to one transformation given by $Y_1 = n(X_{(1)} - \mu)$, $Y_i = (n - i + 1)\,(X_{(i)} - X_{(i-1)})$, $i = 2, 3, \ldots, n$, we can show that $(Y_1, Y_2, \ldots, Y_n)$ are i.i.d. exponentials and

$$g(x_{(1)}, \ldots, x_{(n)}, \mu, \sigma) = \frac{n}{\sigma} \exp \left\{ -\frac{n(x_{(1)} - \mu)}{\sigma} \right\} \frac{1}{\sigma^{n-1}} \exp \{- \sum_{i=2}^{n} y_i / \sigma\}$$

and hence $X_{(1)}$ and $T = \sum_{i=2}^{n} Y_i = \sum_{i=2}^{n} (n - i + 1)\,(X_{(i)} - X_{(i-1)}) = \sum_{i=2}^{n} (X_{(i)} - X_{(1)})$ are independent with pdfs given by

$$g_1(x_{(1)}, \mu, \sigma) = \frac{n}{\sigma} \exp \left\{ -\frac{n(x_{(1)} - \mu)}{\sigma} \right\}, \; x_{(1)} > \mu$$

$$g_2(t, \mu, \sigma) = \frac{1}{\Gamma(n - 1)} \frac{1}{\sigma^n} e^{-t} / \sigma \; t^{n-1}, \; t > 0$$

By factorizability criterion it follows that $(X_{(1)}, T)'$ is sufficient for $(\mu, \sigma)'$ and one can show that it is minimal sufficient and complete. The proof is beyond our scope and we will therefore assume the result.

Suppose we want to obtain M-optimal estimator of $(\mu, \sigma)'$. Then $\frac{T}{n - 1}$ is MVUE of σ. Observing that $E(X_{(1)}) = \mu + \sigma/n$ we have $X_{(1)} - \frac{T}{n\,(n - 1)}$ is MVUE of μ. Therefore the M-optimal estimator of $(\mu, \sigma)'$ is given by $\left(X_{(1)} - \frac{T}{n\,(n - 1)}, \frac{T}{n - 1} \right)'$. The MVUE of $E(X) = \mu + \sigma$ is then given by

$$T_1 = X_{(1)} - \frac{T}{n(n-1)} + \frac{T}{n-1} = X_{(1)} + \frac{T}{n} = [nX_{(1)} + \sum_{2}^{n} (X_{(i)} - X_{(1)})]/n$$

$$T = \frac{1}{n}\sum_{1}^{n} X_{(i)} = \bar{X}$$

Consider the reliability function, the probability that the item will not fail before time t_0

$$R(t_0) = P[X > t_0] = 1 \text{ if } t_0 \leq \mu$$

$$= \exp\left\{-\frac{(t_0 - \mu)}{\sigma}\right\} \text{ if } t_0 > \mu$$

Then Laurent (1963) showed that by Rao-Blackwellizing w.r.t. $(X_{(1)}, T)$, starting with the estimator

$$T_1(X_1) = 1 \quad \text{if } X_1 > t_0$$

$$= 0 \quad \text{otherwise}$$

the MVUE of $R(t_0)$ is given by

$$\tilde{R}(t_0) = 1 \quad \text{if } t_0 \leq x_{(1)}$$

$$= \left(1 - \frac{1}{n}\right)\left[1 - \frac{t_0 - x_{(1)}}{T}\right]^{n-2}, \; x_{(1)} < t_0 < x_{(1)} + \sum_{i=2}^{n} (x_{(i)} - x_{(1)})$$

$$= 0 \qquad t_0 > x_{(1)} + \sum_{i=2}^{n} (x_{(i)} - x_{(1)})$$

We have then the result that $(\tilde{R}(t_1), \ldots, \tilde{R}(t_k))'$ is M-optimal for the vector of reliability functions evaluated at k-points $(t_1, \ldots, t_k)$.

Exercise 4.3.4 Let $(X_1, \ldots, X_n)$ be i.i.d. $N(\mu, \sigma^2)$. Then $(\bar{X}, S^2)'$ is complete sufficient for $(\mu\ \sigma^2)'$ and $(\bar{X}, S^2/(n-1))$ is M-optimal for $(\mu, \sigma^2)'$. Let $\varphi_1(X_1) = 1$ if $X_1 < a_1$ and zero otherwise so that $E[\varphi_1(X_1)] = \Phi((a_1 - \mu)/\sigma)$. Now for any C

$$P[X_1 < C \mid (\bar{X}, S^2)] = P\left[\frac{X_1 - \bar{X}}{S} < \frac{C - \bar{X}}{S} \mid \bar{X}, S^2\right]$$

$$= P[U < u_0 \mid \bar{X}, S^2]$$

where $U = \sqrt{n}(X_1 - \bar{X})/\sqrt{(n-1)S^2}$ and $u_0 = \sqrt{n}(C - \bar{X})/\sqrt{(n-1)S^2}$. The distribution of U is independent of $(\bar{X}, S^2)$, in view of Theorem 3.6.5 (Basu's Theorem) and therefore $P[U < u_0 \mid \bar{X}, S^2] = P[U < u_0]$. The distribution of U is known and its pdf is given by

$$g(u) = \frac{1}{\pi}\frac{\Gamma((n-1)/2)}{\Gamma((n-2)/2)}(1 - u^2)^{(n-4)/2}, \quad |u| < 1$$

from which $P[U < u_0]$ can be obtained by using tables of incomplete beta distribution taking into account the sign of u_0 for given observed value of $(\bar{X}, S^2)$ for the sample. The idea is to use the fact that U^2 is beta, $B(1, (n-2)/2)$ and use the result that

$$P[U < u_0] = \frac{1}{2} + P[U^2 < u_0^2] \quad \text{if } u_0 > 0$$

$$= \frac{1}{2} - P[U^2 < u_0^2] \quad \text{if } u_0 < 0$$

Kolmogorov (1950) considered this problem in the context of quality control. Suppose that specifications indicate that the diameter of head of a screw must be within (1 ± 0.01) mm. The proportion of acceptable screws in a given batch, when the process mean is μ mm, and process variance is σ^2 $(\text{mm})^2$ is given by

$$\pi = \Phi\left(\frac{1.01 - \mu}{\sigma}\right) - \Phi\left(\frac{0.99 - \mu}{\sigma}\right)$$

and we can obtain MVUE of π using the above technique. The batch then is acceptable if $\hat{\pi} \geq \pi_0 = 0.95$ say, i.e. percentage defective < 5%.

We point out here that the abovementioned paper of Kolmogorov defines sufficiency from Bayesian view point and also gives Rao-Blackwell Theorem. In some Russian texts the Rao-Blackwell Theorem is referred to as Kolmogorov-Rao-Blackwell Theorem.

EXAMPLE 4.3.5 Let (X_i, Y_i), $i = 1, 2, \ldots, n$ be i.i.d. bivariate discrete r.v. with p.m.f. given by

$$f(x, y, \lambda, p) = \binom{x}{y} p^y (1 - p)^{x-y} e^{-\lambda} \quad \lambda^x/x!, \; y = 0, 1, 2, \ldots, x, \quad x = 0, 1, \ldots$$

Then as seen in Example 2.5.5 we have $V_1 = \sum X_i$ and $V_2 = \sum Y_i$ are jointly complete and sufficient for $(\lambda, p)'$. The conditional distribution of V_2 given V_1 is binomial and marginal of V_1 is Poisson with mean $n\lambda$. Hence the joint pmf of (V_1, V_2) is

$$g(v_1, v_2, \lambda, p) = \binom{v_1}{v_2} p^{v_2} (1 - p)^{v_1 - v_2} e^{-n\lambda v_1} (n\lambda)^{v_1}/v_1!,$$

$$v_2 = 0, 1, \ldots, v_1, \; v_1 = 0, 1, \ldots$$

As $E(V_2 \mid V_1) = V_1 p$ we have $E(V_2) = pE(V_1) = n\lambda p$. Hence V_2/n is MVUE of λp and V_1/n is MVUE of λ and therefore $(V_1/n, V_2/n)'$ is M-optimal for $(\lambda, \lambda p)'$.

By a straight forward calculation we can show that Y is Poisson with mean λp and hence V_2 is Poisson with mean $n\lambda p$. Consider estimation of $P(Y = 0) = e^{-\lambda p}$. This is the probability that no wrong calls are connected at the exchange if X = the number of total telephone calls received in unit time and Y = the number of calls wrongly connected. Using the results of Example 3.4.2 we have

$$\varphi(v_1, v_2) = \left(\frac{n-1}{n}\right)^{v_2} \quad v_2 = 0, 1, 2, \ldots$$

is the MVUE of $e^{-\lambda p}$.

Exercise 4.3 (1) Let $y_i = \alpha + \beta x_i + \varepsilon_i$, $i = 1, 2, \ldots, n$ be the normal regression model where ε_i are i.i.d. $N(0, \sigma^2)$ and where x_i's are constants with $\sum (x_i - \bar{x})^2 > 0$. As seen in Example 2.5.4 the joint pdf of $(y_1, \ldots, y_n)$ belongs to 3 parameter exponential family. Determine M-optimal estimator of $(\alpha, \beta, \sigma^2)$, if it exists.

(2) Let $(X, Y)'$ be bivariate normal with parameters $(\mu_1, \mu_2, \sigma_1^2, \sigma_2^2, \rho)$. Show that the pdf of $(X, Y)'$ belongs to five parameter exponential family. Find M-optimal estimators of (a) the mean vector $(\mu_1, \mu_2)'$ (b) Variance-Covariance matrix

$$\begin{pmatrix} \sigma_1^2 & \rho\sigma_1\sigma_2 \\ \rho\sigma_1\sigma_2 & \sigma_2^2 \end{pmatrix}$$

(3) Let X_1 and X_2 be independent exponentials with means θ_1 and θ_2. Show that the joint pdf of $(X_1, X_2)'$ belongs to two parameter exponential family and obtain M-optimal estimator of $(\theta_1, \theta_2, P(X < Y))'$.

4.4 Cramer-Rao Inequality

Let X be a random vector with pdf belonging to the class $\{f(x, \theta), \theta \in \Omega \subset R_m\}$ which satisfies the regularity conditions given in Section 2.2 so that the Fisher information matrix $J(\theta)$ is well defined and is pd. Let $\psi = (\psi_1, \ldots, \psi_m)'$ be a vector of parametric functions such that the Jacobian $\left|\dfrac{\partial(\psi_1, \ldots, \psi_m)}{\partial(\theta_1, \ldots, \theta_m)}\right|$ is non-zero or $\theta \leftrightarrow \psi$ is one to one transformation. Let D_ψ denote the matrix $D = (\partial\psi_i/\partial\theta_j) = d_{ij}$, $i = 1, 2, \ldots, m$, $j = 1, 2, \ldots, m$. Let $T \in U_\psi$ be such that

$$\int T_i(x)\, f(x;\, \theta)\, dx = \psi_i, \quad i = 1, 2, \ldots, m$$

can be differentiated under the integral sign w.r.t. θ_j, $j = 1, 2, \ldots, m$ so that

$$\int T_i(x) \frac{\partial \log f}{\partial \theta_j}\, dx = \frac{\partial \psi_i}{\partial \theta_j} = d_{ij}$$

i.e.

$$\text{Cov}\,(T_i, S_j) = d_{ij} \tag{4.4.1}$$

where $S = (S_1, \ldots, S_m)'$ is the vector of score functions $\left(\dfrac{\partial \log f}{\partial \theta_1}, \ldots, \dfrac{\partial \log f}{\partial \theta_m}\right)'$. Thus D is the Variance-Covariance matrix. Let $u, v \in R_m$ and consider $Y = u'S$ and $Z = v'T$. Then $\text{Cov}\,(Y, Z) = E(u'\,ST'v) = u'Dv$ and $\text{Var}\,(Y) = u'Ju$ and $\text{Var}\,(Z) = v'M_Tv$. Then by Cauchy Schwarz inequality

$$(u'D'v)^2 \leq (u'Ju)\,(v'M_Tv) \tag{4.4.2}$$

Let $u = J^{-1}D'v$ then as $(J^{-1})' = J^{-1}$.

$$(v'\,DJ^{-1}D'v)^2 \le (v'DJ^{-1}J\,J^{-1}D'v)\;.\;(v'\,M_T v)$$
$$\le (v'\,DJ^{-1}D'v)\,(v'M_T v).$$

However since J is pd and D is non-singular we have

$$v'DJ^{-1}D'v \ge 0 \text{ and thus } v'\,DJ^{-1}D'v \le v'M_T v$$

or
$$v'(M_T - DJ^{-1}D')v \ge 0 \quad v \in R_m.$$

Thus $M_T - DJ^{-1}D'$ is nnd and we have the theorem,

THEOREM 4.4.1 Under the regularity conditions given above, for $T \in U_\psi$, $M_T - DJ^{-1}D'$ is nnd.

Corollary 4.4.1 For $\psi = (\theta_1, \ldots, \theta_m)'$ we have $D = I$.

$$M_T - J^{-1} \text{ is nnd.}$$

Remark 4.4.1 For the generalized variance of T, defined by $|\,M_T\,|$ we have $|\,M_T\,| \ge |\,D\,|^2/|\,J\,|$ and $Tr(M_T) \ge Tr(DJ^{-1}D')$. For $\psi = (\theta_1, \ldots, \theta_m)$ we have

$|\,M_T\,| \ge 1/|\,J\,|$ and $Tr(M_T) \ge \sum_{i=1}^{k} J^{ii}(\theta)$ where J^{ii} denotes the (i, i)-th diagonal element of the inverse of the Fisher information matrix J. Further for i-th component of T

$$\text{Var}(T_i) \ge \sum_{j=1}^{m}\sum_{l=1}^{m} d_{ij}J^{jl}d_{il}.$$

EXAMPLE 4.4.1 Let $\{X_i\}_1^n$ be i.i.d. $N(\theta_1, \theta_2)$ and let $\psi = (\theta_1, \theta_2)$. Then as observed in Example 3.7.3, $\bar{X}$ is MVUE of θ_1 and $S^2/(n-1)$ is MVUE of θ_2 and $T = (\bar{X}, S^2/(n-1))'$ is M-optimal estimator of $(\theta_1, \theta_2)'$. But $M_T = \begin{pmatrix} \theta_2^2/n & 0 \\ 0 & 2\theta_2^2/(n-1) \end{pmatrix}$ as $\bar{X} \sim N(\theta_1, \theta_2/n)$ and $S^2 \sim \theta_2\chi^2_{n-1}$ and $\bar{X}$ and S^2 are independent. Routine calculations will show that the Fisher information matrix $J(\theta) = \begin{pmatrix} n/\theta_2 & 0 \\ 0 & n/2\theta_2^2 \end{pmatrix}$, so that the CRLB to M_T is given by

$$DJ^{-1}D' = \begin{pmatrix} \theta_2/n & 0 \\ 0 & 2\theta_2^2/n \end{pmatrix} \text{ and } M_T - DJ^{-1}D' = \begin{pmatrix} 0 & 0 \\ 0 & 2\theta_2^2/(n(n-1)) \end{pmatrix}.$$

EXAMPLE 4.4.2 Consider the bivariate discrete distribution given in Example 4.3.5. Routine calculation show that

$$J(\lambda, p) = \begin{pmatrix} n/\lambda & 0 \\ 0 & n\lambda/p(1-p) \end{pmatrix}$$

and for $\psi = (\lambda, p)$ we have $M_T - \begin{pmatrix} \lambda/n & 0 \\ 0 & p(1-p)/n\lambda \end{pmatrix}$ is nnd by Theorem 4.4.1. However in this case we can show that U_ψ is empty as although unbiased estimators of λ and λp exist, there does not exist unbiased estimator of p. To prove this, if possible let $\varphi(v_1, v_2)$ be unbiased for p then for any $p \in (0, 1)$ and $\lambda > 0$ we have

$$\sum_{v_1=0}^{\infty} \sum_{v_2=0}^{v_1} \varphi(v_1, v_2)\, e^{-n\lambda} \frac{(n\lambda)^{v_1}}{v_1!} \binom{v_1}{v_2} p^{v_2}(1-p)^{v_1-v_2} = p$$

or letting $n\lambda = \theta$ we have for any $p \in (0, 1)$ and $\theta > 0$

$$\sum_{v_1=0}^{\infty} \left[\sum_{v_2=0}^{v_1} \binom{v_1}{v_2} \varphi(v_1, v_2)\, p^{v_2}\, (1-p)^{v_1-v_2} \right] \frac{\theta^{v_1}}{v_1!} = p \sum_{v_1=0}^{\infty} \frac{\theta^{v_1}}{v_1!} \tag{4.4.3}$$

Compare the coefficient of θ^0 on both sides of (4.4.3) for a fixed $p \in (0, 1)$. Then we must have for each fixed $p \in (0, 1)$.

$$\varphi(0, 0) = p \tag{4.4.4}$$

This is a contradiction since $\varphi(0, 0)$ must take only a single value in R_1, and therefore U_p is empty.

Remark 4.4.2 As observed in the case $k = 1$, the equality is attained in CRLB iff there exists a matrix $\Lambda(\theta)$ such that $\dfrac{\partial \log L}{\partial \theta} = \Lambda(\theta)\,[T - \psi(\theta)]$ where $T = (T_1, \ldots, T_m)'$ $\psi = (\psi_1, \ldots, \psi_m)'$ and $\dfrac{\partial \log L}{\partial \theta}$ is the vector of score functions $\left(\dfrac{\partial \log L}{\partial \theta_1} \cdots \dfrac{\partial \log L}{\partial \theta_m}\right)'$. One can show that in this case the pdf belongs to an m-parameter exponential family and T is minimal sufficient for θ.

In the above work we have assumed that the dimensionality of the parameter is same as that of the vector of functions that we want to estimate. This need not be the case. Suppose $\theta = (\theta_1, \ldots, \theta_m)'$ and $\psi = (\psi_1(\theta), \ldots, \psi_k(\theta))'$ and let $D = (d_{ij})$ be $k \times m$ matrix of $E(\partial\psi_i/\partial\theta_j)$, $i = 1, 2, \ldots, k$, $j = 1, 2, \ldots, m$. Then CRLB results states that $M_T - DJ^{-1}D'$is nnd $\forall\ \theta \in \Omega \subset R_m$. The proof is exactly similar as given in case of $k = m$. The case $k < m$ is usually of interest which corresponds to say for example, estimating $\theta^{(1)} = (\theta_1, \ldots, \theta_k)'$ only while $(\theta_{k+1}, \ldots, \theta_m)'$ acts as a nuisance parameter. In this case CRLB result states that $M_T - J^{11}$ is nnd where J^{-1} is partitioned as $\begin{pmatrix} J^{11} & J^{12} \\ J^{21} & J^{22} \end{pmatrix}$. Consider Example 4.3.5 and let $\theta_1 = \lambda$ be the parameter of interest and $\theta_2 = p$ be a nuisance parameter. Then $J^{11} = \lambda/n$ and Var $(T) \geq \lambda/n$ for any $T \in U_\lambda$ and CRLB is attained by $\frac{V_1}{n} = \bar{X}$. Similar situation occurs in Example 4.4.1 $N(\theta_1, \theta_2)$ where θ_1 is the parameter of interest. In the same example if θ_2 is the parameter of interest CRLB, J^{22} is given by $2\theta_2^2/n$ and there does not exist $T \in U_{\theta_2}$ which attains CRLB although MVUE of θ_2 is given by $S^2/(n-1)$ as seen before. On the other hand in Example 4.3.5 unbiased estimator of p does not exist and attaining CRLB $p(1-p)/n\lambda$ for estimating p does not arise. In both of these examples J is diagonal matrix. This case is referred to as parameters being orthogonal and we have $J^{11} = 1/J_{11}$. However if J is not orthogonal we have $J^{11} > 1/J_{11}$ and we have a sharper lower bound for variance of $T_1 \in U_{\theta_1}$ when nuisance parameters $(\theta_2, \ldots, \theta_m)'$ are involved. We illustrate this by an example.

EXAMPLE 4.4.6 Let $(X_1, \ldots, X_{n-1})$ be i.i.d. exponential with mean $\theta > 0$ and let X_n be independent exponential with mean $\lambda\theta$ where $\lambda \geq 1$. This model is used to describe a situation in which an identified outlier may be present in the life testing experiment. Now the joint pdf is given by

$$L(x, \theta, \lambda) = \frac{1}{\lambda\theta^n} \exp\left\{-\frac{S_{n-1}}{\theta} + \frac{X_n}{\lambda\theta}\right\}, \text{ where } S_{n-1} = \sum_{i=1}^{n-1} X_i.$$

The joint pdf belongs to two parameter exponential family with $(S_{n-1}, X_n)'$ being complete sufficient for (θ, λ). Routine calculations will show that

$$J = \begin{pmatrix} n/\theta^2 & 1/\lambda\theta \\ 1/\lambda\theta & 1/\lambda^2 \end{pmatrix} \text{ and } J^{-1} = \begin{pmatrix} \theta^2/(n-1) & -\lambda\theta/(n-1) \\ -\lambda\theta/(n-1) & n\lambda^2/(n-1) \end{pmatrix}.$$

Let parameter of interest be θ then $J^{11} = \theta^2/(n-1)$ and $\frac{1}{J_{11}} = \frac{\theta^2}{n} < J^{11} = \frac{\theta^2}{n-1}$. Now we can show that by RBLS Theorem $\frac{S_{n-1}}{n-1}$ is MVUE of θ with Variance $\frac{\theta^2}{n-1} = J^{11}$. On the other hand CRLB for estimating λ is given by $\frac{n\lambda^2}{n-1} = J^{22} > \frac{1}{J_{22}} = \lambda^2$. Using the fact S_{n-1} is Gamma with parameters $(n-1)$ and θ, an unbiased estimator of $\frac{1}{\theta}$ is $\frac{n-2}{S_{n-1}}$. As X_n and S_{n-1} are independent $E\left[\frac{(n-2)X_n}{S_{n-1}}\right] = \lambda$ and $\frac{(n-2)X_n}{S_{n-1}}$ is MVUE of λ and thus $\left(\frac{S_{n-1}}{n-1}, \frac{(n-2)X_n}{S_{n-1}}\right)'$ is M-optimal for $(\theta, \lambda)'$. We calculate M_T and particularly note that $J^{22} = \frac{n\lambda^2}{n-1}$ is smaller than

$$\text{Var}\left(\frac{(n-2)X_n}{S_{n-1}}\right) = \frac{\lambda^2(n-1)}{(n-3)}.$$

Further $$\text{Cov}\left(\frac{S_{n-1}}{n-1}, \frac{(n-2)X_n}{S_{n-1}}\right) = \frac{-\lambda\theta}{n-1} = J^{12}.$$

4.5 Other Optimality Criteria

In Sec. 4.1 we have defined *M, D, T* and *E* optimality criteria and observed that *M* optimality implies *D, T* and *E* optimality. Now corresponding to each of these criterion we have preference order between T_1 and $T_2 \in U_\psi$. For example we prefer T_1 to T_2 as per *D* criterion if $| M_{T_1} | \le | M_{T_2} |, \forall\, \theta \in \Omega$. Now if *A* and *B* are symmetric positive definite matrices then $| A | \le | B |$ does not imply that $(B - A)$ is nnd and therefore if we prefer T_1 to T_2 according to *D*-criterion it does not follow that we should prefer T_1 to T_2 as per *M* criterion. Similar remark holds for *T* and *E* optimality criterion. Thus although *M*-optimality of T^* implies *D, T* and *E* optimality of T^* given that T_1^* is *D*-optimal say, it does not follow that it would be *M, T* or *E*-optimal and in general we may have different optimal estimators according to different optimality criteria.

The following shows that fortunately this does not occur and if T_1^* is *D*-optimal then it is *M*-optimal and therefore it is *T* and *E*-optimal also. Similarly, if T_2^* is *T*-optimal then it is *M*-optimal and hence also *D* and *E*-optimal. We show by an example that similar result does not hold for *E*-optimal estimator and *E*-optimality does not necesarily imply optimality as per other criteria. Thus *M, D*, and *T* criteria are equivalent in the sense that they lead to the same optimal estimator, but *E* criterion is different.

THEOREM 4.5.1 (i) If T_1^* is *T*-optimal then T_1^* is *M*-optimal and hence *D* and *E* optimal also.

(ii) if T_2^* is *D*-optimal then T_1^* is *M*-optimal and hence *T* and *E* optimal also.

Proof Let $u \in U_0^{(1)}$ with Var $(u) = \alpha(\theta) > 0$. Suppose T_1^* is not *M*-optimal. Then it follows that there exists a subscript *i* such that $\text{Cov}(T_{1i}^*, u) \ne 0$ at some $\theta_0 \in \Omega$. Let $\beta_j(\theta) = \text{Cov}\,(T_{1j}^*, u)/\alpha(\theta)$ and let $\beta(\theta) = (\beta_1(\theta), \ldots, \beta_k(\theta))'$.

Then $\beta(\theta_0) \ne 0$. Define $T = T_1^* - \beta(\theta_0)u$ so that $T \in U_\psi$. Now at θ_0

$$M_{T_1^*, \beta(\theta_0)u} = \alpha(\theta_0)\beta(\theta_0)\,\beta'(\theta_0)$$

$$M_T(\theta_0) = M_{T_1^*} - \alpha(\theta_0)\beta(\theta_0)\beta'(\theta_0)$$

and therefore

$$M_{T_1^*}(\theta_0) = M_T(\theta_0) + \alpha(\theta_0)\ \beta(\theta_0)\ \beta'(\theta_0) \tag{4.5.1}$$

which implies that $Tr[M_{T_1^*}(\theta_0)] > Tr[M_T(\theta_0)]$ as $\beta(\theta_0) \neq 0$. This contradicts the datum that T_1^* is T-optimal. Hence by reductio ad absurdum argument T_1^* must be M-optimal and therefore also D-optimal.

To prove (ii) we again argue as above. Suppose T_2^* is not M-optimal then there exists a subscript i such that Cov $(T_{2i}^*, u) \neq 0$ and thus if $T = T_2^* - \beta(\theta_0)u$ we have analoguous to (4.5.1) the result that

$$M_{T_2^*}(\theta_0) = M_T(\theta_0) + \alpha(\theta_0)\beta(\theta_0)\beta'(\theta_0) \tag{4.5.2}$$

i.e. $$A = B + C$$

where A, B are pd matrices and C is nnd. By using the fact that rank of C is unity and that of B is k, one can show that $| B + C | > | B |$ by using simultaneous diagonalization of B and C. Hence $| M_{T_2^*}(\theta_0) | > | M_T(\theta_0) |$ which contradicts the datum that T_2^* is D-optimal and hence the result.

For more details we refer to Kale and Chandrasekar (1983) which contains two distinct proofs of part 2 of the above theorem.

We now show by an example that E-optimality of T_3 does not imply M, T or D optimality of T_3.

EXAMPLE 4.5.1 [Kale and Chandrasekar (1983)]

Let $(X_1, \ldots, X_{10})$ be a random sample of size 10 from $N(\theta_1, \theta_2)'$ where the parameter space is $\Omega_0 = \{(\theta_1, \theta_2)' \mid \theta_1 \in R_1, \theta_2 \geq 5\}$. Then the pdf belongs to two parameter exponential family with $(\bar{X}, S^2)'$ as minimal complete sufficient statistic for $(\theta_1, \theta_2)'$ even though the parameter space Ω_0 is a truncated subset of the natural parameter space $\Omega = R_1 \times (0, \infty)$. Now by RBLS theorem $\bar{X}$ and $S^2/9$ are MVUE of θ_1, θ_2 respectively and $T^* = (\bar{X}, S^2/9)'$ is M-optimal for $(\theta_1, \theta_2)'$ with $M_{T^*} = \text{diag}\left(\frac{\theta_2}{10}, \frac{2\theta_2^2}{9}\right)$. For $\theta \in \Omega_0$, $\lambda_1(M_{T^*}) = 2\theta_2^2/9$. Now consider $T_1 = (X_1, S^2/9)'$ and $M_{T_1} = \text{diag}$ $(\theta_2, 2\theta_2^2/9)$ and $\lambda_1(M_{T_1}) = 2\theta_2^2/9 = \lambda_1(M_{T^*})$ as $\theta_2 \geq 5$. Now we know that T^* is M-optimal and therefore D, T and E optimal and the above argument shows that $T_1$is also E-optimal since $\lambda_1(M_{T_1}) = \lambda_1(M_{T^*}) \leq \lambda_1(M_T)$ for any $T \in U_\psi$. Thus T_1 is E-optimal but not M-optimal. Hence we conclude that although M optimality implies E optimality, the converse is not true.

Remark 4.5.1 We observe that the D, T and E optimality criteria are related to characteristic roots $(\lambda_1, \ldots, \lambda_k)$ of the Variance-Covariance matrix

M_T. Thus D criterion is based on minimizing $\prod_{i=1}^{n} \lambda_i$, T-optimality is based on minimizing $\sum \lambda_i$ and E-optimality is based on minimizing $\max_i \lambda_i$. One could consider other symmetric functions of $(\lambda_i, \ldots, \lambda_k)$ to develop other criteria for optimality for vector unbiased estimator T for ψ. However as observed in Section 4.2, T, D and E optimality have nice geometric interpretations.

5

Consistent Estimators

5.1 Consistency

This chapter deals with the estimation of a real or vector valued parametric function $\psi(\theta)$, based on a random sample of size n, where n is assumed to be large enough such that if T is an estimator of $\psi(\theta)$, its performance can be studied by using large sample or asymptotic distribution of T. Corresponding to the criterion of unbiasedness in Chapter 3 we now consider the criterion of consistency of an estimator T. We assume that T, a real valued statistic, is to be used as an estimator of real parameter θ based on a random sample of size n from $\{f(x, \theta), \theta \in \Omega\}$ where $\Omega \subset R_1$.

Since the behaviour of T is to be studied for large values of n only, we consider the sequence of r.v.s. $\{T_n\}$ and base our study on its convergence properties. There are several types of convergences that can be defined e.g. convergence in probability, convergence in distribution, convergence in quadratic mean among others. We refer to Cramèr (1946) Chapter 20 Sec. 6, and Rao (1973) Chapter 2, Section 2-C for definitions and properties and further details which are adequate for the purpose of this text and it will be assumed that the reader is familiar with the material particularly with the Weak Law of Large Numbers (WLLN) and the Central Limit Theorem (CLT) for i.i.d. r.v.s.

Definition 5.1.1 An estimator T_n is said to be consistent for θ if $T_n \xrightarrow{p} \theta$ for each $\theta \in \Omega$ and the convergence in probability is taken under the distribution indexed by θ.

Recalling the definition of $\xrightarrow{p}$ there are following alternative ways in which consistency can be defined.

Definition 5.1.2 (a) For any $\varepsilon > 0$

$$\lim_{n\to\infty} P_\theta[|T_n - \theta| < \varepsilon] = 1 \;\forall\; \theta \in \Omega \tag{5.1.1}$$

(b)

$$\lim_{n\to\infty} P_\theta[|T_n - \theta| \geq \varepsilon] = 0 \;\forall\; \theta \in \Omega \tag{5.1.2}$$

(c) For any $\varepsilon > 0$ and $\delta > 0$ there exists an $n_0(\varepsilon, \delta, \theta)$ such that for $\forall\; n \geq n_0\, (\varepsilon, \delta, \theta)$

$$P_\theta[|T_n - \theta| < \varepsilon] \geq 1 - \delta, \;\forall\; \theta \in \Omega \tag{5.1.3}$$

or

$$P_\theta[|T_n - \theta| \geq \varepsilon] \leq \delta, \;\forall\; \theta \in \Omega \tag{5.1.4}$$

All these definitions essentially convey the requirement that for large samples, with very high probability the values of the estimator T_n are sufficiently close to the "true value" of the parameter θ. Indeed if T_n is consistent for θ then $P_{\theta_1}[| T_n - \theta_2 | < \varepsilon]$ does not tend to unity or $P_{\theta_1}[| T_n - \theta_2 | > \varepsilon]$ does not tend to zero, as $n \to \infty$. Thus T_n does not converge in probability to θ_2 if $\xrightarrow{p}$ is taken under $\theta_1 \neq \theta_2$. This is a consequence of the property that the probability limit is unique, i.e. if $Y_n \xrightarrow{p} a$ then it cannot converge to another constant b.

A very important property of a consistent estimator is the invariance under continuous transformation, a property not enjoyed by an unbiased estimator. Thus if $\psi(\theta)$ is continuous function and if T is consistent for θ then $\psi(T)$ is consistent for $\psi(\theta)$. As seen earlier $E(T) = \theta, \forall \theta \in \Omega$ does not in general imply that $E(\psi(T)) = \psi(\theta)$ unless ψ is linear in θ.

THEOREM 5.1.1 Let $T \xrightarrow{p} \theta, \forall \theta \in \Omega$ and let ψ be continuous function from Ω to R_1 then $\psi(T) \xrightarrow{p} \psi(\theta)$.

As ψ is continuous, given $\varepsilon > 0$ there exists a δ such that $| \psi(T) - \psi(\theta) | < \varepsilon$ whenever $| T - \theta | < \delta$. Let $E_\varepsilon = \{x \mid | \psi(T(x)) - \psi(\theta) | < \varepsilon\}$ and $F_\delta = \{x \mid | T(x) - \theta | < \delta\}$. Then $F_\delta \subset E_\varepsilon$ and $P_\theta(F_\delta) \le P_\theta(E_\varepsilon)$. But $\lim_{n\to\infty} P_\theta(F_\delta) = 1$, $\forall \theta \in \Omega$ and any $\delta > 0$. Therefore $\lim_{n\to\infty} P_\theta(E_\varepsilon) = 1, \forall \theta \in \Omega$ and $\forall \varepsilon > 0$.

As a consequence of this property of invariance of consistent estimators, for all practical purposes we need consider consistent estimators of θ only.

Now suppose that the parameter θ is vector valued say $\theta = (\theta_1, \ldots, \theta_m)'$. Then a vector valued statistic $T = (T_1, \ldots, T_m)'$ is said to be consistent for θ if

$$T_i \xrightarrow{p} \theta_i, \; i = 1, 2, \ldots, m, \; \forall \theta \in \Omega \tag{5.1.5}$$

This is analogous to defining a vector valued statistic T to be unbiased for vector valued function $\psi(\theta)$ if T is componentwise unbiased. However in a vector space of m dimensions one could also define convergence by introducing a distance function defined by a suitable norm. Thus we could define T to be consistent for θ if

$$\lim_{n\to\infty} P_\theta[\| T - \theta \| < \varepsilon] = 1, \; \forall \theta \in \Omega, \; \forall \varepsilon > 0 \tag{5.1.6}$$

Now we take $\| T - \theta \| = \max_i | T_i - \theta_i |$, i.e. the supnorm which is equivalent to the usual Euclidean distance, namely, $\| T - \theta \| = \left\{\sum_{i=1}^{m} (T_i - \theta_i)^2\right\}^{1/2}$. The definition of consistency induced by (5.1.5) is usually referred to as marginal consistency whereas that induced by (5.1.6) is called as joint consistency. However, one can show that marginal and joint consistency are equivalent.

THEOREM 5.1.2 T is marginal consistent for θ iff T is jointly consistent for θ.

Now, assume that T is jointly consistent so that (5.1.6) holds and let $E_i = \{x \mid |T_i - \theta_i| < \varepsilon\}$ and let $E = \{x \mid \max_i |T_i - \theta_i| < \varepsilon\}$. Then we have $E = \bigcap_{i=1}^{m} E_i$ and therefore $E \subset E_i$ and $P_\theta(E)) \le P_\theta(E_i)$. As T is jointly consistent, for any $\varepsilon > 0$, $P_\theta(E) \to 1$ as $n \to \infty$, $\forall\, \theta \in \Omega$ and therefore for each $i = 1, 2, \ldots, m$, $P_\theta(E_i) \to 1$ as $n \to \infty\ \forall\, \theta \in \Omega$ and each $T_i \xrightarrow{p} \theta_i$ or T is marginally consistent for θ.

Next assume that for each $i = 1, 2, \ldots, m$, $T_i \xrightarrow{p} \theta_i$ then by (5.1.2) $P_\theta(E_i^c) \to 0$ as $n \to \infty$, $\forall\, \theta \in \Omega$. As $E = \bigcap_{i=1}^{m} E_i$, we have $E^c = \bigcup_{i=1}^{m} E_i^c$ and $P_\theta(E^c) \le \sum_{i=1}^{m} P_\theta(E_i^c)$. But this implies that $P_\theta(E^c) \to 0$ as $n \to \infty$, $\forall\, \theta \in \Omega$ and $\forall\, \varepsilon > 0$. Thus T is jointly consistent for θ.

We can now generalize Theorem 5.1.1 to a situation in which T and θ are vector valued and ψ is a continuous function from Ω to say R_k where k is not necessarily equal to m.

THEOREM 5.1.3 Let T be jointly consistent for θ and let ψ be a k-dimensional continuous function from Ω to R_k then $\psi(T)$ is jointly consistent for $\psi(\theta)$.

Since ψ is a continuous function from $\Omega \subset R_m$ to R_k given $\varepsilon > 0$ there exists a $\delta > 0$ such that $\| \psi(T) - \psi(\theta) \| < \varepsilon$ whenever $\| T - \theta \| < \delta$. Therefore $P_\theta[\| \psi(T) - \psi(\theta) \| < \varepsilon] \ge P_\theta[\| T - \theta \| < \delta]$ and as T is jointly consistent, for any $\delta > 0$, and $\forall\, \theta \in \Omega$ we have

$$\lim_{n\to\infty} P_\theta[\| T - \theta \| < \delta] = 1 \text{ which implies that}$$

$$\lim_{n\to\infty} P_\theta[\| \psi(T) - \psi(\theta) \| < \varepsilon] = 1,\ \forall\, \varepsilon > 0 \text{ and } \forall\, \theta \in \Omega.$$

Thus $\psi(T)$ is jointly consistent for $\psi(\theta)$ and therefore each $\psi_i(T)$ is consistent for $\psi_i(\theta)$, $i = 1, 2, \ldots, k$.

We now consider some examples to illustrate the above theory.

EXAMPLE 5.1.1 Let $\{X_i\}_1^n$ be i.i.d. $N(\theta, 1)$ and consider the MVUE of θ given by $\bar{X}$. As $\{X_i\}_1^n$ are i.i.d. with $E(X_i) = \theta$ by using WLLN we have $\bar{X} \xrightarrow{p} \theta$ and thus $\bar{X}$ is consistent for θ. Note that WLLN does not provide us any information about the rate of convergence of $p_n(\varepsilon, \theta) = P_\theta[|\bar{X} - \theta| < \varepsilon] \to 1$. Noting that Var $(\bar{X}) = 1/n$, by Tchebychev's inequality $p_n(\varepsilon, \theta) = P_\theta[|\bar{X} - \theta| < \varepsilon] \geq 1 - \frac{1}{n\varepsilon^2} \to 1 \ \forall \ \theta \in \Omega$. We can then determine $n_0(\varepsilon, \delta, \theta)$ as needed in definition (1.2.c), if we select n_0 such that $1 - \frac{1}{n\varepsilon^2} \geq 1 - \delta$ or $n \geq \frac{1}{\varepsilon^2\delta}$. Thus we can take $n_0 = \left[\frac{1}{\varepsilon^2\delta}\right] + 1$, where $[u]$ denotes the largest integer less than a positive real number u. We observe that this $n_0(\varepsilon, \delta, \theta)$ does not depend on θ and therefore $p_n(\varepsilon, \theta) \to 1$ uniformly in θ. In this case we say that $\bar{X}$ is uniformly consistent for θ or $\bar{X} \xrightarrow{p} \theta$ uniformly for $\theta \in \Omega$. Next since $\bar{X} \sim N\left(\theta, \frac{1}{n}\right)$ we can obtain an exact value for

$$\begin{aligned} p_n(\varepsilon, \theta) &= \Phi(\sqrt{n}\ \varepsilon) - \Phi[-\sqrt{n}\ \varepsilon] \\ &= 2\Phi(\sqrt{n}\ \varepsilon)) - 1 \end{aligned}$$

as $\Phi[\sqrt{n}\ \varepsilon] \to 1$ as $n \to \infty$, $p_n(\varepsilon, \theta) \to 1$ as $n \to \infty$. Further to determine $n_0(\varepsilon, \delta, \theta)$ we must select n_0 such that

$$2\Phi[\sqrt{n}\ \varepsilon] - 1 \geq 1 - \delta \text{ or } \Phi[\sqrt{n}\ \varepsilon] \geq 1 - \delta/2$$

or $$n \geq \frac{1}{\varepsilon^2}\{\Phi^{-1}[1 - \delta/2]\}^2 \text{ or } n_0 = \left[\frac{1}{\varepsilon^2}\{\Phi^{-1}[1 - \delta/2]\}^2\right] + 1.$$

As is expected, the values of n_0 determined by Tchebychev inequality would be much larger than the one determined by using the exact distribution of $\bar{X} \sim N(\theta, 1/n)$. Thus for $\varepsilon = 10^{-1}, 10^{-2}$ and $\delta = 10^{-1}, 10^{-2}$ we get the following data:

n_0 (Tchebychev Inequality)			n_0 (Exact distribution)		
δ/ε	0.1	0.01	δ/ε	0.1	0.01
0.1	$10^3 + 1$	$10^5 + 1$	0.1	385	38417
0.01	$10^4 + 1$	$10^6 + 1$	0.01	790	78962

The practical meaning of $n_0(\varepsilon, \delta, \theta)$ is that it gives the minimum sample size required to attain the level of accuracy specified by (ε, δ) combination. Thus the minimum sample size required to estimate the mean θ in $N(\theta, 1)$ case correct to first place of decimal ($\varepsilon = 10^{-1}$) with probability $p_n(\varepsilon, \theta) \geq .99$ (i.e. $\delta = 0.01$) would be 38417. Note that n_0 determined by using Tchebychev inequality will be $10^5 + 1$ or over hundred thousand.

EXAMPLE 5.1.2 Let $\{X_i\}_1^n$ be i.i.d. $N(\theta_1, \theta_2)$ then by WLLN applied to $\{X_i\}$ and $\{X_i^2\}$ we have $m_1' = \bar{X} \xrightarrow{p} \theta_1$ and $m_2' = \frac{1}{n}\Sigma X_i^2 \xrightarrow{p} \theta_2 + \theta_1^2$. Now by invariance under continuous transformation $\bar{X}^2 \xrightarrow{p} \theta_1^2$ and $\left(\bar{X}^2, \frac{1}{n}\Sigma X_i^2\right)'$ is consistent for $(\theta_1^2, \theta_2 + \theta_1^2)'$ and, therefore,

$\frac{1}{n}\Sigma X_i^2 - \bar{X}^2 = \frac{S^2}{n} = \frac{1}{n}\Sigma(X_i - \bar{X}^2) \xrightarrow{p} \theta_2$. Note that while $\bar{X}$ is MVUE for $\theta_1, \bar{X}^2$ is not MVUE for θ_1^2 and $\frac{S^2}{n}$ is not MVUE for θ_2. The MVUE of θ_2 is $\frac{S^2}{n-1} = \left(\frac{n}{n-1}\right)\frac{S^2}{n} \xrightarrow{p} \theta_2$ as $\frac{S^2}{n} \xrightarrow{p} \theta_2$ and $\frac{n}{n-1} \to 1$. This follows from the property that if $Y_n \to \alpha$ and $\{a_n\}$ and $\{b_n\}$ are real sequences converging to a and b respectively then $a_n Y_n + b_n \xrightarrow{p} a\alpha + b$.

We also note that the standard deviation $\sqrt{\theta_2}$ the positive root of θ_2, is consistently estimated by $\sqrt{S^2/n} = \sqrt{m_2}$. Note that in general taking square root is not a function, but if the domain and range is restricted to positive axes we have infact a continuous function.

Consider now estimating p-th percentile of $N(\theta_1, \theta_2)$ given by $\theta_1 + \xi_p\sqrt{\theta_2}$ which is a continuous function of $(\theta_1, \sqrt{\theta_2})'$. Then $\bar{X} + \xi_p S/\sqrt{n}$ is consistent estimator of 100p% point of $N(\theta_1, \theta_2)$. On the other hand suppose we want to estimate $\psi_a(\theta_1, \theta_2) = P[X \le a] = \Phi\left[\frac{a-\theta_1}{\sqrt{\theta_2}}\right]$ then as Φ is a continuous function of $(\theta_1, \sqrt{\theta_2})'$ we have $\Phi\left[\frac{(a-\bar{X})\sqrt{n}}{\sqrt{S}}\right]$ is consistent for $\psi_a(\theta_1, \theta_2)$. We note the ease with which we can obtain consistent estimators as opposed to MVU estimators of these functions.

EXAMPLE 5.1.3 Let $\{(X_i, Y_i)\}_1^n$ be i.i.d. Bivariate Normal with mean vector $(\mu_1, \mu_2))'$ and covariance matrix $\begin{pmatrix}\sigma_1^2 & \rho\sigma_1\sigma_2 \\ \rho\sigma_1\sigma_2 & \sigma_2^2\end{pmatrix}$. Then $\{X_i\}_1^n$ being i.i.d. $N(\mu_1, \sigma_1^2)$ we have from Example 5.1.2

$$\bar{X} \xrightarrow{p} \mu_1, \frac{S_x^2}{n} = \Sigma(X_i - \bar{X})^2/n \xrightarrow{p} \sigma_1^2.$$

Similarly $\bar{Y} \xrightarrow{p} \mu_2$ and $\frac{1}{n}S_y^2 \xrightarrow{p} \sigma_2^2$. To estimate covariance term let $\frac{1}{n}S_{xy} = \frac{1}{n}\Sigma(X_i - \bar{X})(Y_i - \bar{Y}) = \frac{1}{n}\Sigma X_i Y_i - \bar{X}\bar{Y}$. Now $\frac{1}{n}\Sigma X_i Y_i \xrightarrow{p} E(X_i Y_i)$ and as $\bar{X} \xrightarrow{p} \mu_1$ $\bar{Y} \xrightarrow{p} \mu_2$, $\frac{1}{n}S_{xy} \xrightarrow{p} E(XY) - E(X)E(Y) = \rho\sigma_1\sigma_2$. Now $S_{XY}/S_X S_Y \xrightarrow{p} \rho\sigma_1\sigma_2/\sigma_1\sigma_2 = \rho$ and therefore $S_{XY}/S_X S_Y$ is consistent for ρ.

The reader can easily generalize this example to the case where $(X_1, \ldots, X_k)'$ is multivariate normal with mean vector $(\mu_1, \ldots, \mu_k)'$ and variance covariance matrix Λ. Thus it is easy to show that the sample mean vector $(\bar{X}_1, \ldots, \bar{X}_k)'$ is consistent for $(\mu_1, \ldots, \mu_k)'$ and Λ is consistently estimated by the sample variance covariance matrix S where

$$S_{ij} = \frac{1}{n}\sum_{r=1}^{n} X_{ir}X_{jr} - \bar{X}_i\bar{X}_j,\ i = 1, 2, \ldots, k, j = 1, 2, \ldots, k.$$

Once the above result is established we can conclude that the population multiple correlation coefficient $\rho^2_{1,23\ldots k} = 1 - |\Lambda| / \lambda_{11} |\Lambda^{(1,1)}|$ is estimated consistently by the sample multiple correlation coefficient $R^2_{1,23\ldots k} = 1 - \dfrac{|S|}{S_{11}|S^{(1,1)}|}$. Using similar techniques we can claim that the population partial correlation coefficients and regression coefficients are consistently estimated by their sample counterparts.

EXAMPLE 5.1.4 Let $\{X_i\}_1^n$ be i.i.d. Then we know that $\bar{X} \xrightarrow{p} \mu$ iff $E(X) = \mu$. Thus if we have a random sample of size n from a population for which $E(X)$ does not exist then $\bar{X}$, the sample mean would not be consistent for the population mean. Such a situation obtains for the Cauchy distribution with location parameter μ, having pdf

$$f(x, \mu) = \frac{1}{\pi} \frac{1}{1 + (x - \mu)^2}, \; x \in R_1, \mu \in R_1$$

and $\bar{X}$ does not converge in probability to μ. Indeed we know that $\bar{X}$ has same pdf as that of a single observation and $P[|\bar{X} - \mu| < \varepsilon] = \dfrac{2}{\pi}\tan^{-1}(\varepsilon)$ which does not tend to one as $n \to \infty$.

In case of Pareto distribution with pdf $f(x, \lambda) = \dfrac{\lambda}{x^{\lambda+1}}$, $x \geq 1$ which is often used as a model to describe income distribution, $E(X)$ exists only if $\lambda > 1$ and then it is given by $\dfrac{\lambda}{\lambda - 1}$. Therefore $\bar{X} \xrightarrow{p} \dfrac{\lambda}{\lambda - 1}$ only when $\lambda > 1$ but for $\lambda < 1$, $\bar{X}$ will not converge in probability.

In the next Section we will consider the method of moments in general and the quantile or percentile method of estimation in Section 5.3.

Exercise 5.1

1. Let $\{X_i\}_1^n$ be i.i.d. Gamma, $G(\lambda, \sigma)$ with p.d.f. $f(x, \sigma, \lambda) = \dfrac{1}{\Gamma(\lambda)} \dfrac{e^{-x/\sigma}}{\sigma^\lambda} x^{\lambda-1}$, $x > 0, \sigma > 0, \lambda > 0$. Using WLLN for $\{X_i\}$ and $\{X_i^2\}$ show that $m_1' \xrightarrow{p} \lambda\sigma$ and $m_2' \xrightarrow{p} \lambda(\lambda + 1)\sigma^2$ and therefore $m_2 = m_2' - (m_1')^2 \xrightarrow{p} \lambda\sigma^2$ where m_k' and m_k denote k-th raw and central moments of the sample. Show that $\left(\dfrac{m_2}{m_1'}, \dfrac{m_1'^2}{m_2}\right)'$ is consistent for $(\sigma, \lambda)'$.
2. Let $\{X_i\}_1^n$ be i.i.d. $b(1, \theta)$ show that $\bar{X} \xrightarrow{p} \theta$. Using this fact obtain (i) a consistent estimator of the probability that in future m trials s successes occur and (ii) a consistent estimator of $\theta(1 - \theta)$ the population variance. (iii) $\dfrac{1}{\theta(1 - \theta)} = I_X(\theta)$, the Fisher Information per unit observation.
3. Let $\{X_i\}_1^n$ be i.i.d. exponential with mean σ, so that $\bar{X} \xrightarrow{p} \sigma$. Let $T = \sum_{i=1}^{n} l_i X_i$

be consistent for σ. Show that $E(T-\sigma)^2$ is minimized for $l_1 = l_2 = \dots\ l_n = \frac{1}{n+1}$. Consider $T^* = \frac{n\bar{X}}{n+1}$. Show that $T^* \xrightarrow{p} \sigma$ and MSE $(T^*) <$ MSE $(\bar{X})$ although $\bar{X}$ is MVUE of σ. Obtain consistent estimator of $R(a, \sigma) = e^{-a/\sigma} = P_\sigma(X > a)$.

4. Let $\{X_i\}_1^n$ be i.i.d. Poisson with mean λ so that $\bar{X} \xrightarrow{p} \lambda$ and $e^{-\bar{X}} \xrightarrow{p} e^{-\lambda}$. Obtain $E(e^{-\bar{X}})$ and Var $(e^{-\bar{X}})$ and compare these with corresponding expressions for MVUE of $e^{-\lambda}$ given by $\varphi(T) = \left(\frac{n-1}{n}\right)^T$.. (Hint: $T = \Sigma X_i \sim$ Poisson $(n\lambda)$)
5. We have seen that in $U(0, \theta)$ case, $T = \frac{n+1}{n} X_{(n)}$ is MVUE. Show that T is consistent and so also is $\frac{n}{n+1} T = X_{(n)}$. Compare the MSE($T$) and MSE $\left(\frac{n}{n+1} T\right)$. Show that $X_{(n-1)}$ is consistent for, θ but $X_{(1)}$ is not.
(Hint: To prove that $X_{(1)}$ is not consistent calculate $P_\theta[|X_{(1)} - \theta| < \varepsilon]$ and show that there exists a pair $(\theta_0, \varepsilon_0)$ for which the above probability does not go to one. You must avoid the trap of wrongly claiming inconsistency of $X_{(1)}$ by showing that $E(X_{(1)} - \theta)^2 =$ MSE $(X_{(1)})$ does not tend to 0 as $n \to \infty$. Recall that if Y_n converges in quadratic mean to zero then $Y_n \xrightarrow{p} 0$ but we have situations in which $Y_n \xrightarrow{p} 0$ but $E(Y_n^2)$ does not tend to 0 e.g. take $Y_n = 0$ with probability $1 - \frac{1}{n}$ and n with probability $1/n$ then $Y_n \xrightarrow{p} 0$ but $E(Y_n^2) \to \infty$.
6. Let $X_{ij} = \mu_i + \varepsilon_{ij}$, $i = 1, 2, \dots, k$, $j = 1, 2, \dots, n$, the model used in balanced one way analysis of variance to analyse k sample problem. For $\{\varepsilon_{ij}\}$ i.i.d. $N(0, \sigma^2)$ we have seen in Example 4.3.1, that $T = \bar{X}_1, \dots, \bar{X}_k, S^2/(n-1)k)'$ is M-optimal for $(\mu_1, \dots, \mu_k, \sigma^2)' = \theta$ where $S^2 = \Sigma S_i^2$ and $S_i^2 = \sum_{j=1}^{n} (X_{ij} - \bar{X}_i)^2$. Show that the M-optimal Estimator is consistent. Further even if we only assume that $\{\varepsilon_{ij}\}$ are i.i.d. with $E(\varepsilon_{ij}) = 0$, Var $(\varepsilon_{ij}) = \sigma^2$ without assuming their normality, show that T remains consistent for θ.
7. In Neyman-Scott problem with $X_{ij} = \mu_i + \varepsilon_{ij}$, $i = 1, 2, \dots, k$, $j = 1, 2$ where $\{\varepsilon_{ij}\}$ are i.i.d. $N(0, \sigma^2)$, we can show that $T = (\bar{X}_1, \dots, \bar{X}_k, \sum_{i=1}^{k} S_i^2/k)'$ is M-optimal for θ. Show that when $k \to \infty$, $\bar{X}_i$ is not consistent for μ_i, $i = 1, 2, \dots, k$ although $\Sigma S_i^2/k$ is consistent for σ^2, and thus T is not consistent for θ.

5.2 Method of Moments

The Reader would have noticed from the previous section that using WLLN in case where $\{X_i\}_1^n$ are i.i.d. is one of the simplest method to generate consistent estimators. Let $(X_1, \dots, X_n)'$ be a random sample of size n from $\{f(x, \theta), \theta \in \Omega \subset R_1\}$. Then if $E(X_i) = \mu(\theta)$ then $\bar{X} \xrightarrow{p} \mu(\theta)$ and $\bar{X}$ is consistent for $\mu(\theta)$. If $\mu(\theta)$ is such that μ^{-1} exists and is continuous then by Theorem 5.1.1, property of invariance under continuous transformation, $\mu^{-1}(\bar{X}) \xrightarrow{p} \mu^{-1}(\mu(\theta)) = \theta$ and $\mu^{-1}(\bar{X})$ is consistent for θ. This is the well known method of moments where we obtain an estimator by solving the moment equation $\bar{X} = \mu(\theta)$. A sufficient condition for μ^{-1} to exist and be continuous is that $\frac{d\mu}{d\theta} \neq 0$. Note that we could have used transformed observations $Y_i = U(X_i)$ if $E(Y_i) = \eta(\theta)$ is such that $\frac{d\eta}{d\theta} \neq 0$. to obtain a consistent estimator based on $\{Y_i\}$ given by $\eta^{-1}(\bar{Y})$.

EXAMPLE 5.2.1 Let $\{X_i\}$ be i.i.d. exponential with pdf given $f(x, \theta) = \theta\, e^{-\theta x}$, $\theta > 0$, $x > 0$. Here $E(X) = 1/\theta$ and $\frac{d\mu}{d\theta} = -\frac{1}{\theta^2} \neq 0$. Hence the moment equation based on $\{X_i\}$ gives $\frac{n}{\Sigma X_i} = \frac{1}{\bar{X}}$ as a consistent estimator of θ, the failure rate or the reciprocal of the mean life time of the r.v. X.

EXAMPLE 5.2.2 Consider the Pareto distribution with pdf $f(x, \lambda) = \frac{\lambda}{x^{\lambda+1}}$, $x \geq 1$, $\lambda > 0$, which is used as a model for income distribution. As observed in Example 5.1.4, $E(X)$ does not exist if $0 < \lambda < 1$ and unless apriori we know that $\lambda > 1$, the method of moment will fail. However, suppose $\Omega = (1, \infty)$ then $E(\bar{X}) = \mu(\lambda) = \frac{\lambda}{\lambda - 1}$ and $\frac{d\mu}{d\lambda} = -\frac{1}{(\lambda-1)^2} \neq 0$. Therefore the moment equation $\bar{X} = \frac{\lambda}{\lambda - 1}$, determines the estimator $T = \frac{\bar{X}}{\bar{X} - 1}$ which is consistent for λ. However if $\Omega = (0, \infty)$ then the method of moments can not be applied using original observations themselves. Under the transformation $y = \log x$ we observe that $\{f(y, \lambda), \lambda > 0\}$ is one parameter exponential family with $k(x) = \log x$ and $Y = \log X$ is exponential with pdf $g(y, \lambda) = \lambda e^{-\lambda y}$, $\lambda >$, $y \geq 0$. Now by Example 5.2.1, $\frac{n}{\Sigma Y_i} = \frac{n}{\Sigma \log X_i}$ is consistent for λ.

EXAMPLE 5.2.3 Let X have a Weibull distribution with parameter α so that $f(x, \alpha) = \alpha x^{\alpha-1} \exp\{-x^{\alpha}\}$, $x > 0$, $\alpha > 0$. Thus X is such that X^{α} is standard exponential with mean one. Weibull distribution is quite frequently used as a model for failure time distribution. Here $E(X) = \Gamma\left(\frac{1}{\alpha} + 1\right)$ and the moment equation is given by $\bar{X} = \Gamma\left(\frac{1}{\alpha} + 1\right) = \mu(\alpha)$. Here $\mu(\alpha) = \Gamma\left(\frac{1}{\alpha} + 1\right)$ is such that μ^{-1} does not exist although $\frac{1}{\alpha} + 1$ is monotone decreasing function varying over $(\infty, 1)$ as α varies over $(0, \infty)$. Let $\theta = \frac{1}{\alpha} + 1$ then $\theta\, \varepsilon\, (1, \infty)$. Now $\Gamma(1) = \Gamma(2) = 1$, and as $\Gamma(\theta)$ is continuous and differentiable by Rolle's Theorem there exists a $\theta \in (1, 2)$ such that $\frac{d\Gamma(\theta)}{d\theta} = 0$. Thus there exists $\alpha \in (1, \infty)$ such that $\frac{d\mu}{d\alpha} = 0$ and μ^{-1} does not exist and therefore the method of moments based on X does not work unless a priori we know that $\alpha \in (0, 1)$. To avoid this problem suppose we consider the k-th moment of X then $E(X^k) = \Gamma\left(\frac{k}{\alpha} + 1\right)$ and as α varies over $(0, \infty)$ whatever be the value of k, $\frac{k}{\alpha} + 1$ will vary over $(1, \infty)$ which includes the interval $(1, 2)$ and $\mu_k'(\alpha) = \Gamma\left(\frac{k}{\alpha} + 1\right)$ will be such that $\frac{d\mu_k'}{d\alpha} = 0$ for some $\alpha \in (1, 2)$ and the moment equation will fail to give an estimator. In fact this failure is due to non-uniqueness or non-existence of solution to the moment equation $m_k'(\alpha) = \Gamma\left(\frac{k}{\alpha} + 1\right)$. A detailed study using simulation has been carried out by Paranjape (1994). It is observed that the minimum value of $\Gamma\left(\frac{k}{\alpha} + 1\right) = .8856$ and occurs around $\frac{k}{\alpha} = 0.46$. If observed $m_k' < .8856$ then moment equation has no solution, and if $m_k' \in [.8856, 1]$ then the moment equation would have two roots. If observed $m_k' > 1$ then only the moment equation has the unique solution. To get around this difficulty consider mgf of $\log X$, given by $E(e^{t \log X}) = E(X^t) = \Gamma\left(\frac{t}{\alpha} + 1\right)$. This gives us $E(\log X) = \frac{\Gamma'(1)}{\alpha} = \frac{-\gamma}{\alpha}$ where γ is Euler's constant. For properties of Gamma function $\Gamma(1 + u)$ and its derivatives we refer to Abramowitz and Stegun [1968]. Thus $\frac{1}{n} \Sigma \log X_i \xrightarrow{P} -\frac{\gamma}{\alpha}$ and therefore $\hat{\alpha} = -n\gamma/\Sigma \log X_i$ is consistent for α. There is still some problem with the possibility that $\hat{\alpha}$ may be negative if $\frac{1}{n} \Sigma \log X_i > 0$. Now $\{\log X_i\}_1^n$ are i.i.d. r.v.s with mean $-\frac{\gamma}{\alpha}$ and variance $\frac{\pi^2}{6\alpha^2}$ therefore by CLT as applied to $\{\log X_i\}_1^n$, $\frac{1}{n} \Sigma \log X_i$ for large n is approximately normal with mean $= \frac{-\gamma}{\alpha}$ and the variance $= \frac{\pi^2}{n6\alpha^2}$. Hence, $P\left[\frac{1}{n} \Sigma \log X_i > 0\right] \simeq 1 - \Phi\left[\frac{\gamma}{\pi} \cdot \sqrt{6n}\right] \to 0$ as $n \to \infty$. Thus for large samples there will be in general no problem in obtaining moment estimator based on $\log X$.

In case parameter $\theta = (\theta_1, \ldots, \theta_m)'$, is vector valued statistic and $\{f(x, \theta), \theta \in \Omega \subset R_m\}$ is such that there exist m-functions $\{(T_1(X_i), \ldots, T_m(X_i)\}$ such that $E(T_r(X_i) = \psi_r(\theta)$. $\gamma = 1, 2, \ldots m$. Then by WLLN, $\frac{1}{n} \sum_{i=1}^{n} T_r(x_i) = \bar{T}_r(x)$

is consistent for $\psi_r(\theta)$. If $\{\psi_1, \ldots, \psi_m\}$ are such that $\left|\dfrac{\partial(\psi_1, \ldots, \psi_m)}{\partial(\theta_1, \ldots, \theta_m)}\right| \neq 0$, then the moment equations given by

$$\bar{T}_r(x) = \psi_r(\theta), \quad r = 1, 2, \ldots, m$$

determine the estimator $(\tilde{\theta}_1, \ldots, \tilde{\theta}_m)'$ which is consistent for θ since the implicit function theorem is applicable to the above system of equations. Indeed originally the method of moments was proposed by Pearson to determine the parameters occurring in distributions whose pdf is determined by the differential equation

$$\frac{\partial \log f(x)}{\partial x} = \frac{(x-a)}{b_{00} + b_1 x + b_2 x^2} \; x \in (c, d)$$

K. Pearson showed that under some regularity conditions which guarantees existence of first four moments the constants (a, b, c, d) are determined uniquely by $(\mu_1', \mu_2', \mu_3', \mu_4')$ or equivalently by $(\mu_1', \mu_2, \beta_1, \beta_2)$ i.e. mean, variance, skewness and kurtosis. K. Pearson also prepared tables of (β_1, β_2) using which we first guess the type of the curve and then estimate (a, b, c, d). Thus for a real valued data $(x_1, \ldots, x_n)'$, Pearson technique consisted of: (i) calculating (m_1', m_2', m_3', m_4') and then $(m_1', m_2, b_1, b_2)'$ and (ii) using sample skewness b_1 and sample kurtosis b_2 and using Biometrika tables (first prepared by Pearson) guess the type of the pdf and then estimate (a, b, c, d). This was a standard procedure used for a long time. Our formulation allows us to select appropriate functions $(T_1, \ldots, T_m)$ rather than $(X, X^2, \ldots, X^m)$.

EXAMPLE 5.2.4 Consider Exercise 5.1 (1). Here X is Gamma with pdf $f(x, \sigma, \lambda) = \dfrac{1}{\Gamma(\lambda)} e^{-x/\sigma} \dfrac{1}{\sigma^\lambda} x^{\lambda-1}$, $x > 0$, $\sigma > 0$, $\lambda > 0$. Now $\mu_1' = \lambda\sigma$ and $\mu_2' = \lambda(\lambda+1)\sigma^2$ so that

$$\left|\frac{\partial(\mu_1', \mu_2')}{\partial(\lambda, \sigma)}\right| = \begin{vmatrix} \sigma & \lambda \\ \sigma^2(2\lambda+1) & 2\sigma\lambda(\lambda+1) \end{vmatrix} = \sigma^2\lambda > 0$$

and the moment equations $m_1' = \lambda\sigma$, $m_2' = \lambda(\lambda+1)\sigma^2$ determine $(\tilde{\lambda}, \tilde{\sigma})$ uniquely which are given by $\tilde{\sigma} = \dfrac{m_2' - m_1'^2}{m_1'}$, $\tilde{\lambda} = \dfrac{m_1'^2}{m_2' - m_1'^2}$, which can be shown to be consistent for σ and λ respectively. However, if we do not know that X is Gamma distributed then fitting of frequency curve can be done by using data to calculate $(b_1, b_2)'$ and use these values to guess the type, say Pearson Type III, which is Gamma distribution.

Pearson's method of moments was used by Student (1908) to obtain the sampling distribution of $S^2 = \sum (X_i - \bar{X})^2$ and that of $t_{n-1} = \bar{X}\sqrt{n}/\sqrt{S^2/n-1}$, known as Student's t with $(n - 1)$ degrees of freedom. Assuming that $\{X_i\}_1^n$ are i.i.d. $N(0, 1)$, Student obtained first four moments of S^2 and t_{n-1}, their b_1, b_2 and guessed the Pearson types and then obtained their exact sampling distributions. The usual derivation of distribution of S^2 and t_{n-1} was derived later by Fisher (1922). Student's 1908 paper is regarded as a major breakthrough in Statistics as it introduced the concept of exact sampling distribution theory. It is interesting to point out that Student obtained moments of t_{n-1} by claiming independence of $\bar{X}$ and S^2 on the basis of Cov $(\bar{X}, S^2)$ = 0. Of course Cov $(T_1, T_2) = 0$ does not imply that T_1, T_2 are independent. But in Student's case his conclusion was true (as later verified by Fisher) but his argument was certainly incorrect.

Exercise 5.2 (1) Consider $b(1, p)$ model where $p = P[X = 1]$ is a function of θ, a situation that occurs in bio-assay problem. Suppose that at the dosage level $d > 0$, $P[X = 1] = p(\theta) = \dfrac{e^{\theta d}}{1 + e^{\theta d}}$, $\theta \in R_1$ then obtain the moment estimator of θ based on $\bar{X}$.

(2) Show that the moment estimator of θ based on X in the following three models is same. Compare the variances of the moment estimator under these three models.

(a) $X \sim N(\theta, 1)$, (b) $X \sim U(\theta - 1, \theta + 1)$, (c) X with pdf $f(x, \theta) = \dfrac{1}{2} \exp\{-|x - \theta|\}$, $x \in R_1$, $\theta \in R_1$ (Double Exponential Distribution).

(3) Suppose that apriori it is known that $X \sim N(\theta, 1)$ with $\Omega_0 = \{\theta \mid -1 \le \theta \le 1\}$. Obtain the probability p_θ that the moment equation $\bar{x} = \theta$ would not have a solution in Ω_0. Show that $p_\theta \to 0$ as $n \to \infty$ for $\theta \in (-1, 1)$ but at $\theta = \pm 1$, $p_\theta \to 1/2$.

(4) For truncated Poisson distribution with pmf

$$P[X = r] = e^{-\lambda}\lambda^r/r!\ (1 - e^{-\lambda}),\ r = 1, 2, 3, \ldots .$$

show that moment equation $m_1' = E(X)$ has unique solution.

5.3 Method of Percentiles

Suppose $(X_1, \ldots, X_n)$ is a random sample of size n on a continuous r.v. X from pdf $\{f(x, \theta), \theta \in \Omega \subset R_1\}$ then the 100p% population percentile $\xi_p(\theta)$ is defined as the solution of the equation

$$F(\xi_p(\theta), \theta) = \int_{-\infty}^{\xi_p(\theta)} f(x, \theta)\, dx = p$$

where $0 < p < 1$ or $\xi_p(\theta) = F_\theta^{-1}(p)$. Note that if the r.v. X is assumed to have a non-vanishing pdf at $\xi_p(\theta)$ then $\xi_p(\theta)$ is uniquely determined. Let $r = [np] + 1$, then the r-th order statistic of the sample $X_{(r)}$ can be shown to be consistent estimator of $\xi_p(\theta)$ or $X_{(r)} \xrightarrow{p} \xi_p(\theta)$, $\forall\ \theta \in \Omega$. The method of percentiles consists in obtaining an estimator of θ by solving the "percentile equation"

$$X_{(r)} = \xi_p(\theta)$$

which admits a unique solution if $\dfrac{d\xi_p}{d\theta} \neq 0$.

We note that $X_{(r)}$ can be interpreted as 100p% percentile of the sample. Thus as the method of moments equates sample moments with corresponding population moments, the method of percentiles can be viewed as a method where we equate sample percentiles with corresponding population percentiles.

To see this we note that if $\{X_{(1)}, \ldots, X_{(n)}\}$ is the order statistic (o.s.) of a sample of size n on a continuous r.v. X with pdf f and df F then $U_{(r)} = F(X_{(r)})$ has the same distribution as that of r-th order statistic for a sample of size n from $U(0, 1)$. Hence

$$E(U_{(r)}) = \frac{r}{n+1}, \quad E(U_{(r)}^2) = \frac{r(r+1)}{(n+1)(n+2)}$$

Therefore, $E(U_{(r)} - p)^2 = \dfrac{r(r+1)}{(n+1)(n+2)} - 2p\dfrac{r}{n+1} + p^2$. But as $r = [np] + 1$, $np \le r \le np + 1$ and therefore $\dfrac{r}{n} \to p$ as $n \to \infty$. Thus $E(U_{(r)} - p)^2 \to 0$ as $n \to \infty$ and therefore by Tchebychev inequality $U_{(r)} \xrightarrow{p} p$ as $n \to \infty$ or $F(X_{(r)}) \xrightarrow{p} p$. However as F^{-1} exists and is continuous since $\dfrac{dF}{dx} = f(x) > 0$ at $x_p = \xi_p(\theta)$, we have by invariance property $F^{-1}(F(X_{(r)})) = X_{(r)} \xrightarrow{p} F^{-1}(p) = \xi_p(\theta)$. We note that for $p = 1/4$, $3/4$, we get lower and upper quantiles and $p = 1/2$ we get the median.

EXAMPLE 5.3.1 Consider $(X_1, \ldots, X_n)$ a random sample of size n from $N(\theta, 1)$ then for $p = 1/2$ we get $X_{([n/2]+1)} = M_n$ the sample median as a consistent estimator of θ, the population median. In fact the same result will hold for a random sample of size n from $U(\theta - 1, \theta + 1)$ or Double Exponential (Laplace distribution) with pdf $f(x, \theta) = \frac{1}{2} \exp\{-|x - \theta|\}$, $x \in R_1$, $\theta \in R_1$. As seen in Exercise 5.2.2 in all these three case $\bar{X}$, the sample mean is also a consistent estimator of θ. On the other hand in Example 5.1.4 we have seen that for Cauchy distribution with pdf $f(x, \theta) = \dfrac{1}{\pi}\dfrac{1}{1 + (x - \theta)^2}$, $x \in R_1$, $\bar{X}$ is not consistent for θ. However noting that θ is population median in case of Cauchy distribution, the sample median M_n would be consistent for θ.

EXAMPLE 5.3.2 We have seen in Example 5.2.2 that in case of Pareto distribution, if $\lambda \in (0, 1)$, the method of moments fails. However here $\xi_p(\lambda) = \exp\{-\log(1 - p)/\lambda\}$ and $X_{([np]+1)} \xrightarrow{p} \exp\{-\log(1 - p)/\lambda\}$. As $d\xi_p/d\lambda \ne 0$, ξ_p is a continuous invertible function of λ the percentile equation is given by

$$X_{([np]+1)} = \exp\{-\log(1 - p)/\lambda\}$$

which yields a consistent estimator of λ given by

$$\tilde{\lambda} = [-\log(1-p)]/\log X_{([np]+1)}.$$

Note that the percentile method does not put any restriction on λ such as $\lambda \in (1, \infty)$ as is required for the method of moments.

Recall that if X is Pareto with parameter λ then $Y = \log X$ is exponential with mean $1/\lambda$ or $g(y, \lambda) = \lambda e^{-\lambda y}$, $y \geq 0$. Now 100p% percentile of Y is $\xi_p(\lambda) = -\log(1-p)/\lambda$. Now percentile equation based on Y is $\log X_{([np]+1)} = -\log(1-p)/\lambda$ which leads to the same estimator $\tilde{\lambda}$ given above. On the other hand the method of moments as applied to Y gives the estimator $\dfrac{n}{\Sigma \log X_i}$. Observe that $y = \log x$ is strictly increasing transformation and therefore $(\log X_{(1)}, \ldots, \log X_{(n)})$ is the o.s. of a sample of size n from $g(y, \lambda)$ with $Y_{(i)} = \log X_{(i)}$.

EXAMPLE 5.3.3 We have seen in Example 5.2.3 that the method of moments has various problems when X is Weibull with parameter α. Now observing that X^α is exponential with mean one, we have 100p% percentile $\xi_p(\alpha) = \log[-\log(1-p)] \cdot \dfrac{1}{\alpha}$ and a consistent estimator of α based on percentile method yields

$$\tilde{\alpha} = \log[-\log(1-p)]/\log X_{([np]+1)}$$

Consider in general a location scale parameter family with pdf $f(x, \mu, \sigma) = \dfrac{1}{\sigma} f_0\left(\dfrac{x-\mu}{\sigma}\right)$, $x \in R_1$, $\mu \in R_1$, $\sigma > 0$ where $f_0 = f(x, 0, 1)$, is the standard pdf. Let $0 < p < 1$. Then $\xi_p = \mu + c_p \sigma$ where c_p is the 100p% percentile of the standard pdf given by $p = \int_{-\infty}^{c_p} f_0(x)\, dx = F(c_p)$. Consider $0 < p_1 < p_2 < \ldots < p_k < 1$. Then the corresponding order statistic $X_{([np_r]+1)}$ is consistent for ξ_{p_r}, and we have k equations to determine estimators of μ and σ. These are given by

$$x_{([np_r]+1)} = \mu + c_{p_r} \sigma, \quad r = 1, 2, \ldots, k.$$

We may obtain estimators of $(\mu, \sigma)'$ by the method of least squares which are given by

$$\tilde{\sigma} = \frac{k \sum_{r=1}^{k} c_{p_r} X_{([np_r]+1)} - \sum_{r=1}^{k} c_{p_r} \sum_{r=1}^{k} X_{([np_r]+1)}}{k \sum_{r=1}^{k} c_{p_r}^2 - \left(\sum_{r=1}^{k} c_{p_r}\right)^2}$$

$$\tilde{\mu} = \frac{\sum_{r=1}^{k} X_{([np_r]+1)} - \tilde{\sigma} \sum_{r=1}^{k} c_{p_r}}{k}$$

Noting that $X_{([np_r]+1)} \xrightarrow{p} \mu + c_{p_r}\sigma$ as $n \to \infty$, one can easily demonstrate that $\tilde{\sigma} \xrightarrow{p} \sigma$ and $\tilde{\mu} \xrightarrow{p} \mu$ so that $(\tilde{\mu}, \tilde{\sigma})'$ is consistent for (μ, σ). We emphasize here the fact that k is a fixed integer and the estimators $\tilde{\mu}$ and $\tilde{\sigma}$ are based on fixed number of k selected percentiles.

On the other hand in the range depending on parameter case such as $U(0, \theta)$ considered in Exercise 5.1(5) we have shown that $X_{(n)}$ is consistent for θ, the upper end of the distribution. Similarly we can show that $X_{(1)}$ the first order statistic of the sample from pdf $f(x, \mu) = \exp\{-(x - \mu\}$, $x \geq \mu$, is consistent for μ. This follows from the fact that $E(X_{(1)}) = \mu + 1/n$ and Var $(X_{(1)}) = 1/n^2$ so that $E(X_{(1)} - \mu)^2 \to 0$. Indeed if $(X_{(1)}, \ldots, X_{(n)})$ is the order statistic from a continuous distribution with range (a, b) then one can show that $X_{(1)} \xrightarrow{p} a$ and $X_{(n)} \xrightarrow{p} b$. If the random variable X has range $(-\infty, b)$ then $X_{(n)} \xrightarrow{p} b$ but $X_{(1)}$ will diverge to minus infinity and if the range is (a, ∞) then $X_{(1)} \xrightarrow{p} a$ but $X_{(n)}$ will diverge to plus infinity. The proofs are immediate consequence of the fact that the distribution function of $X_{(n)}$ is given by $[F(u)]^n$ and that of $X_{(1)}$ is given by $1 - (1 - F(u))^n$. We prove this for $X_{(n)}$ and leave the results for $X_{(1)}$ as an exercise to the reader. Now $G_n(u)$ the d.f. of $X_{(n)}$ when the range of X is a bounded interval (a, b) is given by

$$\begin{aligned} G_n(u) &= 0 \quad u < a \\ &= [F(u)]^n \quad a \leq u < b \\ &= 1 \quad u \geq b \end{aligned}$$

Taking limit as $n \to \infty$

$$\begin{aligned} \lim G_n(u) &= 0 \quad \text{if } u < b \\ &= 1 \quad u \geq b \end{aligned}$$

which is the d.f. of the r.v. degenerate at b. Let $\varepsilon > 0$ then

$$\begin{aligned} P[b - \varepsilon < X_{(n)} < b + \varepsilon] &= G_n(b + \varepsilon) - G_n(b - \varepsilon) \\ &= 1 - G_n(b - \varepsilon) \end{aligned}$$

But $G_n(b - \varepsilon) \to 0$ for any $\varepsilon > 0$ and therefore we have

$$P[| X_{(n)} - b | < \varepsilon] \to 1 \text{ and } X_{(n)} \xrightarrow{p} b.$$

On the other hand if range of X is (a, ∞) then $G_n(u) = 0$ if $u < a$ and $G_n(u) = [F(u)]^n$, $a \leq u$ and $G(u) = \lim G_n(u) = 0$ for any $u \in R_1$ and $P[X_{(n)} \leq u] \to 0$ as $n \to \infty$ for any $u \in R_1$. This implies that $P[X_{(n)} > u] \to 1$ as $n \to \infty$ for any $u \in R_1$ or $X_{(n)}$ diverges to plus infinity.

We now consider an example by which the distinction between the behaviour of extreme order statistics and that of central order statistics will be clarified further.

EXAMPLE 5.3.4 Let $(X_1, \ldots, X_{(2n+1)})$ be the order statistic of a random sample of size $(2n + 1)$ from $U(\theta - 1, \theta + 1)$ with pdf $f(x, \theta) = \frac{1}{2}$, $\theta - 1 \leq x \leq \theta + 1$. From what we have seen earlier $X_{(1)} \xrightarrow{p} \theta - 1$ and $X_{(2n+1)} \xrightarrow{p} \theta + 1$. So that $T_1 = X_{(1)} + 1$ and $T_2 = X_{(2n+1)} - 1$ are both consistent for θ. Further the sample median $X_{(n+1)}$ is consistent for θ the population median.

Routine calculations will show that the pdf of $X_{(2n+1)}$ is given by

$$g(x, \theta) = (2n + 1)(x - \theta + 1)^{2n} \frac{1}{2^{2n+1}}, \ \theta - 1 < x < \theta + 1.$$

Let $W = \dfrac{X_{(2n+1)} - \theta + 1}{2}$ then the pdf of W is given by

$$g(w) = (2n + 1)w^{2n}, \ 0 < w < 1.$$

This gives $E(w) = \dfrac{2n + 1}{2n + 2}$ and $E(w^2) = \dfrac{2n + 1}{2n + 3}$

and

$$\text{Var}(w) = \frac{(2n + 1)}{(2n + 2)^2(2n + 3)}.$$

Now

$$E(T_2) = E(X_{(2n+1)} - 1) = 2E(w) + \theta - 2 = \theta - \frac{1}{n + 1}.$$

$$\text{Var}(T_2) = 4 \text{ Var}(w) = \frac{4(2n + 1)}{(2n + 2)^2(2n + 3)}$$

Hence
$$E(T_2 - \theta)^2 = \frac{4(2n + 1)}{(2n + 2)^2(2n + 3)} + \frac{1}{(n + 1)^2}$$

$$= \frac{4}{(n + 1)(2n + 3)}$$

Therefore $T_2 \xrightarrow{p} \theta$ since T_2 converges in quadratic mean to θ, i.e. $E(T_2 - \theta)^2 \to 0$. The distribution theory of T_2 and T_1 is almost identical and the reader can verify that $T_1 \xrightarrow{p} \theta$ and infact $E(T_1 - \theta)^2 = E(T_2 - \theta)^2$.

Now consider median $M_n = X_{(n+1)}$ then its pdf is given by

$$g(y_{(n+1)}, \theta) = \frac{1}{B(n + 1, n + 1)} \frac{(x_{(n+1)} - \theta + 1)^n}{2} \left(1 - \frac{x_{(n+1)} - \theta + 1)}{2}\right)^n \frac{1}{2},$$

$\theta - 1 < x_{(n+1)} < \theta + 1$ or if we use $w = \dfrac{x_{(n+1)} - \theta + 1}{2}$ then

$$g(w) = \frac{1}{B(n + 1, n + 1)} w^n(1 - w)^n \ 0 < w < 1.$$

This gives $E(w) = \dfrac{B(n+2, n+1)}{B(n+1, n+1)} = \dfrac{1}{2}$ so that $E(X_{(n+1)}) = \theta$ and $E(w^2) = \dfrac{B(n+3, n+1)}{B(n+1, n+1)} = \dfrac{n+2}{2(2n+3)}$ and $\text{Var}(w) = \dfrac{1}{4(2n+3)}$ so that $\text{Var}(X_{(n+1)}) = 4\,\text{Var}(w) = \dfrac{1}{2n+3}$. Thus $E(X_{(n+1)} - \theta)^2 = \dfrac{1}{2n+3} \to 0$ as $n \to \infty$. $X_{(n+1)}$ is also consistent for θ. We however observe that the MSE (T_1) = MSE (T_2) tends to zero at the rate of $1/n^2$ but MSE $(X_{(n+1)})$ tends to zero at the rate of $1/n$.

5.4 Choosing Between Consistent Estimators

To choose between two unbiased estimators we have used the criterion of smallness of variance. In a similar way assuming that mean squared error i.e. $E(T - \theta)^2$ exists, we use the criterion of smallness of mean squared error (MSE) to choose between two consistent estimators. Thus if T_1 and T_2 are both consistent for θ then we would prefer T_1 to T_2 if MSE $(T_1) \le$ MSE (T_2), $\forall\ \theta \in \Omega$ and there exists an n_1 such that $\forall\ n \ge n_1$, MSE (T_1) $\le$ MSE (T_2), $\forall\ \theta \in \Omega$. Thus if T_1 is preferred to T_2 then by Tchebychev inequality it follows that $P[|T_1 - \theta| < \varepsilon]$ converges to unity faster than $P[|T_2 - \theta| < \varepsilon] \to 1$ as $n \to \infty$, $\forall\ \theta \in \Omega$ and any $\varepsilon > 0$. Therefore the minimum sample size (see Section 5.1) to achieve the level of accuracy specified by (ε, δ) for T_1 would be smaller than that required for T_2 for every combination of values of $\varepsilon, \delta, \theta$. We consider a few examples to illustrate this approach based on the rate at which MSE $(T_i) \to 0$ as $n \to \infty$.

EXAMPLE 5.4.1 Let $\{X_i\}_1^n$ be i.i.d. $N(\theta, \sigma^2)$ then as seen in Example 5.1.2, S^2/n and $S^2/(n-1)$ are both consistent for σ^2 where $S^2 = \Sigma (X_i - \bar{X})^2 \sim \sigma^2\chi^2_{n-1}$. Now

$$\text{MSE}\left(\frac{S^2}{n-1}\right) = \text{Var}\left(\frac{S^2}{n-1}\right) = \frac{2\sigma^4}{(n-1)}$$

and

$$\text{MSE}\left(\frac{S^2}{n}\right) = \text{Var}\left(\frac{S^2}{n}\right) + \left(\text{bias}\ \frac{S^2}{n}\right)^2$$

$$= \frac{2\sigma^4}{n^2}(n-1) + \frac{1}{n^2}\sigma^4 = \frac{\sigma^4(2n-1)}{n^2}$$

$$\text{MSE}\left(\frac{S^2}{n-1}\right) - \text{MSE}\left(\frac{S^2}{n}\right) = \frac{2\sigma^4}{n-1} - \frac{\sigma^4(2n-1)}{n^2}$$

$$= \frac{\sigma^4(1-3n)}{n^2}$$

$$< 0 \text{ for any } n \ge 2$$

Therefore we prefer $\frac{S^2}{n}$ to $\frac{S^2}{n-1}$. Consider in general $T_1 = a_n \frac{S^2}{n-1}$ where $a_n > 0$ such that $a_n \to 1$ as $n \to \infty$. Then T_1 is consistent for θ, as

$$\text{MSE}(T_1) = \text{Var}\left(a_n \frac{S^2}{n-1}\right) + \left(\text{bias}\ \frac{a_n S^2}{n-1}\right)^2 = \frac{2\sigma^4 a_n^2}{n-1} + (a_n - 1)^2 \sigma^4$$

$$= \sigma^4 \left\{\frac{2a_n^2}{(n-1)} + (a_n - 1)^2\right\}$$

Reader may verify that for $a_n = \frac{n-1}{n}$ we get $T_1 = S^2/n$, and for $a_n = \frac{n-1}{n+1}$, $T_1 = S^2/(n+1)$ with MSE $= 2\sigma^4/(n+1)$ and we have

$$\text{MSE}\left(\frac{S^2}{n+1}\right) < \text{MSE}\left(\frac{S^2}{n}\right) < \text{MSE}\left(\frac{S^2}{n-1}\right).$$

One can show that within the class of consistent estimators given by $\frac{S^2}{n+k}$ for k a fixed real constant, MSE $\left(\frac{S^2}{n+k}\right)$ is minimum when $k = 1$. We leave this as an exercise to the reader.

Yet another way of comparing MSE (T_1) and MSE (T_2) is to expand the expressions in powers of $1/n$ and compare the coefficients. For example consider

$$\text{MSE}\left(\frac{S^2}{n-1}\right) = \frac{2\sigma^4}{n-1} = \frac{2\sigma^4}{n}\left(1 - \frac{1}{n}\right)^{-1} = \frac{2\sigma^4}{n}\left\{1 + \frac{1}{n} + \frac{1}{n^2} + \ldots\right\}$$

$$\text{MSE}\left(\frac{S^2}{n}\right) = \frac{\sigma^4(2n-1)}{n^2} = \sigma^4\left(\frac{2}{n} - \frac{1}{n^2}\right)$$

$$\text{MSE}\left(\frac{S^2}{n+1}\right) = \frac{2\sigma^4}{n+1} = \frac{2\sigma^4}{n}\left(1 + \frac{1}{n}\right)^{-1}$$

$$= \frac{2\sigma^4}{n}\left\{1 - \frac{1}{n} + \frac{1}{n^2} + \ldots\right\}$$

We see that the coefficient of $\frac{1}{n}$ is 2 σ^4 in all the three expansions but coefficients of $\frac{1}{n^2}$ in the three cases are +2, –1 and –2, respectively, with a common factor σ^4 and therefore MSE $\left(\frac{S^2}{n+1}\right) \to 0$ at a rate faster than that of the other two. Note that to define an estimator based on S^2 we need $n \geq 2$.

EXAMPLE 5.4.2 Consider a random sample of size m from $U(\theta - 1, \theta + 1)$ a situation considered in Example 5.3.2 for $m = 2n + 1$ an odd number. Now $E(X_i) = \theta$, Var $(X_i) = 3$ so that the sample mean $\bar{X}_m$ is consistent for θ with MSE $(\bar{X}_m) = 1/3m =$ Var $(\bar{X}_m)$. Let $(X_{(1)}, \ldots, X_{(m)})$ denote the order statistic of the sample. Then $y_{(i)} = \dfrac{X_{(i)} - \theta + 1}{2}$ then $(Y_{(1)}, \ldots, Y_{(m)})$ is the order statistic of a sample of size m from $U(0, 1)$ and the pdfs of $y_{(1)}$ and $y_{(m)}$ are

$$g_1(u) = m(1 - u)^{m-1}, \quad 0 < u < 1$$

$$g_m(v) = mv^{m-1}, \quad 0 < v < 1$$

and the joint pdf of $y_{(1)}, y_{(m)}$ is

$$g_{1,m}(u, v) = m(m - 1)\,(v - u)^{m-2}, \; 0 < u < v < 1.$$

Now routine calculations (which we request the reader to carry out) will show that

$$E(Y_{(1)}) = \frac{1}{m+1}, \quad E(Y_{(m)}) = \frac{m}{m+1}$$

$$\mathrm{Var}(Y_{(1)}) = \mathrm{Var}\,(Y_{(m)}) = \frac{m}{(m+1)^2(m+2)}$$

$$\mathrm{Cov}\,(Y_{(1)}, Y_{(m)}) = \frac{1}{(m+1)^2(m+2)}$$

Therefore $E(X_{(1)} + 1) = \theta + \dfrac{2}{m+1}$, $\mathrm{Var}(X_{(1)} + 1) = \dfrac{4m}{(m+1)^2(m+2)}$

$$E(X_{(m)} - 1) = \theta - \frac{2}{m+1}, \; \mathrm{Var}(X_{(m)} - 1) = \frac{4m}{(m+1)^2(m+2)}$$

$$\mathrm{Cov}\,(X_{(1)} + 1, X_{(m)} - 1) = \frac{1}{(m+1)^2(m+2)}$$

It now follows that $T_1 = (X_{(1)} + 1)$ $T_2 = (X_{(m)} - 1)$ are both consistent for θ with MSE $(T_1) =$ MSE $(T_2) = \dfrac{8}{(m+1)(m+2)}$. Further $T_3 = \dfrac{T_1 + T_2}{2}$ has bias zero and Var $(T_3) = \dfrac{(4m+1)}{2(m+1)^2(m+2)}$ so that T_3 is also consistent for θ. Expanding MSE $(\bar{X}_m)$ and MSE (T_i), $i = 1, 2, 3$ in powers of $1/m$ we observe that

$$\text{MSE}\,(\bar{X}_m) = 1/3m$$

$$\text{MSE}\,(T_1) = \text{MSE}\,(T_2) = \frac{8}{m^2}\left(1 + \frac{1}{m}\right)^{-1}\left(1 + \frac{2}{m}\right)^{-1}$$

$$= \frac{8}{m^2}\left\{1 + \frac{3}{m} + 0\left(\frac{1}{m^2}\right)\right\}$$

$$= \frac{8}{m^2} + \frac{24}{m^3} + 0\left(\frac{1}{m^3}\right)$$

$$\text{MSE}\,(T_3) = \frac{(4m + 1)}{2m^3}\left\{\left(1 + \frac{1}{m}\right)^{-1}\left(1 + \frac{2}{m}\right)^{-1}\right\}$$

$$= \left(\frac{2}{m^2} + \frac{1}{2m^3}\right)\left\{1 + \frac{3}{m} + 0\left(\frac{1}{m}\right)\right\}$$

$$= \frac{2}{m^2} + \frac{13}{2m^3} + 0\left(\frac{1}{m^3}\right).$$

Note that MSE $(\bar{X}_m) \to 0$ at the rate $\frac{1}{3m}$ and MSE (T_1) = MSE $(T_2) \to 0$ at the rate $8/m^2$ whereas MSE $(T_3) \to 0$ at the rate $2/m^2$. Thus fastest convergence here is for MSE (T_3) and T_3 is the best and $\bar{X}_m$ is the worst among these four estimators.

We now consider the estimator $M_m = X_{([m/2]+1)}$ the sample median. Now $M_m = X_{(n+1)}$ if $m = 2n + 1$ or $m = 2n$ but the distributions of M_m in two cases are different. The case $m = 2n + 1$ has been considered in detail in Example 5.3.2 hence we will consider the case $m = 2n$ only. The pdf of M_m in this case is given by

$$g(u) = \frac{1}{B(n + 1, n)}\left(\frac{u - \theta + 1}{2}\right)^n\left(1 - \frac{u - \theta + 1}{2}\right)^{n-1}\frac{1}{2},\quad \theta - 1 < u < \theta + 1.$$

Again substituting $\dfrac{M_m - \theta + 1}{2} = W$ we have pdf of W given by

$$g(w) = \frac{1}{B(n + 1, n)}\, w^n(1 - w)^{n-1},\quad 0 < w < 1.$$

Thus $\quad E(W) = \dfrac{n + 1}{2n + 1}$ and Var $(W) = \dfrac{n}{2(2n + 1)^2}$ so that

$$E(M_m) = \theta + \frac{1}{2n + 1}$$

$$\text{Var}\,(M_m) = 4\,\text{Var}\,(W) = \frac{2n}{(2n + 1)^2}.$$

Thus $$\text{MSE}(M_m) = \frac{1}{m+1} \quad \text{if } m \text{ is even.}$$

$$= \frac{1}{m+2} \quad \text{if } m \text{ is odd.}$$

However, MSE $(M_m) \to 0$ as $m \to \infty$ and M_m is consistent for θ.

$$\text{MSE}(M_m) = \frac{1}{m}\left(1 + \frac{1}{m}\right)^{-1} = \frac{1}{m}\left\{1 - \frac{1}{m} + 0\left(\frac{1}{m}\right)\right\} \quad \text{if } m \text{ is even}$$

$$= \frac{1}{m}\left(1 + \frac{2}{m}\right)^{-1} = \frac{1}{m}\left\{1 - \frac{2}{m} + 0\left(\frac{1}{m}\right)\right\} \quad \text{if } m \text{ is odd.}$$

Thus M_m is least efficient as MSE $(M_m) \to 0$ at the rate of $\frac{1}{m}$ whereas MSE $(\bar{X}_m) \to 0$ at the rate of $\frac{1}{3m}$. The following table gives the minimum sample size m_0 to achieve accuracy level $\varepsilon = 0.1$ and $\delta = 0.1$ for all the five estimators. This table can be obtained by using the fact that m_0 is determined by inequality MSE $(T_i) \le \delta\,\varepsilon^2$ and the expansion of MSE (T_i) in powers of $\frac{1}{m}$.

Estimator	$\bar{X}_m$	M_m	T_1	T_2	T_3
m_0	334	1000	90	90	45

We have already noted in Example 5.1.1, that the number $n_0(\varepsilon, \delta, \theta)$ given by using Tchebychev inequality could be much higher than the number obtained by using the large sample or asymptotic distribution of the estimator. Let T be a consistent estimator of θ then we want to estimate

$$p_n(\varepsilon, \theta) = P_\theta[|\,T - \theta\,| < \varepsilon] = F_n(\theta + \varepsilon) - F_n(\theta - \varepsilon)$$

assuming T to be continuous r.v. with continuous distribution function $F_n(u, \theta) = P_\theta[T \le u]$ where T is the estimator based on sample size n. As $T \xrightarrow{p} \theta$ we have

$$\lim_{n\to\infty} F_n(u, \theta) = 0 \quad \text{if } u < \theta$$

$$= 1 \quad \text{if } u \ge \theta$$

and $\lim_{n\to\infty} p_n(\varepsilon, \theta) = 1$ for any $\varepsilon > 0$. Note that convergence in probability and convergence in distribution are equivalent when the limiting distribution function is degenerate. Thus convergence in probability by itself will not provide us any information about the rate at which $p_n(\varepsilon, \theta) \to 1$ as $n \to \infty$.

To obtain such an information about the rate at which $p_n(\varepsilon, \theta) \to 1$ we must obtain an approximation for $p_n(\varepsilon, \theta)$. Recall that for $a_n > 0$, $P[|T - \theta| < \varepsilon] = p_n(\varepsilon, \theta) = P[a_n |T - \theta| < a_n\varepsilon]$. Thus if we can find a sequence $a_n > 0$, so that limiting distribution of $a_n(T - \theta)$ is non-degenerate or $Y_n = a_n(T - \theta) \xrightarrow{d} Y$ where Y is a non-degenerate r.v. then $p_n(\varepsilon, \theta) \simeq P[|Y| < a_n\varepsilon] = G(a_n\varepsilon) - G(-a_n\varepsilon)$ assuming that Y has its d.f. is continuous. For example in case of $\{X_i\}_1^n$ i.i.d. with $E(X_i) = \theta$ and Var $(X_i) = \sigma^2$ by WLLN $\bar{X} \xrightarrow{p} \theta$ and by CLT $\frac{\sqrt{n}(\bar{X} - \theta)}{\sigma} \xrightarrow{d} N(0, 1)$ or if $a_n = \sqrt{n}$, $p_n(\varepsilon, \theta) \simeq \Phi(\sqrt{n}\ \varepsilon/\sigma) - \Phi(-\sqrt{n}\ \varepsilon/\sigma)$ and the rate at which $p_n(\varepsilon, \theta) \to 1$ can now be studied by known results about the rate at which $\Phi(\sqrt{n}\ \varepsilon) \to 1$ and $\Phi(-\sqrt{n}\ \varepsilon) \to 0$. For example, for large a, it is known that $1 - \Phi(a) \simeq \frac{1}{a}\varphi(a) = \frac{1}{a}\frac{e^{-a^2/2}}{\sqrt{2\pi}}$. Therefore in above case $p_n(\varepsilon, \theta) \simeq 2\Phi(\sqrt{n}\ \varepsilon/\sigma) - 1$ and $p_n(\varepsilon, \theta) \to 1$ at the rate of $\frac{2\sigma}{\sqrt{n\varepsilon 2\pi}} e^{-n\varepsilon^2/2\sigma^2}$.

Observe that by considering $\sqrt{n}(\bar{X} - \theta)$ rather than $(\bar{X} - \theta)$ we have a non-degenerate limiting distribution of the sequence of r.v.s. under study. Multiplication by the factor a_n in general and $\sqrt{n}$ in particular blows the infinitesimal r.v. $(\bar{X} - \theta)$ to $\sqrt{n}(\bar{X} - \theta)$ a non-infinitesimal r.v. This is similar to studying a phenomenon that cannot be studied by naked eyes only with the help of a microscope or studying a map obtained by an airborne camera by enlarging. Note that we have to blow the picture just enough. Thus in this case $a_n = \sqrt{n}$ is just right as $n^{\delta}(\bar{X} - \theta) \xrightarrow{p} 0$ if $0 < \delta < 1/2$ and $n^{\delta}(\bar{X} - \theta)$ will diverge to infinity if $\delta > 1/2$. It is only for $\delta = 1/2$ the r.v. $(\bar{X} - \theta)$ is blown just enough so that $n^{\delta}(\bar{X} - \theta) \xrightarrow{d} N(0, \sigma^2)$, a proper non-degenerate r.v. We illustrate this magnifying operation by a few examples and in the next chapter consider this approach in detail.

EXAMPLE 5.4.3 Consider a r.s. of size m from $U(\theta - 1, \theta + 1)$ and the five estimators considered in Example 5.4.2. We now consider the asymptotic distributions of each one of these i.e. use the limiting distribution of $b_m^{(1)}(\bar{X}_m - \theta)$, $b_m^{(2)}(X_{([m/2]+1)} - \theta)$, and $a_m^{(i)}(T_i - \theta)$, $i = 1, 2, 3$ where the constants $a_m^{(i)}$ and $b_m^{(1)}, b_m^{(2)}$ are chosen appropriately. A quick rule of thumb for the choice of such norming constants is to use them as proportional to the square root of the MSE. Thus for $\bar{X}_m$ and $X_{([m/2]+1)}$ we use $\sqrt{m}$. Now by CLT $(\bar{X}_m - \theta)/\sqrt{3m} \xrightarrow{d} N(0, 1)$ i.e. $\bar{X}_m$ is Asymptotic Normal with mean θ and variance $1/3m$ a fact denoted by $\bar{X}_m \sim AN(\theta, 1/3m)$. Similarly the theorem on asymptotic distribution of 100p% sample percentile states that provided $f(\xi_p) \neq 0$ $X_{([mp]+1)} \sim AN\left(\xi_p, \frac{p(1-p)}{m[f(\xi_p)]^2}\right)$. For proof refer to David

(1981). For $p = 1/2$ we have $M_m = X_{([m/2]+1)} \sim AN(\theta, 1/m)$ as $f(\xi_{1/2}) = 1/2$ in $U(\theta - 1, \theta + 1)$ case. Thus by using asymptotic distributions of $\bar{X}_m$, M_m, we get $m_0(\varepsilon, \delta, \theta)$ as $[[\Phi^{-1}(1 - \delta/2)]^2/3\varepsilon^2] + 1$ and $[[\Phi^{-1}(1 - \delta/2)]^2/\varepsilon^2] + 1$ respectively. For $\varepsilon = 0.1$, $\delta = .1$ we have $m_0(\bar{X}_m) = 91$ and $m_0(M_m) = 271$. For T_1, T_2 we can use the exact distribution for each m. Consider $T_1 - \theta = (X_{(1)} + 1 - \theta)$. Observe that $T_1 - \theta = 2y_{(1)}$ where $Y_{(1)} = (X_{(1)} + 1 - \theta)/2$ has pdf $g_1(y_{(1)}) = m(1 - y_{(1)})^{m-1}$, $0 < y_{(1)} < 1$. Since $T_1 = X_{(1)} + 1 \geq \theta$, we have $P[| T_1 - \theta | < \varepsilon] = P[0 < y_{(1)} < \varepsilon/2] = 1 - (1 - \varepsilon/2)^m$ if $\theta < \varepsilon < 1/2$ and one otherwise. Thus $m_0(T_1)$ is such that

$$1 - (1 - \varepsilon/2)^m \geq 1 - \delta \text{ or } m_0(T_1) = \left[\frac{\log(\delta)}{\log(1 - \varepsilon/2)}\right] + 1 \text{ if } 0 < \varepsilon < 1/2$$

otherwise $m_0(T_1) = 1$. For $\delta = \varepsilon = .1$ we obtain $m_0(T_1) = 45$. Similarly for $T_2 - \theta = X_{(m)} - 1 - \theta$ since $T_2 - \theta < 0$ we have $| T_2 - \theta | = \theta - T_2 = \theta - (X_{(m)} - 1) = \theta + 1 - X_m$ and

$$\begin{aligned} P[| T_2 - \theta | < \varepsilon] &= P[\theta - (X_{(m)} - 1) < \varepsilon] \\ &= P[X_{(m)} > \theta + 1 - \varepsilon] \\ &= P[2Y_{(m)} + \theta - 1 > \theta + 1 - \varepsilon] \\ &= P[Y_{(m)} > 1 - \varepsilon/2] \\ &= 1 - (1 - \varepsilon/2)^m \quad \text{if } 0 < \varepsilon \leq 1/2 \\ &= 1 \qquad \text{otherwise} \end{aligned}$$

If $\varepsilon < 1/2$, $m_0(T_2)$ is given by $1 - (1 - \varepsilon/2)^m \geq 1 - \delta$ or $m_0(T_2) = \left[\frac{\log(\delta)}{\log(1 - \varepsilon/2)}\right] + 1$, which for $\varepsilon = 0.1$, $\delta = 0.1$ gives $m_0(T_2) = 45$, as one would expect. For $\varepsilon \geq 1/2$ we have $m_0(T_2) = 1$. It would be interesting to consider the asymptotic distribution of $a_m(T_1 - \theta)$ and $a_m(T_2 - \theta)$ where a_m is chosen such that the limiting distributions are non-degenerate. Since $E(T_1 - \theta)^2$ is of the order $1/m^2$ one could consider $a_m = m$ and indeed if $Z_m = m(T_1 - \theta)$ then $0 < Z_m < 2m$ and the d.f. of Z_m,

$$\begin{aligned} G_m(u) &= 1 - (1 - u/2m)^m \quad \text{if } 0 < u < 2m \\ &= 1 \qquad \text{if } u \geq 2m \end{aligned}$$

To consider $\lim_{m\to\infty} G_m(u)$ for fixed u, we observe that for given u there exists an m' such that $0 < u < 2m$ and $G_m(u) = 1 - (1 - u/2m)^m$ for $m \geq m'$ and therefore $\lim_{m\to\infty} G_m(u) = 1 - e^{-u/2}$, $u \geq 0$. Thus $m(T_1 - \theta)$ is asymptotically exponential with mean 2. Hence $P[| (T_1 - \theta) | < \varepsilon] = P[0 < Z_m < m\,\varepsilon]$ is approximately equal to $1 - e^{-m\varepsilon/2}$. Thus m_0 based on the asymptotic distribution of $m(T_1 - \theta)$ is determined by the condition that $1 - e^{-m\varepsilon/2} \geq 1 - \delta$ or $m_0 = [-2 \log \delta/\varepsilon] + 1$. For $\varepsilon = .1$ and $\delta = .1$ we obtain $m_0'(T_1) = 47$ which is quite close to $m_0(T_1) = 45$ obtained by using the exact distribution of T_1. Note that as $\log(1 - \varepsilon/2) = -\varepsilon/2 + 0(\varepsilon/2)$, the values of $m_0(T_1)$ and $m_0'(T_1)$ are bound to be quite close.

Now consider $T_3 = \dfrac{T_1 + T_2}{2} = \dfrac{X_{(1)} + 1 + X_{(m)} - 1}{2}$ so that $T_3 = \dfrac{X_{(1)} + X_{(m)}}{2}$ and $T_3 - \theta = \dfrac{X_{(1)} - \theta + X_{(m)} - \theta}{2}$. In terms of $Y_{(i)} = \dfrac{X_{(i)} - \theta + 1}{2}$ we have $T_3 - \theta = [y_{(1)} + y_{(m)} - 1]$ and

$$P[|\,T_3 - \theta\,| < \varepsilon] = P[1 - \varepsilon < y_{(1)} + y_{(m)} < 1 + \varepsilon]$$

$$= \iint m(m-1)\,(y_{(m)} - y_{(1)})^{m-2}\,dy_{(1)}\,dy_{(m)}$$

where A is the shaded region. Therefore,

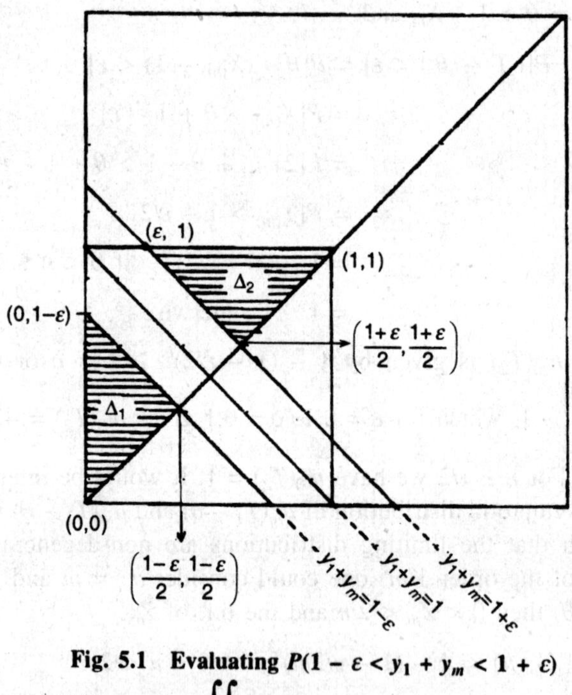

Fig. 5.1 Evaluating $P(1 - \varepsilon < y_1 + y_m < 1 + \varepsilon)$

$$\text{Required probability} = 1 - \iint_{\Delta_1} g(y_{(1)}, y_{(m)})\,dy_{(1)}\,dy_{(m)}$$

$$- \iint_{\Delta_2} g(y_{(1)}, y_{(m)})\,dy_{(1)}\,dy_{(m)}$$

where Δ_1 is the triangle formed by (0, 0), $\left(\dfrac{1-\varepsilon}{2}, \dfrac{1-\varepsilon}{2}\right)$ and $(0, 1-\varepsilon)$ and Δ_2 the triangle formed by (1, 1) $\left(\dfrac{1+\varepsilon}{2}, \dfrac{1+\varepsilon}{2}\right)$ and $(\varepsilon, 1)$. Now

$$P(\Delta_1) = \frac{(1-\varepsilon)^m}{2} = P(\Delta_2)$$

as can be seen by integrating the pdf over Δ_1 and Δ_2. Therefore

$$P[1 - \varepsilon < y_{(1)} + y_{(m)} < 1 + \varepsilon] = 1 - (1-\varepsilon)^m.$$

Thus $m_0(T_3)$ is given by $1 - (1-\varepsilon)^m \geq 1 - \delta$ or $m_0(T_3) = \left[\frac{\log \delta}{\log (1-\varepsilon)}\right] + 1$.
For $\varepsilon = 0.1$ and $\delta = 0.1$, $m_0(T_3) = 22$.

To obtain asymptotic distribution of $m(T_3 - \theta)$ we see that

$$P[|\, m(T_1 - \theta)\, | < \varepsilon] = P[|\, T_3 - \theta\, | < \varepsilon/m] = 1 - (1 - \varepsilon/m)^m$$

and $\lim_{m\to\infty} P[m \,|\, T_1 - \theta \,|< u] = 1 - e^{-u}$. Therefore for large m, $P[|\, T_3 - \theta \,| < \varepsilon] = P[|\, m(T_3 - \theta)\,| < m\varepsilon] \simeq 1 - e^{-m\varepsilon}$ and $m_0'(T_3)$ based on asymptotic distribution of T_3 is given by $m_0'(T_3) = \left[\frac{-\log \delta}{\varepsilon}\right] + 1$ for $\varepsilon = 0.1$, $\delta = 0.1$ we have $m_0'(T_3) = 24$. This is quite close to the exact value of $m_0(T_3) = 22$ as we have $\log (1-\varepsilon) = -\varepsilon + 0(\varepsilon)$.

The other interesting way to obtain this result is to observe that $my_{(1)} \xrightarrow{d} Z_1$ and $m(1 - y_{(m)}) \xrightarrow{d} Z_2$ where Z_1, Z_2 are standard exponentials. Further $y_{(1)}$ and $y_{(m)}$ can be shown to be asymptotically independent. Hence $P[|\, y_{(1)} + y_{(m)} - 1\,| < \varepsilon] \simeq P[|\, Z_1 - Z_2 \,| < m\varepsilon]$ where Z_1, Z_2 are independent exponentials (see David, 1981). Now $P[|\, Z_1 - Z_2\,| < m\varepsilon]$ can be shown to be equal to $1 - e^{-m\epsilon}$ and $m_0'(T_3) = \left[\frac{-\log \delta}{\varepsilon}\right] + 1$. Table 2 summarizes the results for all the five estimators for $\varepsilon = .1$ and $\delta = .1$.

Table 2

Min sample size \ Estimator	$\bar{X}_m$	M_m	T_1	T_2	T_3
Based on MSE	314	1000	90	90	45
Based on exact distribution	NA	NA	45	45	22
Based on asymptotic distribution	91	274	47	47	24

Exercise 5.4 (1) Let $(X_1, \ldots, X_n)$ be i.i.d. $U(0, \theta)$. Carry out similar exercise as in Example 5.4.3 for consistent estimation of θ using $2\bar{X}_m$, M_m, $T_2 = X_{(m)}$, $T_3 = \frac{m+1}{m} X_{(m)}$. Note that here the minimum number m_0 depends on θ in each case. Calculate m_0 for $\theta = 1/4, 1/2, 1, 2, 4$.

(2) Let $(X_1, \ldots, X_n)$ be i.i.d. $N(\theta, \sigma^2)$ and let $S^2 = \Sigma(X_1 - \bar{X})^2 \sim \sigma^2\chi^2_{n-1}$. Obtain a general expression for $Q = \text{MSE}(c_n S^2)$ and show that Min Q occurs at $c_n = \dfrac{1}{n+1}$ and $\text{MSE}\left(\dfrac{S^2}{n+1}\right) < \text{MSE}\left(\dfrac{S^2}{n}\right) < \text{MSE}\left(\dfrac{S^2}{n-1}\right)$.

(3) Use the well known normal approximation to χ^2_k given by $\chi^2_k \sim AN(k, 2k)$ to determine $n_0(\varepsilon, \delta, \sigma^2)$ for

$$T_1 = \frac{S^2}{n-1}, T_2 = \frac{S^2}{n}, T_3 = \frac{S^2}{n+1} \text{ as estimators of } \sigma^2.$$

(4) For the double exponential distribution with Mean = Median = θ, use CLT to obtain asymptotic distribution of $\bar{X}_n$ and the theorem quoted in Example 5.4.3 to obtain asymptotic distribution of M_n. Determine $n_0(\varepsilon, \delta, \theta)$ for $\bar{X}_n$ and M_n.

6

Consistent Asymptotically Normal Estimators

6.1 A Basic Result

Let T be consistent for a real parameter θ so that $(T - \theta) \xrightarrow{p} 0$, $\forall\ \theta \in \Omega$. We have seen that in order to obtain an approximation to $P[\,|T - \theta| < \varepsilon]$ for large n, we blow up infinitesimal r.v. $(T - \theta)$ by a factor a_n so that $a_n(T - \theta) \xrightarrow{d} Z$ where Z is a non-degenerate r.v. We saw by way of examples that for random sample of size n or $\{X_i\}_1^n$ i.i.d., $a_n = \sqrt{n}$ is adequate for estimators obtained by method of moments or percentiles and Z is $N(0,\ \sigma_T^2(\theta))$ so that $T \sim AN(\theta,\ \sigma_T^2(\theta)/n)$. For extremes such as $X_{(n)}$ or $X_{(1)}$ the blow up factor was n and limiting distribution was non-normal, typically exponential with mean $\sigma(\theta)$. When limiting distribution is normal, the minimum sample size $n_0(\varepsilon,\ \delta,\ \theta)$ required for consistent estimator T to achieve degree of accuracy specified by (ε, δ) is given by smallest n such that $a_n \geq \sigma_T(\theta)\ \Phi^{-1}(1 - \delta/2)/\varepsilon$. If T' is any other consistent estimator such that $T' \sim AN(\theta,\ \sigma_{T'}^2(\theta)/a_n^2)$, i.e. T' has the same blow up factor a_n, then the corresponding number $n_0'(\varepsilon,\ \delta,\ \theta)$ is given by smallest n such that $a_n \geq \sigma_{T'}(\theta)\ (\Phi^{-1}(1 - \delta/2))/\varepsilon$. We then would prefer T' to T if $\sigma_{T'}^2(\theta) \leq \sigma_T^2(\theta)$, $\forall\ \theta \in \Omega$. The number $\sigma_{T'}^2(\theta)/\sigma_T^2(\theta)$ is called as the asymptotic relative efficiency (ARE) of T with respect to T'. If ARE $(T, T') = \sigma_{T'}^2(\theta)/\sigma_T^2(\theta) \leq 1\ \forall\ \theta \in \Omega$ then we prefer T' to T. If the sequence of blow up factor is not the same, i.e. $T \sim AN(\theta,\ \sigma_T^2(\theta)/a_n^2)$ and $T' \sim AN(\theta,\ \sigma_{T'}^2(\theta)/b_n^2)$ then we choose T' to T if $n_0'(\varepsilon,\ \delta,\ \theta) \leq n_0(\varepsilon,\ \delta,\ \theta)$, $\forall\ \theta \in \Omega$ and this would happen if $\sigma_T^2(\theta)/a_n^2 \geq \sigma_{T'}^2(\theta)/b_n^2$. The ARE of T w.r.t. T' is then defined as limit of the ratio $\sigma_{T'}^2(\theta)b_n^2/\sigma_T^2(\theta)a_n^2$. If the limiting distributions after the use of appropriate blow up factors are normal, the choice between consistent estimators is governed by their asymptotic variances. We therefore first concentrate on those consistent estimators which are such that their blown up versions $a_n(T - \theta) \xrightarrow{d} N(0,\ \sigma_T^2(\theta))$ or $T \sim AN(\theta,\ \sigma_T^2(\theta)/a_n^2)$. Such an estimator is called as Consistent Asymptotically Normal or CAN estimator.

We have seen in the previous chapter that if T is consistent and if ψ is a continuous function then $\psi(T)$ is consistent for $\psi(\theta)$. Similarly we have an invariance property for CAN estimators under a continuous differentiable transformation ψ.

THEOREM 6.1.1 Let T be CAN for θ so that $T \sim AN(\theta, \sigma_T^2(\theta)/a_n^2)$ and let ψ be a differentiable function such that $\frac{d\psi}{d\theta}$ is continuous and nonvanishing then $\psi(T)$ is CAN for $\psi(\theta)$ and

$$\psi(T) \sim AN\left(\psi(\theta), \sigma_T^2(\theta)\left(\frac{d\psi}{d\theta}\right)^2 / a_n^2\right).$$

Since ψ is differentiable, it is continuous and $\psi(T)$ is consistent for $\psi(\theta)$. Now by mean value theorem

$$\psi(T) - \psi(\theta) = (T - \theta)\frac{d\psi}{d\theta} + R$$

where the remainder R is such that $|R| \leq M|T - \theta|^{1+\delta}$ for some $\delta > 0$.

Now $a_n(\psi(T) - \psi(\theta)) = a_n(T - \theta)\frac{d\psi}{d\theta} + Ra_n$ and the theorem is proved if we show that $|Ra_n| \xrightarrow{P} 0$. Note that $a_n(T - \theta)\frac{d\psi}{d\theta} \xrightarrow{d} N\left(0, \sigma_T^2(\theta)\left(\frac{d\psi}{d\theta}\right)^2\right)$ since $a_n(T - \theta) \xrightarrow{d} N(0, \sigma_T^2(\theta))$. Now $|a_n R| \leq M|a_n(T - \theta)||T - \theta|^{\delta}$ and $a_n(T - \theta) \rightarrow N(0, \sigma_T^2(\theta))$, a proper r.v. and $|T - \theta|^{\delta} \xrightarrow{P} 0$ so that $|a_n R| \xrightarrow{P} 0$. Here we are using the result that if $U_n \xrightarrow{d} U$ and $V_n \xrightarrow{P} 0$ then $U_nV_n \xrightarrow{P} 0$. For proof we refer to Rao (1973) or Cramer (1946).

We now consider a few examples to illustrate the invariance of CAN estimators.

EXAMPLE 6.1.1 Let $(X_1, ..., X_n)$ be i.i.d. Poisson with mean λ then by CLT, $\bar{X} \sim AN(\lambda, \lambda/n)$ and thus MVUE $\bar{X}$ is CAN for λ. Consider estimation of $\psi(\lambda) = e^{-\lambda} = P(X = 0)$. Then $\frac{d\psi}{d\lambda} = -e^{-\lambda} < 0, \forall \lambda > 0$ and is continuous. Hence by the invariance property $\psi(\bar{X}) = e^{-\bar{X}} \sim AN\left(e^{-\lambda}, e^{-2\lambda} \cdot \frac{\lambda}{n}\right)$. Note that the $AV(e^{-\bar{X}}) = e^{-2\lambda}\frac{\lambda}{n}$ = CRLB for estimating $\psi(\lambda) = e^{-\lambda}$. We have already seen that the MVUE of $e^{-\lambda}$ given by $\varphi_0(\bar{X}) = \left(\frac{n-1}{n}\right)^{n\bar{X}}$ has variance strictly greater than the CRLB. Next consider estimating $\psi(\lambda) = \lambda e^{-\lambda} = P[X = 1]$. Here $\frac{d\psi}{d\lambda} = e^{-\lambda}(1 - \lambda) \neq 0, \lambda \neq 1$. Thus $\psi(\bar{X}) = \bar{X}e^{-\bar{X}}$ is CAN for $\lambda e^{-\lambda}$ with asymptotic variance $\frac{\lambda}{n}e^{-2\lambda}(1 - \lambda)^2$ for any $\lambda \neq 1$. At $\lambda = 1$ the CAN ness of $\psi(\bar{X})$ may not hold.

EXAMPLE 6.1.2 Let $(X_1, ..., X_n)$ be i.i.d. $N(\mu, 1)$ then $\bar{X} \sim N\left(\mu, \frac{1}{n}\right)$ and is

CAN for μ. Consider $\psi(\mu) = \mu^2$. Then $\frac{d\psi}{d\lambda} = 2\mu \neq 0$ for any μ except $\mu = 0$. Thus $\bar{X}^2$ is CAN for μ^2 for all $\mu \neq 0$ and $AV(\bar{X}^2) = \frac{4\mu^2}{n}$ = CRLB for estimating μ^2. Consider $\sqrt{n}(\bar{X}^2 - \mu^2)$ at $\mu = 0$ i.e. $Z = \sqrt{n}\ \bar{X}^2$. Then as $\bar{X} \sim N(0, 1/n)$ under $\mu = 0$, $n\bar{X}^2 \sim \chi_1^2\ \forall\ n \geq 1$ and thus $n\bar{X}^2 \xrightarrow{d} \chi_1^2$ at $\mu = 0$. Therefore $\sqrt{n}(\bar{X}^2 - \mu^2)$ at $\mu = 0$ behaves like $\chi_1^2/\sqrt{n}$ which converges in probability to 0. Therefore here the inflating factor would be $a_n = n$ rather than $a_n = \sqrt{n}$ and the asymptotic distribution of $n(\bar{X}^2 - \mu^2)$ is χ_1^2 at $\mu = 0$ which is same as the exact distribution. This example illustrates that the sequence a_n need not always be $\sqrt{n}$ and the limiting distribution is not always normal.

The phenomenon observed in the above example can be generalized and stated as a theorem as follows.

THEOREM 6.1.2 Assume the frame work of Theorem 6.1.1 but suppose that $\frac{d\psi}{d\theta} = 0$ and $\frac{d^2\psi}{d\theta^2} \neq 0$ at $\theta = \theta_0$ and is continuous then $a_n^2(\psi(T) - \psi(\theta_0))$ $\xrightarrow{d} \frac{\sigma_T^2(\theta_0)}{2}\left(\frac{d^2\psi}{d\theta^2}\right)_{\theta=\theta_0} \chi_1^2$.

By Taylor series expansion of $\psi(T)$ around $\psi(\theta_0)$ we have

$$a_n^2\ (\psi(T) - \psi(\theta_0)) = \frac{a_n^2(T - \theta_0)^2}{2!}\left(\frac{d^2\psi}{d\theta^2}\right)_{\theta=\theta_0} + a_n^2 R$$

where $|\ a_n^2\ R\ | \leq M\ a_n^2\ (T - \theta_0)^2\ |\ T - \theta_0\ |^{\delta}$ for $\delta > 0$.

Now $a_n(T - \theta_0) \xrightarrow{d} N(0, \sigma_T^2(\theta_0))$ therefore

$$\frac{a_n^2(T - \theta_0)^2}{2} \xrightarrow{d} \frac{\sigma_T^2(\theta_0)}{2}\chi_1^2 \text{ and } a_n^2 R \xrightarrow{P} 0 \text{ since } |\ T - \theta_0\ |^{\delta} \xrightarrow{P} 0.$$

Therefore, $\quad a_n^2(\psi(T) - \psi(\theta_0)) \xrightarrow{d} \frac{\sigma_T^2(\theta_0)}{2}\left(\frac{d^2\psi}{d\theta^2}\right)_{\theta=\theta_0} \chi_1^2.$

EXAMPLE 6.1.1 (*Contd.*)

For $\psi(\lambda) = \lambda e^{-\lambda}$ we have for any $\lambda \neq 1$ $\sqrt{n}(\bar{X}e^{-\bar{X}} - \lambda e^{-\lambda}) \xrightarrow{d} N(0, \lambda(1-\lambda)^2 e^{-2\lambda})$ as $\frac{d\psi}{d\lambda} = e^{-\lambda}(1 - \lambda)$. However at $\lambda = 1$ $\frac{d\psi}{d\lambda} = 0$ but $\left(\frac{d^2\psi}{d\lambda^2}\right)_{\lambda=1} = -e^{-1}$. Therefore, at $\lambda = 1$, $n(\bar{X}e^{-\bar{X}} - e^{-1}) \xrightarrow{d} \frac{1}{2}(-e^{-1})\chi_1^2$.

The selection of norming constant a_n so that $a_n(T - \theta) \xrightarrow{d} N(0, \sigma_{T'}^2(\theta))$ is a tricky job. We have already seen that for $(X_1, \ldots, X_n)$ i.i.d. $U(0, \theta)$,

$T = X_{(n)}$ is a consistent estimator for θ, and $n(X_{(n)} - \theta) \xrightarrow{d} -Y$ where Y is exponential with mean θ. However there is no choice of a norming sequence a_n for which $a_n(X_{(n)} - \theta)$ would converge to a normal r.v. This follows from the observation that no matter how we choose a_n, $P[a_n(X_{(n)} - \theta) \le u] = 1$ for $u \ge 0$ and the limiting distribution of $a_n(X_{(n)} - \theta)$, say $G(u)$ is such that $G(u) = 1$ for $u \ge 0$ and $G(u) \ne \Phi\left(\frac{u}{\sigma_T(\theta)}\right)$ for $u \in (0, \infty)$ for any a_n. In case observations are on r.v.s which are non i.i.d., behaviour of limiting distribution of $a_n(T - \theta)$ can be very complex. We illustrate this phenomenon by way of an example where observations are independent but not identically distributed. When dependence is introduced in the sequence $\{X_i\}_1^\infty$ the behaviour of $a_n(T - \theta)$ can be still more complex. However we will not pursue this problem in this introductory course.

EXAMPLE 6.1.3 Let $(X_1, \ldots, X_n)$ be independent r.v.s where $X_i = \theta + \varepsilon_i$ where $\varepsilon_i \sim N(0, i^{2\alpha})$ where α is a known constant. Now $E(X_i) = \theta$, and $E(\bar{X}_n) = \theta$ and $\bar{X} \sim N\left(\theta, \frac{1}{n^2}\sum_{i=1}^n i^{2\alpha}\right)$. Let $b_n^2 = \frac{1}{n^2}\sum_{i=1}^n i^{2\alpha}$. Now $P[\,|\bar{X} - \theta| < \varepsilon] = P\left[\frac{|\bar{X} - \theta|}{b_n} < \frac{\varepsilon}{b_n}\right] = 2\Phi\left(\frac{\varepsilon}{b_n}\right) - 1$, where $b_n > 0$ is the positive square root of b_n^2. Now

$$b_n^2 = \frac{1}{n^2}\sum_{i=1}^n i^{2\alpha} \simeq \frac{1}{n}\sum_{i=1}^n (i/n)^{2\alpha} \cdot \frac{n^{2\alpha+1}}{n^2} \simeq n^{2\alpha-1}\frac{1}{2\alpha+1} \quad \text{for large } n,$$

as $\lim_{n\to\infty} \frac{1}{n}\sum_{i=1}^n (i/n)^{2\alpha} = \int_0^1 x^{2\alpha}\,dx = \frac{1}{2\alpha+1}$. Thus for $\alpha < \frac{1}{2}$, $b_n \to 0$ and $P[\,|\bar{X} - \theta| < \varepsilon] \to 1$ as $n \to \infty$. For $\alpha = 1/2$, $\lim_{n\to\infty} P[\,|\bar{X} - \theta| < \varepsilon]$ $= \lim_{n\to\infty} 2\Phi(\varepsilon/b_n) - 1 = 2\Phi(\sqrt{2}\varepsilon) - 1$ as $\lim_{n\to\infty} b_n = \frac{1}{\sqrt{2}}$. For $\alpha > 1/2$, $b_n \to \infty$ as $n \to \infty$ and $2\Phi\left(\frac{\varepsilon}{b_n}\right) - 1 \to 0$ as $n \to \infty$ and $\lim_{n\to\infty} P[\,|\bar{X} - \theta| < \varepsilon] = 0$.

Thus $\bar{X}$ is CAN for θ only when $\alpha < \frac{1}{2}$. For $\alpha \ge \frac{1}{2}$, $\bar{X}$ is $N(\theta, b_n^2)$ but $\bar{X}$ is not consistent. This is because in case $\alpha > \frac{1}{2}$ the entire probability mass of the distribution of $\bar{X}$ escapes to infinity and sequence of d.f.s of $\bar{X}$ does not converge to a d.f. For $\alpha = 1/2$, $\bar{X}\ N\left(\theta, \frac{1}{2} + \frac{1}{n}\right)$ and $\bar{X} \xrightarrow{d} N\left(\theta, \frac{1}{2}\right)$. Note that $(\bar{X} - \theta)/b_n$ for each $n \ge 1$ is $N(0, 1)$ and therefore $(\bar{X} - \theta)/b_n \xrightarrow{d} N(0, 1)$ for any α as $n \to \infty$ but the behaviour of $P[\,|\bar{X} - \theta| < \varepsilon]$ is different for different values of α and depends on whether $b_n \to 0$ or $b_n \to 1$ or $b_n \to \infty$ according as $\alpha < \frac{1}{2}$, $\alpha = \frac{1}{2}$ and $\alpha > \frac{1}{2}$ respectively.

In this course we will mostly be concerned with the case of random samples i.e. $(X_1, \ldots, X_n)$ are i.i.d. and the choice of $a_n = \sqrt{n}$ would be appropriate in most situations. We next consider the method of moments and percentiles to generate CAN estimators.

Exercise 6.1.5 (1) Let $(X_1, \ldots, X_n)$ be i.i.d. exponential with mean θ. Using CLT show that $\bar{X}$ is CAN for θ and obtain a CAN estimator for $P[X > t] = e^{-t/\theta}$ and its asymptotic variance.

(2) Let $(X_1, \ldots, X_n)$ be i.i.d. $b(1, \theta)$. Show that $\bar{X}$ is CAN for θ. Let $\psi(\theta) = \theta(1-\theta)$. Show that $\bar{X}(1-\bar{X})$ is CAN for $\theta(1-\theta)$ for all values of θ except $\theta = 1/2$. What is the asymptotic distribution of $\bar{X}(1-\bar{X})$ at $\theta = \frac{1}{2}$?

6.2 Method of Moments and Percentiles

Let $(X_1, \ldots, X_n)$ be i.i.d. r.v.s with $E(X_i) = \mu(\theta)$. In Chapter 5 we have seen that if μ^{-1} exists and is continuous then $\mu^{-1}(\bar{X})$ is consistent for θ. A sufficient condition for μ^{-1} to exist and be continuous and differentiable is $\frac{d\mu}{d\theta} \neq 0$ and is continuous. Now suppose that $(X_i, \ldots, X_n)$ are i.i.d. such that $E(X_i) = \mu(\theta)$ and Var $(X_i) = \sigma^2(\theta)$. Then by CLT $\sqrt{n}(\bar{X} - \mu(\theta)) \xrightarrow{d} N(0, \sigma^2(\theta))$ or $\bar{X}$ is CAN for $\mu(\theta)$ with asymptotic variance $\frac{\sigma^2(\theta)}{n}$. Now suppose that $\frac{d\mu}{d\theta} \neq 0$ and continuous. We can claim by Theorem 6.1.1 that $\mu^{-1}(\bar{X})$ is CAN. Reparametrizing the problem by the transformation $\phi = \mu(\theta)$ so that $\theta = \mu^{-1}(\phi)$ we have $\bar{X} \sim AN\left(\phi, \frac{\sigma^2(\mu^{-1}(\phi))}{n}\right)$. Since $\frac{d\phi}{d\theta}\frac{d\theta}{d\phi} = 1$ we have $\frac{d\theta}{d\phi} = \frac{d(\mu^{-1}(\phi))}{d\phi} = \frac{1}{d\phi/d\theta} = \frac{1}{(d\mu/d\theta)}$. For Theorem 6.1.1 to be applicable we need that $\frac{d\mu^{-1}(\phi)}{d\phi}$ exists and is continuous or equivalently $\left(\frac{d\mu}{d\theta}\right)^{-1}$ exists and is continuous then $\mu^{-1}(\bar{X}) \sim AN\left(\mu^{-1}(\phi), \frac{\sigma^2(\mu^{-1}(\phi))}{n}\left(\frac{d\mu^{-1}(\phi)}{d\phi}\right)^2\right)$ or $\mu^{-1}(\bar{X}) \sim AN\left(\theta, \frac{\sigma^2(\theta)}{n} \Big/ \left(\frac{d\mu}{d\theta}\right)^2\right)$. We thus have the following theorem.

THEOREM 6.2.1 Let $\{X_i\}_1^n$ be i.i.d. with $E(X_i)$, $= \mu(\theta)$ and Var $(X_i) = \sigma^2(\theta)$ such that $[d\mu/d\theta]^{-1}$ is non vanishing and continuous then $\mu^{-1}(\bar{X})$ is CAN for θ with

$$AV(\mu^{-1}(\bar{X})) = \frac{\sigma^2(\theta)}{n} \Bigg/ \left(\frac{d\mu}{d\theta}\right)^2 .$$

In a similar way if $\xi_p(\theta)$ denotes the p-th percentile of X for $0 < p < 1$ then assuming $f(\xi_p) > 0$, $X_{([np]+1)} \sim AN\left(\xi_p(\theta), \frac{p(1-p)}{n[f(\xi_p)]^2}\right)$ or $X_{([np+1])}$ is CAN for $\xi_p(\theta)$ (David, 1981). Again if $\xi_p(\theta)$ is such that $\left(\frac{d\xi_p}{d\theta}\right)^{-1}$ is nonvanishing and continuous then

$$\xi_p^{-1}(X_{([np]+1)}) \sim AN\left(\theta, \frac{p(1-p)}{n[f(\xi_p)]^2} \Bigg/ \left(\frac{d\xi_p}{d\theta}\right)^2\right).$$

Note that $\mu^{-1}(\bar{X})$ and $\xi_p^{-1}(X_{([np]+1)})$ are moment and percentile estimators based on X. We could also consider transformed variables $Y_i = U(X_i)$ such that $\frac{dU}{dx} \neq 0$ and continuous. Then moment and percentile estimators based on $(Y_1, \ldots, Y_n)$, would also have similar properties if $\left(\frac{dE(y)}{d\theta}\right)^{-1}$ and $\left(\frac{d\xi_p(y)}{d\theta}\right)^{-1}$ are non-vanishing and continuous. We now consider some examples to illustrate the above technique of obtaining moment and percentile estimators.

EXAMPLE 6.2.1 Let $(X_1, \ldots, X_n)$ be i.i.d. with pdf $f(x, \theta) = \theta x^{\theta-1}$, $0 < x < 1$, $\theta > 0$. Then $\mu(\theta) = \frac{\theta}{\theta+1}$ and $\sigma^2(\theta) = \frac{\theta}{(\theta+1)^2(\theta+2)}$. Now $\frac{d\mu}{d\theta} = \frac{1}{(\theta+1)^2}$ and $\left(\frac{d\mu}{d\theta}\right)^{-1} = (\theta+1)^2 > 0$ and is continuous. Hence the moment estimator $\mu^{-1}(\bar{X}) = \frac{\bar{X}}{1-\bar{X}}$ is the solution of the moment equation

$$\bar{X} = \frac{\theta}{\theta+1} \text{ and } \frac{\bar{X}}{1-\bar{X}} \sim AN\left(\theta, \frac{\theta(\theta+1)^2}{n(\theta+2)}\right).$$

As can be proved $\{f(x, \theta), \theta > 0\}$ as given above is one-parameter exponential family with $u(\theta) = (\theta - 1)$, $k(x) = \log x$ and if $Y_i = -\log X_i$ then $g(y, \theta) = \theta e^{-\theta y}$, $\theta > 0$, $y > 0$. Now $E(y) = \frac{1}{\theta} = \mu_y(\theta)$ and $\frac{d\mu_y}{d\theta} = -\frac{1}{\theta^2}$ is such that $\left(\frac{d\mu_y}{d\theta}\right)^{-1} = -\theta^2 < 0$ and is continuous. Hence the moment estimator based on $\log X_i$ is given by $\hat{\theta} = \frac{n}{-\Sigma \log X_i}$ and is CAN for θ. Now Var $(Y) = 1/\theta^2$.

Hence $AV(\hat{\theta}) \simeq \frac{1}{n\theta^2} \Big/ (1/\theta^2)^2 = \frac{\theta^2}{n}$ and therefore $\hat{\theta} \sim AN(\theta, \theta^2/n)$. It can be easily cheacked that $AV\left[\frac{\bar{X}}{(1-\bar{X})}\right] > AV(\hat{\theta})$ and $AV(\hat{\theta}) =$ CRLB for estimating θ.

EXAMPLE 6.2.2 Let $(X_1, \ldots, X_n)$ be a random sample from Weibull distribution with pdf $f(x, \alpha) = \alpha x^{\alpha-1} e^{-x^\alpha}$, $x > 0$, $\alpha > 0$. Then as seen in the previous chapter for $0 < p < 1$ we have $\log \xi_p(\alpha) = \log[-\log(1-p)] \frac{1}{\alpha} = c_p/\alpha$. Now

$\left(\frac{d\xi_p}{d\alpha}\right)^{-1} = \frac{\alpha^2}{-c_p \xi_p(\alpha)}$ is non-vanishing and continuous. Therefore Theorem

6.2.1 is applicable. As $X_{([np]+1)} \sim AN\left(\xi_p(\alpha), \frac{p(1-p)}{[f(\xi_p)]^2 n}\right)$ we have

$\hat{\alpha} = \frac{c_p}{\log[X_{([np]+1)}]}$ the p-th percentile estimator is CAN for α with

$$AV(\hat{\alpha}) = \frac{p(1-p)\alpha^2}{n[f(\xi_p)]^2 c_p^2 \xi_p^2}.$$

We again point out here that the method of moments is not applicable in this example as $E(X) = \Gamma\left(\frac{1}{\alpha} + 1\right) = \mu(\alpha)$ is such that the condition $\frac{d\mu}{d\alpha} \neq 0$ does not hold and the moment equation may have no solution or two solutions for several observed values of $\bar{X}$, as noted in Example 5.2.3.

EXAMPLE 6.2.3 Let $X_i = \theta + \varepsilon_i$, $i = 1, 2, \ldots, n$ be i.i.d. with error distribution $f(\varepsilon) = |\varepsilon|$, $-1 < \varepsilon < 1$. Here $E(\varepsilon) = 0$ and Var $(\varepsilon) = \frac{1}{2}$. By CLT, $\bar{X} \sim AN\left(\theta, \frac{1}{2n}\right)$. As the error distribution is symmetric around zero, $\xi_{1/2} = \theta$ and $X_{([n/2]+1)}$ is consistent for θ as per results of Chapter 5. However as pdf of X vanishes at $\xi_{1/2} = \theta$ the theorem quoted above is not applicable. On the other hand if we take $p = 3/4$ then $X_{([3n/4]+1)}$ is CAN for $\xi_{3/4} = \theta + \frac{1}{\sqrt{2}}$ and as $\left(\frac{d\xi_p}{d\theta}\right)^{-1} = 1$ and $f(\xi_{3/4}) = \frac{1}{\sqrt{2}} > 0$, we have $X_{([3n/4]+1)} - \frac{1}{\sqrt{2}} \sim AN\left(\theta, \frac{3}{8n}\right)$.

Similarly we have $X_{([n/4]+1)} + \frac{1}{\sqrt{2}} \sim AN\left(\theta, \frac{3}{8n}\right)$ What about CANness of $T_1 = \frac{X_{([3n/4]+1)} + X_{([n/4]+1)}}{2}$? We can study this if we know the joint asymptotic distribution of $X_{([np_1]+1)}$ and $X_{([np_2]+1)}$ for $0 < p_1 < p_2 < 1$. This we shall consider a little later. Before that we consider a generalization of Example 6.2.1 to the case of one parameter exponential family, and in the next section consider CANness of a vector valued estimators for a vector valued parameter.

Let $(X_1, \ldots, X_n)$ be i.i.d. such that $f(x, \theta) = \exp\{u(\theta) K(x) + v(\theta) + w(x)\}$, $x \in S$, $\theta \in \Omega \subset R_1$ which belongs to one-parameter exponential family. Then as observed in Chapter 3, $K(x)$ is complete sufficient statistic for the family. Now $\frac{\partial \log f}{\partial \theta} = u' K(x) + v'$ and $\frac{E(\partial \log f)}{\partial \theta} = 0$ implies that $E(K(x)) = \frac{-v'}{u'} = \psi(\theta)$ and $\frac{d\psi}{d\theta} = \frac{u''v' - v''u'}{u'^2}$.

$$\text{Var}\left(\frac{\partial \log f}{\partial \theta}\right) = I(\theta) = E(u'K(x) + v'$$

$$= u'^2 E(K(x) - (-v'/u'))^2$$

$$= u'^2[\text{Var } K(x)].$$

Hence $\sigma_K^2(\theta) = \frac{I(\theta)}{u'^2}$. But

$$I(\theta) = E\left(\frac{-\partial^2 \log f}{\partial \theta^2}\right) = E\{-u'' K(x) - v''\}$$

$$= -u''\left(-\frac{v'}{u'}\right) - v'' = \frac{u''v' - v''u'}{u'}$$

$$= \left[\frac{d}{d\theta}\left(\frac{-v'}{u'}\right)\right] u'$$

or $$\frac{d\psi}{d\theta} = \frac{I(\theta)}{u'}.$$

Let $T = \frac{1}{n} \Sigma K(x_i)$. Then by CLT, $T \sim AN\left(\psi(\theta), \frac{\sigma_K^2(\theta)}{n}\right)$. Further $\left(\frac{d\psi}{d\theta}\right)^{-1}$ is continuous and non-zero. The conditions of Theorem 6.2.1 are satisfied. Hence, $\hat{\theta} = \psi^{-1}(T)$ is asymptotically normal with the mean θ and variance $\sigma_K^2(\theta) \Big/ \left(\frac{d\psi}{d\theta}\right)^2$. But $\sigma_K^2(\theta) = \frac{I(\theta)}{(u')^2}$ and $\frac{d\psi}{d\theta} = \frac{I(\theta)}{u'}$, and therefore $\hat{\theta} = \psi^{-1}(T) \sim AN\left(\theta, \frac{1}{nI(\theta)}\right)$. We thus have the theorem,

THEOREM 6.2.2 If $(X_1, \ldots, X_n)$ is a random sample of size n from one parameter exponential family then the moment estimator $\hat{\theta}$ based on sufficient statistic is CAN for θ with $AV(\hat{\theta}) = \dfrac{1}{nI(\theta)}$ or $\hat{\theta} \sim AN\left(\theta, \dfrac{1}{nI(\theta)}\right)$.

We observe that $AV(\hat{\theta})$ = CRLB for unbiased estimation of θ. Thus even if we have an exponential family in which there does not exist an unbiased estimator of θ which attains CRLB for finite n, the moment estimator based on sufficient statistic is CAN and its asymptotic variance attains the CRLB. We illustrate this phenomenon by an example.

EXAMPLE 6.2.4 Let $(X_1, \ldots, X_n)$ be i.i.d. with $f(x, \theta) = \theta x^{\theta-1}$ $0 < x < 1$, $\theta > 0$. Then as observed in Example 6.2.1, $\hat{\theta} = \dfrac{n}{-\Sigma \log X_i}$ is the moment estimator based on sufficient statistic $k(x) = \log x$. Note that $\log f(x, \theta) = \log \theta + (\theta - 1)\log x$ and $\dfrac{\partial \log f}{\partial \theta} = \dfrac{1}{\theta} + \log x$ and $\dfrac{\partial^2 \log f}{\partial \theta^2} = -\dfrac{1}{\theta^2}$ so that $I(\theta) = \dfrac{1}{\theta^2}$ and CRLB for estimating θ is $\dfrac{1}{nI(\theta)} = \dfrac{\theta^2}{n}$. Now $\dfrac{\partial \log L}{\partial \theta} = \dfrac{n}{\theta} + \Sigma \log x_i$ and therefore there does not exist an unbiased estimator of θ which attains CRLB. But $AV(\hat{\theta})$ = CRLB for estimating θ. By using RBLS theorem. Reader should verify that MVUE of θ exists for $n \geq 1$ and its variance is larger than CRLB for estimating θ.

Exercise 6.2 (1) Let $(X_1, \ldots, X_n)$ be i.i.d. Pareto with pdf $f(x, \lambda) = \dfrac{\lambda}{x^{\lambda+1}}$, $x > 1$, $\lambda > 0$. Show that the pdf belongs to one parameter exponential family and the moment estimator of λ based on $K(x) = \log x$ has asymptotic variance equal to CRLB. Obtain MVUE of λ and compare its variance with CRLB.

(2) (a) Let $f(x, \theta) = \dfrac{1}{2\theta} e^{-|x|/\theta}$, $x \in R_1$, $\theta > 0$. Show that the pdf belongs to exponential family and the moment estimator of θ based on $K(x) = |x|$ is also MVUE of θ whose variance attains CRLB for each $n \geq 1$.

(b) Show that $\bar{X}$ the sample mean is not consistent for θ though it is asymptotically normal.

(c) Using method of moments based on $U(x) = x^2$ obtain CAN estimator of θ and compare its asymptotic variance with that of $\dfrac{1}{n}\Sigma |X_i|$.

3. (a) Variance Stabilizing Transformation was introduced by Bartlett (1937) to analyze the data in which the variance is a function of mean, for example as in case of the exponential, binomial or Poisson distribution. Let X_i be i.i.d. Poisson then by CLT, $\bar{X} \sim AN\left(\lambda, \dfrac{\lambda}{n}\right)$. Bartlett suggested considering $\psi(\bar{X})$ such that $\psi(\bar{X}) \sim AN\left(\psi(\lambda), \dfrac{c^2}{n}\right)$ where c^2 does not depend on λ.. Now by Theorem 6.1.1, $AV(\psi(\bar{X})) = \dfrac{\lambda}{n}\left(\dfrac{d\psi}{d\lambda}\right)^2$ so that $\lambda\left(\dfrac{d\psi}{d\lambda}\right)^2 = c^2$ or $\dfrac{d\psi}{d\lambda} = \dfrac{c}{\sqrt{\lambda}}$ or one can take $\psi(\lambda) = \sqrt{\lambda}$. Now $\dfrac{d\psi}{d\lambda} = \dfrac{1}{2\sqrt{\lambda}} > 0$ and

continuous and therefore $\sqrt{X} \sim AN\left(\sqrt{\lambda}, \frac{1}{4n}\right)$. We will see later that such transformations are very useful in constructing large sample tests or confidence intervals for λ.

(b) Determine variance stabilizing transformation for $(X_1, \ldots, X_n)$ i.i.d. $b(1, \theta)$ where $\bar{X} \sim AN\left(\theta, \frac{\theta(1-\theta)}{n}\right)$.

(c) Carry out similar exercise for $(X_1, \ldots, X_n)$ i.i.d. exponential with mean θ where $\bar{X} \sim AN\left(\theta, \frac{\theta^2}{n}\right)$.

6.3 CAN Estimators: Multiparameter Case

Let $T = (T_1, \ldots, T_m)'$ be a vector valued estimator which is consistent for a vector parameter $\theta = (\theta_1, \ldots, \theta_m)'$ as defined in Chapter 5. If there exists a sequence $a_n > 0$ and $a_n \to \infty$ such that the vector valued r.v. $a_n(T - \theta)$ has in the limit a non-singular proper m-dimensional normal distribution then we say that T is CAN for θ. Thus if $a_n(T - \theta) \xrightarrow{d} N^{(m)}(0, \Lambda(\theta))$ where $\Lambda(\theta)$ is a symmetric positive definite matrix then T is said to have asymptotic normal distribution with mean vector θ and variance covariance matrix $\frac{\Lambda(\theta)}{a_n^2}$. This is denoted by $T \sim AN^{(m)}\left(\theta, \frac{\Lambda(\theta)}{a_n^2}\right)$. As in case of $m = 1$, usually the choice of $a_n = \sqrt{n}$ will suffice.

One consequence of the above definition is that each T_i is CAN for θ_i with $AV(T_i) = \frac{\lambda_{ii}(\theta)}{n}$ and any linear combination $T' = \sum_{i=1}^{m} l_i T_i$ is CAN for $\sum_{i=1}^{m} l_i \theta_i$ with $AV(T') = \frac{1}{n} l' \Lambda(\theta) l$.

As univariate CLT for i.i.d.r.v. with finite variance is helpful in constructing CAN estimators for a real parameter θ, in a similar way multivariate CLT (MVCLT) for i.i.d.r.v. with finite variance-covariances is useful to obtain CAN estimators for $\theta = (\theta_1, \ldots, \theta_m)'$. Let $Z = (Z_1, \ldots, Z_m)'$ be such that $E(Z) = \mu$ and $M_z = \Lambda$ which we assume to be pd. Let $\{Z_i\}_1^n$ be i.i.d.r.v.s distributed as Z and let $\bar{Z} = (\bar{Z}_1, \ldots, \bar{Z}_m)'$ be the mean vector of the sample i.e. $\bar{Z}_r = \frac{1}{n} \sum_{i=1}^{n} Z_{ri}$, $r = 1, 2, \ldots, m$. Then MVCLT asserts that $\sqrt{n}(\bar{Z} - \mu) \xrightarrow{d} N^{(m)}(0, \Lambda)$ or $\bar{Z} \sim AN^{(m)}\left(\mu, \frac{1}{n}\Lambda\right)$. We now illustrate this technique by way of a few examples.

EXAMPLE 6.3.1 Let $(X_1, \ldots, X_n)$ be i.i.d. $N(\mu, \sigma^2)$. Then $E(X_i) = \mu$, $E(X_i^2) = \mu^2 + \sigma^2$. Let $Z = (Z_1, Z_2)'$ where $Z_1 = X_1$ and $Z_2 = X_1^2$. Then Λ in general is given by $\begin{pmatrix} \mu_2' - \mu_1'^2, & \mu_3' - \mu_1'\mu_2' \\ \mu_3' - \mu_1'\mu_2' & \mu_4' - \mu_2'^2 \end{pmatrix}$ and for the normal distribution

$\Lambda = \begin{pmatrix} \sigma^2 & 2\mu\sigma^2 \\ 2\mu\sigma^2 & 2\sigma^4 + 4\sigma^2\mu^2 \end{pmatrix}$. Therefore by MVCLT, we have for samples from $N(\mu, \sigma^2)$,

$$\begin{pmatrix} m_1' \\ m_2' \end{pmatrix} \sim AN^{(2)}\left[\begin{pmatrix} \mu \\ \mu^2 + \sigma^2 \end{pmatrix}, \frac{1}{n}\begin{pmatrix} \sigma^2 & 2\mu\sigma^2 \\ 2\mu\sigma^2 & 2\sigma^4 + 4\sigma^2\mu^2 \end{pmatrix}\right]$$

where m_1' and m_2' are first two raw sample moments.

EXAMPLE 6.3.2 Let the pmf of (X, Y) be given by

$$f(x, y, \lambda, p) = \binom{x}{y} p^y (1-p)^{x-y} e^{-\lambda} \frac{\lambda^x}{x!} \quad 0 < p < 1,\ 0 < \lambda,$$

$$y = 0, 1, 2, \ldots, x,\ x = 0, 1, 2, \ldots.$$

Then $E(X) = \lambda$, $E(Y) = \lambda p$ and $M_z = \begin{pmatrix} \lambda & \lambda p \\ \lambda p & \lambda p \end{pmatrix}$. This follows from the fact that $X \sim P(\lambda)$, $Y \sim P(\lambda p)$ and $E(XY) = E\{XE(Y \mid X)\} = E(X\,Xp) = E(X^2\,p) = (\lambda + \lambda^2)p$. Therefore

$$\begin{pmatrix} \overline{X} \\ \overline{Y} \end{pmatrix} \sim AN^{(2)}\left[\begin{pmatrix} \lambda \\ \lambda p \end{pmatrix}, \frac{1}{n}\begin{pmatrix} \lambda & \lambda p \\ \lambda p & \lambda p \end{pmatrix}\right].$$

EXAMPLE 6.3.3 Let $Z = (Z_1, \ldots, Z_k)'$ be a multinomial r.v. in k-cells with $f(z_1, \ldots, z_k) = z_1^{p_1} z_2^{p_2} \ldots z_k^{p_k}$, $z_i = 0$ or 1, $\sum_{i=1}^{k} z_i = 1$ and $0 < p_i < 1$, $\sum p_i = 1$. Then $E(Z_i) = p_i$ and $V(Z_i) = \lambda_{ii} = p_i(1 - p_i)$ and Cov $(Z_i\, Z_j) = -\, p_i p_j$. Thus

$$M_z = \begin{pmatrix} p_1(1-p_1) & \cdots & -p_1 p_k \\ \cdots & & \cdots \\ -p_k p_1 & \cdots & p_k(1-p_k) \end{pmatrix}$$

However as $\sum z_i = 1$, M_z is singular and has rank $(k - 1)$. We therefore consider only $(k - 1)$ of z_i's say $(z_1, \ldots, z_{k-1})$ so that $z_k = (1 - z_1, \ldots, - z_{k-1})$, and $p_k = (1 - p_1 - p_2, \ldots, p_{k-1})$. Then by MVCLT as applied to $\overline{Z} = (\overline{Z}_1, \ldots, \overline{Z}_{k-1})'$ the vector of $(k - 1)$ relative cell frequencies in $(k - 1)$ cells will have the asymptotic normal distribution with mean vector $(p_1, \ldots, p_{k-1})'$ and asymptotic variance covariance matrix $\frac{1}{n}\Lambda$ where $\lambda_{ii} = p_i(1 - p_i)$ and $\lambda_{ij} = -\,p_i p_j$, $i = 1, 2, \ldots, k - 1$, $j = 1, 2, \ldots, k - 1$.

Similar to Theorem 6.1.1 which establishes invariance of the CAN property under continuous differentiable transformation we can show that if $T \sim AN^{(m)}\ (\theta, \Lambda(\theta)/a_n^2)$ and if $\psi = (\psi_1, \ldots, \psi_k)'$ are such that $\dfrac{\partial \psi_i}{\partial \theta_j}$ are continuous for $i = 1, 2, \ldots, k$ and $j = 1, 2, \ldots, m$ then $\psi(T) \sim AN^{(k)} \left(\psi(\theta), \dfrac{G\Lambda G'}{a_n^2}\right)$ where

$$G = \begin{pmatrix} \dfrac{\partial \psi_1}{\partial \theta_1} & \cdots & \dfrac{\partial \psi_1}{\partial \theta_m} \\ \vdots & & \vdots \\ \dfrac{\partial \psi_k}{\partial \theta_1} & & \dfrac{\partial \psi_k}{\partial \theta_m} \end{pmatrix}$$

provided $G \Lambda G'$ is positive definite. The proof is similar to the proof of the Theorem 6.1.1 and uses Taylor series expansion of each component of the vector ψ. Thus $a_n(\psi(T) - \psi(\theta)) = G\, a_n(T - \theta) + a_n R$ where $a_n R \xrightarrow{p} 0$ and as $a_n(T - \theta) \xrightarrow{d} N^{(m)}\ (0, \Lambda)$, $G\, a_n(T - \theta) \xrightarrow{d} N^{(k)}\ (0, G \Lambda G')$. Hence we have Theorem 6.3.1. Let $T \sim AN^{(m)}\ (\theta, \Lambda(\theta)/a_n^2)$ and $\psi = (\psi_1, \ldots, \psi_k)'$ and $G = \left(\left(\dfrac{\partial \psi_r}{\partial \theta_s}\right)\right)$ such that $G\Lambda G'$ is pd. Then $\psi(T) \sim AN^{(k)}\ (\psi(\theta), G\Lambda G'/a_n^2)$.

EXAMPLE 6.3.3 In Example 6.3.1 showed that $(m_1', m_2')'$ is $AN^{(2)}\ ((\mu_1', \mu_2')'$, Λ/n. Note that here $(\theta_1, \theta_2)' = (\mu_1', \mu_2')' = (\mu, \sigma^2 + \mu^2)' \neq (\mu, \sigma^2)'$.

Now suppose we want to obtain CAN estimators of μ and σ^2. Then we take $\psi_1 = \mu_1' = \mu$, $\psi_2 = \mu_2' - \mu_1'^2 = \sigma^2$. Then

$$G = \begin{pmatrix} 1 & 0 \\ -2\mu_1' & 1 \end{pmatrix} = \begin{pmatrix} 1 & 0 \\ -2\mu & 1 \end{pmatrix}.$$

Therefore $\psi_1(m_1', m_2') = m_1' = \bar{X}$ and $\psi_2(m_1', m_2') = m_2' - m_1'^2 = \dfrac{S^2}{n} = m_2$ the second central moment of the sample and we have

$$\begin{pmatrix} \bar{X} \\ S^2/n \end{pmatrix} \sim AN^{(2)} \left[\begin{pmatrix} \mu \\ \sigma^2 \end{pmatrix}, \frac{G\Lambda G'}{n} \right].$$

By straightforward calculations we can show that

$$G\Lambda G' = \begin{pmatrix} \sigma^2 & 0 \\ 0 & 2\sigma^4 \end{pmatrix}.$$

If we use the general expression for Λ given in Example 6.3.1 namely,

$$\Lambda = \begin{pmatrix} \mu_2' - \mu_1'^2 & \mu_3' - \mu_2'\mu_2' \\ \mu_3' - \mu_1'\mu_2' & \mu_4' - \mu_2'^2 \end{pmatrix}$$

and $$G = \begin{pmatrix} 1 & 0 \\ -2\mu_1' & 1 \end{pmatrix} \text{ then } G\Lambda G' = \begin{pmatrix} \mu_2 & \mu_3 \\ \mu_3 & \mu_4 - \mu_2^2 \end{pmatrix}$$

where μ_2, μ_3, μ_4 are central moments of the pdf $\{f(x, \theta), \theta \in \Omega\}$. Note that if the pdf is symmetric about mean or even if just $\mu_3 = 0$ we have $G\Lambda G' =$ diag $(\mu_2, \mu_4 - \mu_2^2)$ and $\bar{X}$ and $\frac{S^2}{n}$ would be asymptotically normal with off diagonal elements zero or equivalently $\bar{X}$, and $\frac{S^2}{n}$ would be asymptotically independent. Further observe that $\left(\bar{X}, \frac{S^2}{n}\right)'$ is a solution of the moment equations $m_1' = \mu$ and $m_2' = \mu^2 + \sigma^2$ and $\frac{\partial(\mu, \mu^2 + \sigma^2)}{\partial(\mu, \sigma^2)} = \begin{pmatrix} 1 & 0 \\ 2\mu & 1 \end{pmatrix} = G^{-1}$. In Section 6.4 we will consider the method of moments which generalizes these results as well as the method of percentiles.

6.4 Method of Moments and Method of Percentiles

Let $(X_1, \ldots, X_n)$ be a random sample of the size n with pdf belonging to $\{f(x, \theta), \theta \in \Omega \subset R_m\}$ and let $\{U_1(X), \ldots, U_m(X)\}$ be m r.v.s such that $E(U_r(X)) = \psi_r(\theta), r = 1, 2, \ldots, m$ and Cov $(U_r(X), U_s(X)) = \lambda_{rs}(\theta)$, where $\Lambda(\theta)$ is pd. Then $\{U(X_i)\}_{i=1}^n$ are i.i.d. r.v.s distributed as $U(X) = (U_1(X), \ldots, U_m(X))'$ for which MVCLT holds and the mean vector of sample $\bar{U}(X) = (\bar{U}_1, \ldots, \bar{U}_m)$ where $\bar{U}_r = \frac{1}{n}\sum_{i=1}^{n} U_r(X_i)$, is CAN for $\psi = (\psi_1, \ldots, \psi_m)'$ with asymptotic variance covariance matrix $\frac{1}{n}\Lambda(\theta)$. The method of moments consists in equating $\bar{U}_r = \psi_r(\theta)$, $r = 1, 2, \ldots, m$ and solving these m equations for $(\theta_1, \ldots, \theta_m)'$. The above moment equations admit unique solution provided $\left|\frac{\partial(\psi_1, \ldots, \psi_m)}{\partial(\theta_1, \ldots, \theta_m)}\right| \neq 0$ and then $\tilde{\theta}_r = h_r(\bar{U}_1, \bar{U}_2, \ldots, \bar{U}_m)$ where $h = (h_1, \ldots, h_m)'$ is the inverse function of the ψ. Let $\frac{\partial(\psi_1, \ldots, \psi_m)}{\partial(\theta_1, \ldots, \theta_m)} = G$ then $H = \frac{\partial(h_1, \ldots, h_m)}{\partial(\theta_1, \ldots, \theta_m)} = G^{-1}$. Note that since G is assumed to be non-singular we can consider $\phi_r = \psi_r(\theta)$, $r = 1, 2, \ldots, m$ as a one-one transformation with unique inverse given by $\theta_r = h_r(\phi_1, \ldots, \phi_m)$ with Jacobian of transformation as $H = G^{-1}$ or $GH = HG = I$ the identity matrix. By Theorem 6.3.1 we have $\tilde{\theta} = (\tilde{\theta}_1, \ldots, \tilde{\theta}_m) \sim AN^{(m)}(\theta, H\Lambda H'/n)$ or equivalently $\tilde{\theta} \sim AN^{(m)}(\theta, G^{-1}\Lambda(G^{-1})'/n)$.

We now consider application of method of moments in multiparameter exponential family. Let $(X_1, \ldots, X_n)$ be i.i.d. with pdf $\{f(x, \theta), \theta \in \Omega \subset R_m\}$ which belongs to m-parameter exponential family so that

$$\log f(x, \theta) = \sum_{r=1}^{m} u_r(\theta)\, K_r(x) + v(\theta) + w(x). \qquad (6.4.1)$$

Since $(K_1(x), \ldots, K_m(x))'$ is minimal sufficient for θ, we consider the method of moments based on $(K_1(x), \ldots, K_m(x))'$. Now as $\left|\dfrac{\partial(u_1, \ldots, u_m)}{\partial(\theta_1, \ldots, \theta_m)}\right| \neq 0$, we consider first a parametric transformation $\phi_r = u_r(\theta)$, $r = 1, \ldots, m$ which is 1:1 and determines the unique inverse transformation $\theta_r = q_r(\phi)$, $r = 1, 2, \ldots, m$. Then in the new parametric form we have

$$\log f(x, q(\phi)) = \sum \phi_r K_r(x) + v_1(\phi) + w(x)$$

where $v_1(\phi) = v(q(\phi))$ and $q(\phi) = (q_1(\phi), \ldots, q_m(\phi))'$. Now

$$\frac{\partial \log f}{\partial \phi_r} = K_r(x) + \frac{\partial v_1}{\partial \phi_r}$$

$$\frac{\partial^2 \log f}{\partial \phi_r \partial \phi_s} = \frac{\partial^2 v_1}{\partial \phi_r \partial \phi_s}$$

and as $E\left(\dfrac{\partial \log f}{\partial \phi_r}\right) = 0$, $r = 1, 2, \ldots, m$, $E(K_r(x)) = -\dfrac{\partial v_1}{\partial \phi_r} = \eta_r$.

Applying method of moments to $(K_1(x), \ldots, K_m(x))'$ the moment equations are

$$\frac{1}{n} \sum_{i=1}^{n} K_r(x_i) = \bar{Z}_r = \eta_r(\phi), \quad r = 1, 2, \ldots, m \qquad (6.5.1)$$

The system of equations given by (6.5.1) will give the estimator $\tilde{\phi} = (\tilde{\phi}_1, \ldots, \tilde{\phi}_m)$ if

$$\left|\frac{\partial(\eta_1, \ldots, \eta_m)}{\partial(\phi_1, \ldots, \phi_m)}\right| \neq 0. \text{ Now } \frac{\partial \eta_r}{\partial \phi_s} = \frac{-\partial^2 \log f}{\partial \phi_r \partial \phi_s} = J_{rs}(\phi)$$

where $J_{rs}(\phi)$ denotes the (r, s)-th element of the Fisher information matrix $J(\phi)$. Thus $\left|\dfrac{\partial(\eta_1, \ldots, \eta_m)}{\partial(\phi_1, \ldots, \phi_m)}\right| = |\, J(\phi)\, | \neq 0$ which implies that $\tilde{\phi}$ is uniquely determined.

Now

$$E\left(\frac{-\partial^2 \log f}{\partial \phi_r \partial \phi_s}\right) = E\left(\frac{\partial \log f}{\partial \phi_r} \cdot \frac{\partial \log f}{\partial \phi_s}\right)$$

$$= E[(K_r(x) - \eta_r)\,(K_s(x) - \eta_s)]$$

$$= \text{Cov}\,(K_r(x), K_s(x)).$$

Thus $(K_1(x), \ldots, K_m(x))'$ are i.i.d. r.v.s with mean vector $\eta(\phi)$ and variance covariance matrix $J(\phi)$, the Fisher information matrix of ϕ. By MVCLT we have

$$\bar{Z} = (\bar{Z}_1, \ldots, \bar{Z}_m)' \sim AN^{(m)}(\eta(\phi), J(\phi)/n)$$

and $\bar{Z}$ is CAN for $\eta(\phi)$. Therefore $\eta^{-1}(\bar{Z})$ is CAN for ϕ and its asymptotic variance covariance matrix is given by $\frac{1}{n} G^{-1}J(G^{-1})'$ where $G = \frac{\partial(\eta_1, \ldots, \eta_m)}{\partial(\phi_1, \ldots, \phi_m)}$ $= J(\phi)$. Therefore $\tilde{\phi} \sim AN^{(m)}\left(\phi, \frac{J^{-1}(\phi)}{n}\right)$. Now $\theta_r = q_r(\phi_1, \ldots, \phi_m)$, $r = 1, 2, \ldots, m$ where

$$Q = \frac{\partial(\theta_1, \ldots, \theta_m)}{\partial(\phi_1, \ldots, \phi_m)} = \left[\frac{\partial(u_1, \ldots, u_m)}{\partial(\theta_1, \ldots, \theta_m)}\right]^{-1}.$$

Again using Theorem 6.5.1 we have

$$\tilde{\theta} = (\tilde{\theta}_1, \ldots, \tilde{\theta}_m)' = (q_1(\tilde{\phi}), \ldots, q_m(\tilde{\phi}))$$

is asymptotically normal with mean vector θ and variance covariance matrix $\frac{QJ^{-1}(\phi)Q'}{n}$ when evaluated in terms of θ by using $\phi_r = u_r(\theta)$, $r = 1, 2, \ldots, m$. This can be shown to be equal to $\frac{1}{n} J^{-1}(\theta)$.

Now $J_{rs}(\theta) = E\left(\frac{\partial \log f}{\partial \theta_r} \cdot \frac{\partial \log f}{\partial \theta_s}\right)$.

But
$$\frac{\partial \log f}{\partial \theta_r} = \sum_{i=1}^{m} \frac{\partial \log f}{\partial \phi_i} \frac{\partial \phi_i}{\partial \theta_r}, \quad r = 1, 2, \ldots, m.$$

Hence
$$J_{rs}(\theta) = \sum_i \sum_j E\left(\frac{\partial \log f}{\partial \theta_i} \cdot \frac{\partial \log f}{\partial \phi_j}\right) \frac{\partial \phi_i}{\partial \theta_r} \frac{\partial \phi_i}{\partial \theta_s}$$

$$= \sum_{i,j} J_{ij}(\phi) \frac{\partial \phi_i}{\partial \theta_r} \frac{\partial \phi_i}{\partial \theta_s}$$

Therefore
$$J(\theta) = Q^{-1} J(\phi) (Q^{-1})'$$
or
$$QJ^{-1}(\phi) Q' = J^{-1}(\theta).$$

Therefore $\tilde{\theta} \sim AN\left(\theta, \frac{J^{-1}(\theta)}{n}\right)$. Hence, we have the theorem,

THEOREM 6.4.1 If $\{f(x, \theta), \theta \in \Omega \subset R_m\}$ is m-parameter exponential family then the moment estimator of θ based on minimal sufficient statistic is CAN for θ with

$$\tilde{\theta} \sim AN^{(m)}\,(\theta, J^{-1}(\theta)/n)$$

where $J(\theta)$ denotes the Fisher information matrix.

EXAMPLE 6.4.1 Let $(X_1, \ldots, X_n)$ be i.i.d. $N(\theta_1, \theta_2)$ then

$$\log f(x, \theta_1, \theta_2) = -\frac{1}{2}\log 2\pi - \frac{1}{2}\log \theta_2 - \frac{(x-\theta_1)^2}{2\theta_2}$$

and as seen earlier pdf of X belongs to 2 parameter exponential family with $K_1(x) = x$, $K_2(x) = x^2$. Thus moment estimators based on minimal sufficient statistics are solutions of moment equations $m_1' = \bar{x} = \theta_1$ and $m_2' = \frac{1}{n}\Sigma x_1^2 = \theta_1^2 + \theta_2$. The solutions are $\tilde{\theta}_1 = \bar{x}$ and $\tilde{\theta}_2 = \frac{1}{n}\Sigma x_1^2 - x^{-2}$. Then

$$(\tilde{\theta}_1, \tilde{\theta}_2) \sim AN^{(2)}\left[(\theta_1, \theta_2)' \begin{pmatrix} \theta_2/n & 0 \\ 0 & 2\theta_2^2/n \end{pmatrix}\right] \quad \text{as}$$

$J(\theta) = \begin{pmatrix} 1/\theta_2 & 0 \\ 0 & 1/2\theta_2^2 \end{pmatrix}$. Earlier this result was obtained in Example 6.3.4 using different techniques.

EXAMPLE 6.4.2 Let $f(x, y, \lambda, p) = \binom{x}{y} p^y (1-p)^{x-y} e^{-\lambda} \frac{\lambda x}{x!}$, $0 < p < 1$, $\lambda > 0$, $y = 0, 1, 2, \ldots, x$, $x = 0, 1, 2, \ldots$. We have already seen that the pdf belongs to two parameter exponential family with

$$J(\lambda, p) = \begin{pmatrix} 1/\lambda & 0 \\ 0 & \lambda/(p(1-p)) \end{pmatrix} \text{ and } K_1 = y,\ K_2 = x.$$

Therefore moment estimators are solution of the moment equations $\bar{x} = \lambda$, $\bar{y} = \lambda p$ or $\tilde{\lambda} = \bar{x}$, $\tilde{p} = \bar{y}/\bar{x}$ and $(\tilde{\lambda}, \tilde{p})'$ is CAN with mean vector $(\lambda, p)'$ and asymptotic variance covariance matrix diag $\left(\frac{\lambda}{n}, \frac{p(1-p)}{n\lambda}\right)$. Note that if $\bar{x} = 0$ then $\bar{y} = 0$ and $\tilde{\lambda} = \bar{y}/\bar{x}$ is technically undefined. In this case moment equations are $\lambda = 0$ and $\lambda p = 0$ which do not admit unique solution for p and $\tilde{\lambda} = 0$ is a solution outside parameter space $\Omega = \{(\lambda, p)' \mid 0 < \lambda, 0 < p < 1\}$. However $P(\bar{X} = 0) = e^{-n\lambda} \to 0$ as $n \to \infty$ for any $\lambda > 0$ and such samples would rarely occur in practice if n is sufficiently large.

EXAMPLE 6.4.3 Consider multinomial distribution in k-cells as given in Example 6.3.3. One can show that the pmf of $(Z_1, \ldots, Z_{k-1})'$ belongs to $(k-1)$ parameter exponential family with $K_i(z) = z_i$, $i = 1, 2, \ldots, k-1$ so that moment equations give $\tilde{p}_i = \bar{z}_i$ = relative frequency in the i-th cell,

$i = 1, 2, \ldots, k-1$. The Fisher information matrix is given by $J(p)$– such that for $r = 1, 2, \ldots, k-1$ $s = 1, 2, \ldots, k-1$, $J_{rr}(p) = \frac{1}{p_r} + \frac{1}{p_k}$, $J_{rs}(p) = \frac{1}{p_k}$ where $p_k = (1 - p_1 - p_2 - \ldots - p_{k-1})$. Now we can show that $J^{-1}(p) = (J^{rs})$ is such that $J^{rr}(p) = p_r(1 - p_r)$ and $J^{rs}(p) = -p_r p_s$ so that we have the same result as in Example 6.3.3.

We now consider the percentile method to generate CAN estimators for vector valued parameters. For this purpose we use the following theorem.

THEOREM 6.4.2 Let $(X_1, \ldots, Z_n)$ be a random sample of size n from $\{f(x, \theta), \theta \in \Omega \subset R_m\}$. Let $0 < p_1 < p_2 < \ldots < p_m < 1$ and $r_i = [np_i] + 1$. Let $(\xi_{p_1}(\theta), \ldots, \xi_{p_m}(\theta))'$ be the m percentiles and $(X_{(r_1)}, \ldots, X_{(r_m)})'$ the corresponding sample percentiles. Then $(X_{(r_1)}, \ldots, X_{(r_m)})'$ is asymptotically normal with mean vector $(\xi_{p_1}(\theta), \ldots, \xi_{p_m}(\theta))'$ and variance covariance matrix Λ with $\text{Var}(X_{(r_i)}) = \dfrac{p_i(1-p_i)}{n[f(\xi_{p_i})]^2} = \lambda_{ii}$

$$\text{Cov}(X_{(r_i)}, X_{(r_j)}) = \frac{p_i(1-p_j)}{nf(\xi_{p_i})f(\xi_{p_j})} = \lambda_{ij}, \quad i < j$$

provided $f(\xi_{p_i}) \neq 0$, $i = 1, 2, \ldots, m$ (David, 1981).

The percentile method consists of setting up equations $X_{(r_i)} = \xi_{p_i}(\theta)$, $i = 1, 2, \ldots, m$ and then solving these for $\theta = (\theta_1, \ldots, \theta_m)'$. A sufficient condition for the unique solution is $|G| = \left|\dfrac{\partial(\xi_{p_1}, \ldots, \xi_{p_m})}{\partial(\theta_1, \ldots, \theta_m)}\right| \neq 0$. Under this condition the percentile estimator $\tilde{\theta} = (\tilde{\theta}_1, \ldots, \tilde{\theta}_m)$ is given by $\tilde{\theta}_s = h_s(X_{(r_1)}, \ldots, X_{(r_m)})$, $s = 1, 2, \ldots, m$ and

$$\tilde{\theta} = (\tilde{\theta}_1, \ldots, \tilde{\theta}_m)' \sim AN^{(m)}(\theta, G^{-1}\Lambda(G^{-1})'/n).$$

We now consider a few examples of applications of percentile method in the multiparameter case.

EXAMPLE 6.4.4 Let $(X_1, \ldots, X_n)'$ be a random sample of size n from Cauchy distribution with location μ and scale σ so that the pdf is given by

$$f(x, \mu, \sigma) = \frac{1}{\pi}\frac{1}{1 + ((x-\mu)/\sigma)^2}\frac{1}{\sigma}$$

$$= \frac{\sigma}{\pi}\frac{1}{\sigma^2 + (x-\mu)^2}, \; x \in R_1, \; \mu \in R_1, \; \sigma > 0.$$

Now $\xi_p = \mu + c_p\sigma$, where $\frac{1}{\pi}\tan^{-1}(c_p) + 1/2 = p$ or $c_p = \tan(\pi p - \pi/2)$ for $0 < p < 1$.

Now as there are two parameters we consider $0 < p_1 < p_2 < 1$ and solve the percentile equations $X_{(r_1)} = \xi_{p_1} = \mu + c_{p_1}\sigma$ and $X_{(r_2)} = \mu + c_{p_2}\sigma$ giving CAN estimators

$$\frac{[X_{(r_2)} - X_{(r_1)}]}{c_{p_2} - c_{p_1}} = \tilde{\sigma} \text{ and } \frac{c_{p_1}X_{(r_2)} - c_{p_2}X_{(r_1)}}{c_{p_1} - c_{p_2}} = \tilde{\mu}$$

which are linear functions of components of order statistic. Now asymptotic variance covariance matrix of $(X_{(r_1)}, X_{(r_2)})'$ is given by

$$\frac{n}{\sigma^2} AV(X_{(r_i)}) = p_i(1-p_i)/[f(\xi_{p_i})]^2 = \pi^2(1+c_{p_i}^2)^2\, p_i(1-p_i),\ i = 1, 2$$

and $$\frac{n}{\sigma^2} ACV(X_{(r_1)}, X_{(r_2)}) = p_1(1-p_2)\,\pi^2\,(1+c_{p_1}^2)(1+c_{p_2}^2).$$

Many choices are possible for $0 < p_1 < p_2 < 1$. Using the fact that the pdf is symmetric about μ the median, one can take $p_1 = 1/4$ and $p_2 = 3/4$ giving $c_{p_1} = -1$ and $c_{p_2} = +1$ so that

$$\tilde{\mu}_1 = \frac{X_{(r_2)} + X_{(r_1)}}{2} \text{ and } \tilde{\sigma}_1 = \frac{X_{(r_2)} - X_{(r_1)}}{2}.$$

But $$AV(X_{(r_i)}) = \frac{3\pi^2}{4n}\sigma^2,\ i = 1, 2 \text{ and } ACV(X_{(r_1)}, X_{(r_2)}) = \frac{\pi^2}{4n}\sigma^2.$$

Therefore $AV(\tilde{\mu}_1) = \dfrac{\pi^2\sigma^2}{2n}, AV(\tilde{\sigma}_1) = \dfrac{\pi^2\sigma^2}{4n}$ and $ACV(\tilde{\mu}_1, \tilde{\sigma}_1) = 0$.

On the other hand noting that μ itself is median we can take $p_1 = 1/2$ and $p_2 = 3/4$ so that $c_{p_1} = 0, c_{p_2} = +1$. The percentile estimators are now $\tilde{\mu}_2 = (X_{([n/2]+1)}) =$ the sample median and $\tilde{\sigma}_2 = (X_{(r_2)} - \tilde{\mu}_2)$. Again using Theorem 6.4.2 we have $(\tilde{\mu}_2, \tilde{\sigma}_2)'$ is CAN for $(\mu, \sigma)'$ with

$$AV(\tilde{\mu}_2) = \frac{\pi^2}{4n}\sigma^2,\ AV(\tilde{\sigma}_2) = \frac{\pi^2\sigma^2}{2n}$$

$$ACV(\tilde{\mu}_2, \tilde{\sigma}_2) = 0.$$

EXAMPLE 6.4.5 We consider a generalization of Example 5.4.3 and Example 6.2.3. Let $f(x, \mu, \sigma) = \dfrac{|x-\mu|}{\sigma}$ for $\mu - \sigma < x < \mu + \sigma$. Let $0 < p_1 < 1/2 < p_2 < 1$ such that $p_1 = \dfrac{1}{2} - \delta$ and $p_2 = \dfrac{1}{2} + \delta$. Here $0 < \delta < 1/2$. Then $\xi_{p_1} = \mu - c\sigma$ and $\xi_{p_2} = \mu + c\sigma$, where $c^2 = 2\delta$. The percentile estimators are given by $\tilde{\mu} = \dfrac{X_{(r_2)} + X_{(r_1)}}{2}$ and $\tilde{\sigma} = \dfrac{X_{(r_2)} - X_{(r_1)}}{2c}$. Now $AV(X_{(r_1)}) = AV(X_{(r_2)}) = \dfrac{\sigma^2}{c^2 n}\left(\dfrac{1}{4} - \delta^2\right)$ and $ACV(X_{(r_1)}, X_{(r_2)}) = \left(\dfrac{1}{2} - \delta\right)^2 \dfrac{\sigma^2}{c^2 n}$. This leads to $AV(\tilde{\mu}) = \dfrac{\sigma^2}{2c^2 n}\left(\dfrac{1}{2} - \delta\right) = \sigma^2 (1 - 2\delta)/8\delta n = AV(\tilde{\sigma})$ and $ACV(\tilde{\mu}, \tilde{\sigma}) = 0$.

Observe that as $f(\mu) = 0$, the pdf vanishes at $\xi_{1/2}$ and the Theorem 6.4.2 is not applicable. Further as observed in Chapter 5 and in Example 6.2.3, $X_{(1)}$ and $X_{(n)}$ would be consistent for $\mu - \sigma$ and $\mu + \sigma$ respectively and $T_1 = \dfrac{X_{(1)} + X_{(n)}}{2}$ and $T_2 = \dfrac{X_{(n)} - X_{(1)}}{2}$ would be consistent for μ and σ respectively. However one can show that the asymptotic distribution $(T_1, T_2)'$ is not normal as that of $(X_{(1)}, X_{(n)})'$ is not normal. We refer an enterprising reader to David (1981) which gives asymptotic distribution theory of extreme order statistics in detail.

Exercise 6.4 (1) (a) Let $(X_1, \ldots, X_n)'$ be i.i.d. with pdf $f(x, \mu, \sigma) = \dfrac{1}{2\sigma} \exp\{-|x - \mu|/\sigma\}$, $x \in R_1$, $\mu \in R_1$, $\sigma > 0$. Obtain moment estimator of $(\mu, \sigma)'$ and its asymptotic variance covariance matrix.

(b) Let $0 < p_1 < \dfrac{1}{2} < p_2 < 1$ and $p_1 = \dfrac{1}{2} - \delta$, $p_2 = \dfrac{1}{2} + \delta$.

Show that $\xi_{p_1} = \mu - c\sigma$ and $\xi_{p_2} = \mu + c\sigma$ and the percentile estimators are $\dfrac{X_{(r_1)} + X_{(r_2)}}{2} = \tilde{\mu}$ and $\dfrac{X_{(r_2)} + X_{(r_1)}}{2c} = \tilde{\sigma}$. Obtain the asymptotic variance covariance matrix of $(\tilde{\mu}, \tilde{\sigma})'$.

(c) Noting that $\xi_{1/2} = \mu$ we have $\tilde{\mu}_1 = X_{([n/2]+1)}$. Obtain asymptotic variance covariance matrix of $\tilde{\mu}_1$ and

$$\tilde{\sigma}_1 = \frac{X_{(r_2)} - X_{([n/2]+1)}}{c}.$$

(2) (a) Let $(X_1, \ldots, X_n)'$ be a random sample of size n from pdf.

$$f(x, \mu, \sigma) = \frac{1}{\sigma} \exp\left\{-\frac{(x - \mu)}{\sigma}\right\}, \quad x \geq \mu,\ \sigma > 0.$$

Let $(X_{(1)}, \ldots, X_{(n)})$ be the order statistic and define $Y_i = (n - i + 1)(X_{(i)} - X_{(i-1)})$, $i = 1, 2, \ldots, n$ with $X_{(0)} = \mu$. Then we have seen that $(Y_1, \ldots, Y_n)$ are i.i.d. exponential with mean σ. Let $T_1 = X_{(1)}$ and $T_2 = \sum_{i=2}^{n} (X_{(i)} - X_{(1)}) = \sum_{i=2}^{n} Y_i$. Then T_1 and T_2 are independent. Show that $T_2/(n - 1)$ is CAN for σ and T_1 is consistent for μ but not CAN. Show that $n(T_1 - \mu) \xrightarrow{d} Y_1$ and $\dfrac{T_2}{n-1} \xrightarrow{d} Z_2 \sim N(\sigma, \sigma^2/(n - 1))$. This is a mixed situation in which one component of vector statistic is CAN but the other is not.

(b) Obtain moment estimator of $(\mu, \sigma)'$ and its asymptotic variance covariance matrix.

(3) (a) Let $X = (X_1, \ldots, X_m)'$ be an m-variate normal r.v. with mean vector $(\mu_1, \ldots, \mu_m)'$ and variance covariance matrix $\sigma^2 I_k$ where I_k is $(k \times k)$ identity matrix. Show that the pdf of X belongs to $(m + 1)$ parameter exponential family. Use the result on method of moments as applied to a r.s. of size n on the vector X to obtain CAN estimator of $(\mu_1, \ldots, \mu_m, \sigma^2)'$ and its asymptotic variance covariance matrix. Obtain CAN estimator of $\Sigma\, l_i \mu_i$ and also that of $(\Sigma\, l_i \mu_i)/\sigma$. Obtain their asymptotic variances.

(b) Use the above result to obtain CAN estimator of dose effect d when $m = 2$ and $\mu_2 = \mu_1 + d$. This corresponds to the situation in clinical trials where X_1 is response to placebo and X_2 is a response to treatment and d is the treatment effect and we have two independent normal samples of size n each.

(c) If in (b) above we have paired data i.e. X_1 is the initial measurement and X_2 is the measurement on the same individual after the treatment, then the assumption of independence of (X_1, X_2) is not realistic and we instead consider $Y_i = (X_{2i} - X_{1i})$, $i = 1, 2, \ldots, n$, the difference between the measurements before and after the treatment on n patients. Then $(Y_1, \ldots, Y_n)'$ can be regarded as a random sample of size n from $N(d, \sigma^2)$ where d measures the treatment effect. Obtain CAN estimator of d/σ and its asymptotic variance.

6.5 Gauss-Legendre-Boscovich Revisited

Chapter 1 outlined how the statistical methods were originally proposed in the context of direct and indirect measurements of "magnitudes of interest" arising in connection with problems in astronomy and geodessy.

First consider the problem of direct measurement where the simplest model is $X_i = \theta + \varepsilon_i$, where ε_i are i.i.d. with $E(\varepsilon_i) = 0$ and Var $(\varepsilon_i) = \sigma^2$. Then the Least Square Estimator (LSE) of θ is $\bar{X}$ and by CLT we have $\bar{X} \sim AN(\theta, \sigma^2/n)$. Thus $\bar{X}$ is CAN for θ. We have also $\tilde{\sigma}^2 = \dfrac{\Sigma (X_i - \bar{X})^2}{n} = \dfrac{S^2}{n}$ is consistent for σ^2 and $\tilde{\sigma}^2$ will be CAN for σ^2 provided $E(X - \theta)^4 < \infty$. As seen in Example 6.3.1, $(\bar{X}, S^2/n)'$ is CAN for $(\theta, \sigma^2)'$.

The above result can be easily extended to the case $\theta = (\theta_1, \ldots, \theta_m)'$ and the model is $X_{ij} = \theta_i + \varepsilon_{ij}$, $i = 1, 2, \ldots, m$ and $j = 1, 2, \ldots, n$ where $\{\varepsilon_{ij}\}$ are i.i.d. for fixed i and $j = 1, 2, \ldots, n$ with covariance matrix Λ. Then LSE of θ is vector of sample means $\bar{X}$ and by MVCLT $\bar{X} \sim AN^{(m)}(\theta, \Lambda/n)$. The CAN estimator of Λ can also be obtained and is given by (S_{ij}/n) where $S_{ij} = \sum_{r=1}^{n} (X_{ir} - \bar{X}_i)(X_{jr} - \bar{X}_j)$. The CANness of S_{ij}/n can be proved assuming that fourth order moments exist.

Next we consider the simplest indirect measurement problem where the model is $y_i = \beta_0 + \beta_1 x_i + \varepsilon_i$, $i = 1, 2, \ldots, n$ where ε_i are i.i.d. with $E(\varepsilon_i) = 0$, Var $(\varepsilon_i) = \sigma^2$. Note that this is also the standard linear regression problem. Then LSE of $(\beta_0, \beta_1)'$ are $\hat{\beta}_1 = \dfrac{\Sigma (x_i - \bar{x})(y_i - \bar{y})}{S_x^2}$ and $\hat{\beta}_0 = \bar{y} - \hat{\beta}_1 \bar{x}$. Note that here $(x_1, \ldots, x_n)$ are constants and $\hat{\beta}_1$ is defined only when $S_x^2 \neq 0$ i.e. there are at least two distinct x_i's, which we will assume to be the case. Note that $E(y_i) = \alpha + \beta x_i$ and $E(\bar{y}) = \alpha + \beta \bar{x}$ and therefore $E(\hat{\beta}_1) = \beta_1$ and therefore $E(\hat{\beta}_0) = \beta_0$ i.e. $(\hat{\beta}_0, \hat{\beta}_1)'$ is unbiased for (β_0, β_1). Now suppose that the errors are normal i.e. $\{\varepsilon_i\}_1^n$ are i.i.d. $N(0, \sigma^2)$. Then as seen in Example 2.5.4 the joint pdf of $(Y_1, \ldots, Y_n)$ is a three parameter exponential family with $(\Sigma x_i y_i, \Sigma y_i, \Sigma y_i^2)$ as a minimal complete sufficient statistic for $(\beta_0, \beta_1, \sigma^2)'$. Therefore it follows that $(\hat{\beta}_0, \hat{\beta}_1)'$ is M-optimal for $(\beta_0, \beta_1)'$ when errors are i.i.d. normal. One can show that $\hat{\sigma}^2 = \dfrac{\Sigma(y_i - \hat{\beta}_0 - \hat{\beta}_1 x_i)^2}{n-2}$ is unbiased for σ^2 and being a function of minimal complete sufficient statistic, it is MVUE for σ^2. Therefore from results of Chapter 4, $(\hat{\beta}_0, \hat{\beta}, \hat{\sigma}^2)'$ is M-optimal for $(\beta_0, \beta_1, \sigma^2)'$. The variance-covariance matrix of $(\hat{\beta}_0, \hat{\beta}_1)$ is given by $M_\beta = \dfrac{\sigma^2}{S_x^2}\begin{pmatrix} \frac{1}{n}\Sigma x_i^2 & -\bar{x} \\ -\bar{x} & 1 \end{pmatrix}$.

Observe that under normal errors, $\hat{\beta}$ would be bivariate normal with mean β and variance covariance matrix M_β for each n as $\hat{\beta}_0$ and $\hat{\beta}_1$ are both linear functions of $(Y_1, \ldots, Y_n')$ which has n-dimensional normal distribution with mean vector $(\beta_0 + \beta_1 x_1, \ldots, \beta_0, + \beta_1 x_n)'$ and variance-covariance matrix $\sigma^2 I_{n\times n}$. Consider $\hat{\beta}_1$, the regression coefficient. Then even under normal errors, $\hat{\beta}_1 \sim N(\beta, \sigma^2/S_x^2)$ and would be CAN for β_1 if and only if $1/S_x^2 \to 0$ as $n \to \infty$. Similarly $\hat{\beta}_0 \sim N\left(\beta, \dfrac{\sigma^2}{S_x^2}\dfrac{\Sigma x_i^2}{n}\right)$ is CAN for β_0 if and only if $\dfrac{\Sigma x_i^2}{nS_x^2} \to 0$ as $n \to \infty$. Thus $(\hat{\beta}_0, \hat{\beta}_1)$ would be CAN for $(\hat{\beta}_0, \hat{\beta}_1)$ only when the constants $(x_1, \ldots, x_n)$ satisfy some conditions. In the usual regression situation $(x_1, \ldots, x_n)$ are fixed values of so called 'independent' or 'control' variable and $(y_1, \ldots, y_n)$ are values of 'dependent' or response variable and usually in large experiments $(n \to \infty)$, there are replications at a few distinct values of x_i where each distinct value x_i occurs n_i times with $\Sigma\, n_i = n$. We will not go into further details. The above results can be extended to the case when $y_i = \beta_0 + \beta_1 x_{1i} + \ldots + \beta_k x_{ki} + \varepsilon_i$ where $\{\varepsilon_i\}_1^n$ are i.i.d. $N(0, \sigma^2)$.

As is well known in this case we have $\hat{\beta} = (X'X)^{-1} X'Y$ and $M_\beta = \sigma^2(X'X)^{-1}$ where $X' = \begin{pmatrix} 1 & \ldots & 1 \\ x_{11} & \ldots & x_{k1} \\ \vdots & & \\ x_{1n} & \ldots & x_{kn} \end{pmatrix}$. Note that under normal errors model $\hat{\beta} \sim N^{(k)}(\beta, M_\beta)$ and $\hat{\beta}$ would be CAN for β if every diagonal element of $(X'X)^{-1} \to 0$.

For the non-normal but independent errors with $E(\varepsilon_i) = 0$ and $\text{Var}(\varepsilon_i) = \sigma^2$ we will need additional conditions on existence of fourth order moments of ε_i in order to claim asymptotic normality of $\hat{\beta}$. The theory of MVUE or CAN estimators of β in more general regression setup is available in standard texts on regression analysis such as Draper and Smith (1981) among others. However we will not go into further details here.

One could consider the method of least absolute deviations (LAD) where in the direct measurement problem, we minimize $\sum_{i=1}^{n} | X_i - \theta |$ which leads to the estimator $\tilde{\theta} = X_{([n/2]+1)}$ the median of the sample which is CAN if the distribution is such that $\xi_{1/2}(F) = \theta$ and the pdf $f(x, \theta)$ does not vanish at $x = \theta$. The situation in case of indirect measurement problem is much more complex but we will briefly outline the solution.

Consider the model $y_i = \beta_0 + \beta_1 x_i + \varepsilon_i$, $i = 1, 2, \ldots, n$. Then following Boscovitch, we determine (β_0, β_1) such that $\sum (y_i - \beta_0 - \beta_1 x_i) = 0$ or $\bar{y} = \beta_0 + \beta_1 \bar{x}$ or $\beta_0 = \bar{y} - \beta_1 \bar{x}$ and subject to this constraint minimize $\sum | y_i - \beta_0 - \beta_1 x_i |$, i.e. β_1 is estimated by minimizing $S = \sum_{i=1}^{n} | y_i - \bar{y} - \beta_1 (x_i - \bar{x}) |$. We will now describe an algorithm given by Laplace which minimizes S and yields $\tilde{\beta}$ the estimate of β having least absolute deviation. First observe that those x_i's which are exactly equal to $\bar{x}$ will not contribute to the process of minimization and therefore we consider only those observations for which $x_i - \bar{x} \neq 0$. Suppose there are m observations such that $x_i \neq \bar{x}$, then we calculate $b_i = \dfrac{y_i - \bar{y}}{x_i - \bar{x}}$, for such observations and arrange b_i in decreasing order of magnitude

$$b_{(1)} \geq b_{(2)} \ldots \geq b_{(m)}, \ (m \leq n). \tag{6.5.1}$$

Corresponding to the above arrangement arrange $| x_i - \bar{x} |$ in a similar way

$$| x_{r_1} - \bar{x} | \geq | x_{r_2} - \bar{x} | \geq \ldots \geq | x_{r_m} - \bar{x} | \tag{6.5.2}$$

i.e. $b_{(1)}$ the largest value among b_i, corresponds to $i = r_1$ etc.

Let $D = \sum_{i=1}^{m} | x_{r_i} - \bar{x} |$, then S would be minimum for $\beta_1 = b_{(k)}$ the k-th term of the sequence (6.5.1) where k is the smallest integer for which the partial sum $S_k = \sum_{j=1}^{k} | x_j - \bar{x} | \geq \frac{1}{2} D$. We take $\tilde{\beta}_1 = b_{(k)} = (y_{(k)} - \bar{y})/(x_{(r)} - \bar{x})$ where k is obtained as above. If $S_k > \frac{1}{2} D$ then $b_{(k)}$ is uniquely determined but if $S_k = \frac{1}{2} D$ then $\tilde{\beta}_1$ is not unique and $\tilde{\beta}_1$ can be taken as any value between $[b_{(r)}, b_{(r+1)}]$. Having determined $\tilde{\beta}_1$, we take $\tilde{\beta}_0 = \bar{y} - \tilde{\beta}_1 \bar{x}$. Boscovitch had $n = 5$ observations and he obtained $(\tilde{\beta}_0, \tilde{\beta}_1)$ using geometric argument. Laplace gave the above algorithm for the general case. Prior to Boscovitch the traditional approach was to calculate $\binom{n}{2}$ slopes $b_{ij} = \dfrac{(y_i - y_j)}{(x_i - x_j)}$ and take $\tilde{\beta}_2 = \dfrac{1}{\binom{n}{2}} \sum\sum_{i \neq j} b_{ij}$.. Euler suggested to obtain estimates of (β_0, β_1) by obtaining two equations in two unknowns in the following manner. The data (x_i, y_i), $i = 1, 2, \ldots, n$ was subdivided by choosing a point x_0 and then solving

$$\left.\begin{aligned}\sum_{x_i \geq x_0} (y_i - \beta_0 - \beta_1 x_i) = 0\\ \sum_{x_i < x_0} (y_i - \beta_0 - \beta_1 x_i) = 0\end{aligned}\right\} \tag{6.5.3}$$

However the estimate $\tilde{\beta}$ would then depend on the choice of x_0. On the other hand Boscovitch method is quite objective. Laplace used it in his 1789 memoir on the Figure of Earth and later in his famous treatise on Celestial Mechanics published in 1799. However the method of least squares advocated by Gauss-Legendre was equally objective and was much more easy to use. Therefore Boscovitch method was given a back seat by astronomers and other scientists.

However currently the Boscovitch method based on L_1-norm i.e. the norm $\Sigma \mid d_i \mid$, for a vector $(d_1, \ldots, d_n)' \in R_n$ is making a strong come back over the Gauss-Legendre method based on L_2-norm i.e. $\Sigma\, d_j^2$ primarily due to the fact that estimates based on L_1-norm are more stable than those based on L_2-norm. The reader may recall the discussion in Fisher's paper (1920) comparing estimators of σ based on $\Sigma \mid x_i - \bar{x} \mid$ and $\Sigma\, (x_i - \bar{x})^2$ referred in Chapter 1. Even here Edington in a footnote to Fisher's paper observed that estimate based on $\Sigma \mid x_i - \bar{x} \mid$ is preferred by astronomers than that based on $\Sigma\, (x_i - \bar{x})^2$ particularly when observations with large deviations are rejected as outliers. Estimation of parameters when outliers are present in the data is a fascinating area and we refer interested readers to an excellent text by Barnett and Lewis (1984).

In the next chapter we consider the method of maximum likelihood—one of the most commonly used method of estimation to generate CAN estimators.

7

Method of Maximum Likelihood

7.1 Introduction

Suppose that a random sample of size n yields a data $X_1 = x_1, \ldots, X_n = x_n$ under the model specified by the distribution of X belonging to the class $\{f(x, \theta), \theta \in \Omega\}$. When X is discrete we can compute the probability under each $\theta \in \Omega$ given by

$$P[X_1 = x_1, \ldots, X_n = x_n \mid \theta] = \prod_{i=1}^{n} f(x_i, \theta) \qquad (7.1.1)$$

We have seen in Chapter 2, while defining Fisher Information that the variation in $\prod_{i=1}^{n} f(x_i, \theta)$ for fixed x as θ varies over Ω provides us information about θ. The RHS of (7.1.1) for fixed x as function of $\theta \in \Omega$ is called as the likelihood function or likelihood of θ. In case X is a continuous r.v. then $P[X_1 = x_1, \ldots, X_n = x_n \mid \theta] = 0$ and we define the likelihood of θ as follows: Consider a neighbourhood of each x_i, given by $x_i \pm \dfrac{\delta x_i}{2}$, then

$$P\left[x_i - \frac{\delta x_i}{2} < X_i < x_i + \frac{\delta x_i}{2}, i = 1, 2, \ldots, n \mid \theta\right]$$

$$= \prod_{i=1}^{n}\left[\left(F\left(x_i + \frac{\delta x_i}{2}, \theta\right)\right) - F\left(x_i - \frac{\delta x_i}{2}, \theta\right)\right] \qquad (7.1.2)$$

By mean value theorem

$$F\left(x_i + \frac{\delta x_i}{2}, \theta\right) - F\left(x_i - \frac{\delta x_i}{2}, \theta\right) = f(x_i, \theta)\, \delta x_i + 0(\delta x_i)$$

and RHS of (7.1.2) reduces to $\left[\prod_{i=1}^{n} f(x_i, \theta)\right] \delta v + 0(\delta v)$, where δv is the volume element $(\delta x_1, \ldots \delta x_n)$. Then the likelihood of the observed sample $(x_1, \ldots, x_n)$ is defined as $\prod_{i=1}^{n} f(x_i, \theta)$. Thus in either case $\prod_{i=1}^{n} f(x_i, \theta)$ for fixed x and variations in θ is defined as the likelihood of the sample. For discrete X it is actually the probability of the data x and for continuous X it is an approximation for the probability of $X = x$ divided by the volume element δv. We note that since the pdf $f(x, \theta)$ can exceed one in many models at many points x, the likelihood function is not a probability function. Even in discrete case, where the likelihood of $X = x$ is numerically same as probability $P[X = x \mid \theta]$ the likelihood function is a point function of θ and there is no meaning to the likelihood of θ_1 or θ_2 and it does not satisfy the axioms of probability which is in fact a set function. The likelihood function of a random sample of size n on X with pdf $\{f(x, \theta), \theta \in \Omega\}$ is denoted by

$$L(x, \theta) = \prod_{i=1}^{n} f(x_i, \theta) \text{ for fixed } (x_1, \ldots, x_n)' \qquad (7.1.3)$$

and is a function of $\theta \in \Omega$. Here θ, the argument of $L(x, \theta)$, can be real or vector valued but $L(x, \theta)$ is always a real number. Following Fisher (1912) we define θ_1 to be more likely, equally likely or less likely than θ_2 according as $L(x, \theta_1) > L(x, \theta_2)$, $L(x, \theta_1) = L(x, \theta_2)$ and $L(x, \theta_1) < L(x, \theta_2)$ respectively. The method of maximum likelihood prescribes estimating θ by $\hat{\theta}$ where $\hat{\theta}$ is the maximal element defined by the above order relation and is the most likely value of θ for the given data x. The above method of maximum likelihood was claimed by Fisher to be more objective than the method of least squares or the method of moments. For the method of moments as noted in previous chapter the choice of the basic random variable $U(X)$ to generate moment equations is indeed arbitrary although it was a convention to use $U(X) = X$. Similarly in the method of least squares as applied to indirect measurement problem the relationship, between y and $(x_1, \ldots, x_k)$ could be better represented by relationship between transformed variables $u(y)$ and transformed functions $(g_1(x_1), \ldots, g_k(x_k))$ where Jacobian $\dfrac{\partial(g_1 \ldots g_k)}{\partial(x_1 \ldots x_k)}$ is non-singular. Further the use of L_2-norm as prescribed by method of least squares is also a matter of choice. Note that Boscovitch had already used L_1-norm and one could use in general L_p-norm given by $\Sigma \mid d_i \mid^p$ for $0 < p < \infty$, or sup-norm given by Max $\mid d_i \mid$, where $(d_1, \ldots, d_k) \in R_k$. On the other hand Fisher pointed out that once the model $\{f(x, \theta), \theta \in \Omega\}$ is specified the method of maximum likelihood is completely objective. It must be pointed out here that the method of maximum likelihood can be applied only when the pdf $f(x, \theta)$ is fully specified except for the indexing parameter θ. On the other hand the method of moments and least square method are applicable without such specification. As noted before in case of problem of direct measurement $X_i = \theta + \varepsilon_i$, where $\{\varepsilon_i\}_1^n$ are i.i.d. with $E(\varepsilon_i) = 0$, Var $(\varepsilon_i) = \sigma^2$, $\bar{X}$ is the moment estimator and the least square estimator for distributions such as normal, double exponential, uniform among others.

Fisher (1912) derived maximum likelihood estimators for $(\theta, \sigma^2)'$ in the normal model and he was careful to point out that likelihood function does not satisfy the laws (axioms) of probability. It is important to mention that Fisher was barely 21 years old and had just obtained his Maths Tripos (Batchelor's degree) from Cambridge when he wrote this path breaking paper presenting the method of maximum likelihood leading to maximum likelihood estimators (MLE) of the indexing parameter θ in the model specified by $\{f(x, \theta), \theta \in \Omega\}$.

Before going into the theoretical study of MLE we consider a few examples.

EXAMPLE 7.1.1 Consider the direct repeated measurement problem where $X_i = \theta + \varepsilon_i$ where $\{\varepsilon_i\}_1^n$ are i.i.d. $N(0, \sigma^2)$ then the likelihood is given by

$$L(x, \theta, \sigma^2) = \left(\frac{1}{\sqrt{2\pi}\sigma}\right)^n \exp\left\{-\frac{\Sigma(x_i - \theta)^2}{2\sigma^2}\right\}, \theta \in R_1, \sigma^2 > 0$$

or

$$\log L(x, \theta, \sigma^2) = -\frac{n}{2}\log 2\pi - \frac{n}{2}\log\sigma^2 - \frac{\Sigma(x_i - \theta)^2}{2\sigma^2} \tag{7.1.4}$$

To maximize likelihood or log likelihood, we consider two stage maximization of (7.1.4). Consider σ^2 fixed then RHS of (7.1.4) is maximized for minimum value of $\Sigma(x_i - \theta)^2$ which occurs at $\hat{\theta} = \bar{x}$ and therefore

$$\sup_{\theta \in R_1} \log L(x, \theta, \sigma^2) = -\frac{n}{2}\log 2\pi - \frac{n}{2}\log\sigma^2 - \frac{S^2}{2\sigma^2} \tag{7.1.5}$$

where $S^2 = \Sigma(x_i - \bar{x})^2$. Next, consider maximization w.r.t. σ^2. Taking the derivative w.r.t. σ^2 of RHS of (7.1.5), we have $\hat{\sigma}^2$—a solution of $\frac{-n}{2\sigma^2} + \frac{S^2}{2\sigma^4} = 0$ or $\hat{\sigma}^2 = S^2/n$. Now the second derivative w.r.t. σ^2 of RHS of (7.1.5) at $\hat{\sigma}^2$ is given by $\frac{-n}{2\hat{\sigma}^4} < 0$. Therefore, the maximum of $\log L$ and hence of L occurs at $\hat{\theta} = \bar{x}$, $\hat{\sigma}^2 = S^2/n$. The MLE of $(\theta, \sigma^2)'$ is then $\hat{\theta} = \bar{x}$, $\hat{\sigma}^2 = \frac{S^2}{n}$. We observe that MLEs are functions of minimal sufficient statistic $(\bar{x}, S^2)'$. The MLE could have been obtained by setting up likelihood equations, given by

$$\frac{\partial \log L}{\partial \theta} = \frac{-2\,\Sigma(x_i - \theta)}{2\sigma^2} = 0$$

and
$$\frac{\partial \log L}{\partial \sigma^2} = \frac{-n}{2\sigma^2} + \frac{\Sigma(x_i - \theta)^2}{2\sigma^4} = 0$$

yielding unique solution $\hat{\theta} = \bar{x}$, $\hat{\sigma}^2 = \frac{S^2}{n}$. Note that the matrix of second derivative is given by

$$\begin{pmatrix} \dfrac{\partial^2 \log L}{\partial \theta^2} & \dfrac{\partial^2 \log L}{\partial \sigma^2 \partial \theta} \\ \dfrac{\partial^2 \log L}{\partial \theta\, \partial \sigma^2} & \dfrac{\partial^2 \log L}{\partial (\sigma^4)^2} \end{pmatrix}_{(\hat{\theta}, \hat{\sigma}^2)} = \begin{pmatrix} -\dfrac{n}{\sigma^2} & 0 \\ 0 & \dfrac{-n}{2\sigma^4} \end{pmatrix}_{(\hat{\theta}, \hat{\sigma}^2)}$$

which ensures that the likelihood is maximum at $(\hat{\theta}, \hat{\sigma}^2)'$. Observe that as seen earlier $(\hat{\theta}, \hat{\sigma}^2)'$ is CAN for $(\theta, \sigma^2)'$ with asymptotic variance covariance matrix $\frac{1}{n} J^{-1}(\theta, \sigma^2)$ where J is the Fisher Information matrix.

EXAMPLE 7.1.2 Let $(X_1, \ldots, X_n)$ be i.i.d. with pdf given by $f(x, \theta, \sigma) = \frac{1}{2\sigma} \exp\left\{\frac{-|x-\theta|}{\sigma}\right\}, x \in R_1, \theta \in R_1, \sigma > 0$. Then the likelihood of a sample is given by

$$\log L(x, \theta, \sigma) = -n \log 2 - n \log \sigma - \sum_{i=1}^{n} |x_i - \theta|/\sigma \qquad (7.1.6)$$

The model corresponds to repeated direct measurements where errors are i.i.d. double exponential (Laplace) with scale σ. Again we adopt two stage maximization, first fix σ then maximizing $\log L$ for variations in θ which is equivalent to minimizing $\Sigma |x_i - \theta| = \Sigma |x_{(i)} - \theta|$ for $\theta \in R_1$. One can show that $\sum_{i=1}^{n} |x_{(i)} - \theta|$ is minimized for

$$\begin{aligned} \hat{\theta} &= x_{(m+1)} \text{ if } n = 2m+1 \\ &= \alpha x_{(m)} + (1-\alpha)\, x_{(m+1)} \text{ if } n = 2m, \text{ for any } \alpha, 0 \le \alpha \le 1 \end{aligned}$$

We note that for $n = 2m$, $\hat{\theta}$ is not uniquely determined and we can take $\hat{\theta} = x_{(m+1)}$ in this case also. Hence we define $\hat{\theta} = x_{([n/2]+1)}$ = Sample Median. Then we maximize $\log L(x, \hat{\theta}, \sigma) = -n \log 2 - n \log s - \sum_{i=1}^{n} \frac{|x_i - \hat{\theta}|}{\sigma}$. By usual differentiation technique we can show that $\hat{\sigma} = \frac{1}{n} \sum_{i=1}^{n} |x_i - \hat{\theta}|$. Here also the MLE $(\hat{\theta}, \hat{\sigma})$ is a function of minimal sufficient statistic which in this case is order statistic. Here $(\hat{\theta}, \hat{\sigma})'$ can be shown to be CAN by using the asymptotic distribution theory of linear functions of order statistics. For details we refer to David (1981).

EXAMPLE 7.1.3 Let $\{X_i\}_1^n$ be i.i.d. Bernoulli with $P[X_i = 1] = \theta$, $P[X_i = 0] = 1 - \theta$. Then the log likelihood is given by $\log L(x, \theta) = \Sigma x_i \log \frac{\theta}{1-\theta} + n \log (1-\theta)$. If $\Sigma x_i \neq 0$ or $\Sigma x_i \neq n$ then using differentiation technique we can show that $\hat{\theta} = \bar{x}$. If $\Sigma x_i = 0$ then $\log L(x, \theta) = n \log (1-\theta)$ and is a monotone decreasing function of θ. Its maximum is attained at $\theta = 0$, but this value of θ is on the boundary of the parameter space $\Omega = (0, 1)$. Similarly if $\Sigma x_i = n$ then $\log L(x, \theta) = n \log \theta$ and its maximum is attained at $\theta = 1$, a value on the boundary of the parameter space. Recall that in earlier chapters in the discussion of Cramer-Rao inequality, Fisher Information etc. we have assumed $\Omega = (0, 1)$ an open interval. Thus there is some problem in defining MLE of θ for certain samples as the maximum is attained on the boundary of Ω. However observe that $P(\Sigma X_i = 0 \mid \theta) = (1-\theta)^n$ is very small for large n as $0 < \theta < 1$ and similar remark holds for other case namely $P(\Sigma X_i = n \mid \theta) = \theta^n$. By convention we take the MLE of θ as given by $\hat{\theta} = \bar{x}$ in all cases. Note that $\hat{\theta}$ is CAN for θ with asymptotic variance $\dfrac{\theta(1-\theta)}{n} = \dfrac{1}{nI(\theta)}$ where $I(\theta)$ is the Fisher information. Another way to tackle the problem of defining MLE of θ when $\Sigma x_i = 0$ or $\Sigma x_i = n$ is to define MLE as

$$\begin{aligned}\hat{\theta}_1 &= \varepsilon_1 \quad \text{if } \Sigma x_i = 0\\ &= \bar{x} \quad \text{if } 0 < \Sigma x_i < n\\ &= 1 - \varepsilon_2 \quad \text{if } \Sigma x_i = n\end{aligned}$$

where ε_1, ε_2 are arbitrary positive numbers close to zero. We show that asymptotic distribution of $\hat{\theta}_1$ and $\hat{\theta} = \bar{X}$ is identical. Now observe that

$$\sqrt{n}(\hat{\theta}_1 - \theta) = \sqrt{n}(\bar{X} - \theta) + \sqrt{n}(\hat{\theta}_1 - \bar{X})$$

Now by CLT, $\sqrt{n}(\bar{X} - \theta) \xrightarrow{d} N(0, \theta(1-\theta))$. Further

$$P[\,|\sqrt{n}(\hat{\theta}_1 - \bar{X})| < \varepsilon] \geq P[\hat{\theta}_1 = \bar{X}] = 1 - \theta^n - (1-\theta)^n$$

But for $0 < \theta < 1$, $1 - \theta^n - (1-\theta)^n \to 1$ as $n \to \infty$. Thus $\sqrt{n}(\hat{\theta}_1 - \bar{X}) \xrightarrow{p} 0$ and therefore $\sqrt{n}(\hat{\theta}_1 - \theta) \xrightarrow{p} N(0, \theta(1-\theta))$ and there is no difference between the asymptotic distribution of $\hat{\theta}_1$ and $\hat{\theta} = \bar{X}$.

We close this section with noting some general properties of MLE. First suppose that $\{f(x, \theta), \theta \in \Omega\}$ is such that there is a minimal sufficient statistic T then by Neyman Factorizability criterion

$$L(x, \theta) = \prod_{i=1}^{n} f(x_i; \theta) = g(T(x), \theta)\, h(x),\ \forall\, x \in S_0^n,\ \forall\, \theta \in \Omega.$$

Since $\hat{\theta}$ the MLE is obtained by maximizing $L(x, \theta)$ for fixed x and for variations in θ, it is equivalent to maximizing $g(T(x), \theta)$ for fixed x or for fixed $T(x) = t$ and for variations in θ. Thus $\hat{\theta}$ depends on x only through $T(x)$ and the MLE is a function of minimal sufficient statistic T. Next suppose that $\theta \to \phi(\theta)$ is an one-to-one nonsingular transformation. Now, if $\hat{\theta}$ is MLE of θ, then $\hat{\phi} = \phi(\hat{\theta})$ would be MLE of ϕ, the new indexing parameter. We illustrate this by an example.

EXAMPLE 7.1.4 Let $(X_1, \ldots, X_n)$ be i.i.d. Poisson with mean λ then

$$L(x, \lambda) = e^{-n\lambda} \frac{\lambda^{\Sigma x_i}}{\pi x_i\,!} = e^{-n\lambda} \frac{(n\lambda)^t}{t!} \frac{t!}{\prod_{i=1}^{n} x_i\,!} \frac{1}{n^t}, \ t = \sum_{i=1}^{n} x_i$$

Maximizing $L(x, \lambda)$ for $\lambda > 0$ is equivalent to maximing the pmf of the minimal sufficient statistic $T = \Sigma X_i$, given by $g(t, \lambda) = e^{-n\lambda} \frac{(n\lambda)^t}{t!}$, $t = 0, 1, 2, \ldots$ for fixed t. By differentiation technique we can show that the MLE $\hat{\lambda} = \frac{t}{n} = \bar{x}$.

Now consider the Poisson distribution with mean λ given by the new indexing parameter $\theta = \log \lambda$ or $\lambda = e^{\theta}$. Then as $\hat{\lambda} = t/n$ we would have $\hat{\theta} = \log (t/n)$ which can be verified to provide maximum of

$$L_1(x, \theta) = L(x, e^{\theta}) = \exp\{-ne^{\theta}\} \frac{(ne^{\theta})^t}{t!} \cdot \frac{t!}{\prod_{i=1}^{n} x_i!} \frac{1}{n^t}$$

Now $\log L_1 = -ne^{\theta} + t\theta + h(x)$, $\frac{\partial \log L_1}{\partial \theta} = -ne^{\theta} + t$ and as $\frac{\partial^2 \log L_1}{\partial \theta^2} = -ne^{\theta} < 0$ the unique solution of likelihood equation given by $\hat{\theta} = \log \frac{t}{n}$, provides the maximum of $L_1(x, \theta)$.

In the next section we will first consider the problem of determining MLE for the situation where $\{f(x, \theta), \theta \in \Omega\}$ is an exponential family. Recall that in defining exponential family we have considered Ω, the natural parameter space as an open set and thus excluding the boundary points. Observe that the boundary points of Ω as in case of Example 7.1.3 correspond to degenerate distributions with all probability mass concentrated at a single point. This is a delicate mathematical point and we would not go in further details in this course.

Exercise 7.1 (1) Let $(X_1, \ldots, X_n)$ be i.i.d. exponential with location θ and sale σ. The pdf is given by $f(x, \theta, \sigma) = \frac{1}{\sigma} \exp\left\{-\frac{(x-\theta)}{\sigma}\right\}$, $x \geq \theta$, $\theta \in R_1$, $\sigma > 0$. Show that $\hat{\theta} = X_{(1)}$ and $\hat{\sigma} = \frac{1}{n}\sum_{i=1}^{n} (X_i - X_{(1)}) = \frac{1}{n}\sum_{i=2}^{n} [X_{(i)} - X_{(1)}]$. We have already seen that $(\hat{\theta}, \hat{\sigma})'$ is consistent but not CAN for $(\theta, \sigma)'$.

(2) For $(X_1, \ldots, X_n)$ i.i.d. $U(0, \theta)$, $\theta > 0$ show that $\hat{\theta} = X_{(n)}$. Again $\hat{\theta}$ is consistent but not CAN for θ.

(3) Consider Example 7.1.1 and reparametrize by $\phi_2 = -\frac{1}{2\sigma^2}$ and $\phi_2 = \frac{\theta}{\sigma^2}$ and verify that $\hat{\phi}_1 = -\frac{1}{2\hat{\sigma}^2}$ and $\hat{\phi}_2 = \frac{\hat{\theta}}{\hat{\sigma}^2}$.

7.2 MLE in Exponential Family

Let $(X_1, \ldots, X_n)$ be a random sample size n on X where the pdf of X belongs to one parameter exponential family so that $\log f(x, \theta) = u(\theta)\, K(x) + v(\theta) + w(x)$, $x \in S$, $\theta \in \Omega$. Then the likelihood $L(x, \theta) = u(\theta) \sum K(x_i) + nv(\theta) + \sum w(x_i)$ and the likelihood equation is $\dfrac{\partial \log L}{d\theta} = u'(\theta) \sum K(x_i) + nv'(\theta) = 0$. This is same as the moment equation based on sufficient statistic, given by $\dfrac{1}{n} \sum K(x_i) = -\dfrac{v'}{u'} = \eta(\theta) = E(K(x))$. Further $\dfrac{\partial^2 \log L}{d\theta^2} = u'' \sum K(x_i) + nv''(\theta)$ and

$$\left(\frac{\partial^2 \log L}{d\theta^2}\right)_{\theta=\hat{\theta}} = u''(\hat{\theta})\, n \left[\frac{-v'(\hat{\theta})}{u'(\hat{\theta})}\right] + nv''(\hat{\theta})$$

$$= \left(\frac{n[u'v'' - v'u'']}{u'}\right)_{\theta=\hat{\theta}} = -nI(\hat{\theta})$$

where $I(\theta)$ is the Fisher information per unit observation. Therefore, $\left(\dfrac{\partial^2 \log f}{\partial\theta^2}\right)_{\hat{\theta}} < 0$ and $\hat{\theta}$ the unique solution of the likelihood equation is MLE of θ. Note that as already seen in Sec. 6.2, $\dfrac{\partial \log f}{d\theta} = u'(\theta)\, K(x) + v'(\theta)$ and $E\left(\dfrac{\partial \log f}{d\theta}\right) = 0$ implies that $E(K(x)) = \dfrac{-v'}{u'}$. Further $\dfrac{\partial^2 \log f}{\partial\theta^2} = u''(\theta)\, K(x)\, v''(\theta)$ and

$$I(\hat{\theta}) = E\left(-\frac{\partial^2 \log f}{\partial\theta^2}\right)_{\hat{\theta}} = -\left[\frac{u'v'' - v'u''}{u'}\right]_{\hat{\theta}}$$

We note that $\hat{\theta}$ the MLE here is same as the moment estimator based on sufficient statistic and by Theorem 6.2.2, $\hat{\theta}$ is CAN for θ with asymptotic variance $\dfrac{1}{nI(\theta)}$.

THEOREM 7.2.1 For distribution belonging to one parameter exponential family the MLE $\hat{\theta}$ is CAN for θ with asymptotic variance $\dfrac{1}{nI(\theta)}$.

The above result can easily be extended to m-parameter exponential family to show that the MLE $\hat{\theta}$ is CAN for θ with asymptotic variance co-variance matrix $\dfrac{1}{n} J^{-1}(\theta)$. Further the MLE is same as the moment estimator based on sufficient statistic $T = (T_1, \ldots, T_m)'$ where $T_r = \dfrac{1}{n} \sum\limits_{j=1}^{n} K_r(X_i)$.

Let $(X_1, \ldots, X_n)$ be a random sample of size n from an m-parameter exponential family with pdf

$$\log f(x, \theta) = \sum_{r=1}^{m} u_r(\theta)\, K_r(x) + v(\theta) + w(x)$$

where $\theta = (\theta_1, \ldots, \theta_m)'$. Consider the parametric transformation $\varphi_r = u_r(\theta)$ so that $\phi = (\phi_1, \ldots, \phi_m)'$ is the canonical parameter. Then the likelihood of ϕ is given by

$$\log L(x, \phi) = n \sum_{r=1}^{m} \phi_r\, T_r(x) + nv_1(\phi) + \sum w(x_i)$$

The likelihood equations are

$$\frac{\partial \log L}{\partial \phi_r} = n\left(T_r(x) + \frac{dv_1}{d\phi_1}\right) = 0,\ r = 1, 2, \ldots, m$$

and

$$\frac{\partial^2 \log L}{\partial \phi_r \partial \phi_s} = n\, \frac{\partial^2 v_1}{\partial \phi_r \partial \phi_s} = -\, n\, J_{rs}(\phi)$$

The MLE $\hat{\phi}$ is now the solution of moment equations given by $T_r(x) = -\dfrac{dv_1}{\partial \phi_r} = E(T_r(x)),\ r = 1, 2, \ldots, m$. As $J = J_{rs}(\phi)$, the Fisher information matrix about ϕ is pd, it follow that $\hat{\phi}$ provides the maximum of the likelihood of ϕ. Again retransforming to $\theta_r = h_r(\phi_1, \ldots, \phi_m),\ r = 1, 2, \ldots, m$ where H^{-1} is the inverse transformation and from results of Section 6.4, it follows that $\hat{\theta}$ is CAN for θ with asymptotic variance covariance matrix $\frac{1}{n} J^{-1}(\theta)$. From the invariance property of MLE under 1:1 transformation discussed in Sec. 7.1, we note that $\hat{\theta}$ is indeed MLE of θ.

In the above discussion it has been tacitly assumed that $\hat{\theta}$ the MLE, the unique solution of the likelihood equations belongs to the parameter space Ω. Recall that as per the regularity conditions for exponential family, the natural parameter space Ω is an open set. Now $\log L(x, \theta)$ for fixed x as a function of θ varying over Ω is continuous and as Ω is an open set the maximum of $L(x, \theta)$ or $\log L(x, \theta)$ may not be attained at an interior point of Ω but may be only on the boundary of Ω. We will allow this to occur and still regard $\hat{\theta}$ which may be on the boundary of Ω as MLE of θ although strictly speaking MLE of θ is not defined in this situation. We illustrate this point by an example

EXAMPLE 7.2.1 Let $(X_1, \ldots, X_n)$ be i.i.d. $b(1, \theta)$ where $\Omega = (0, 1)$.

Then $L(x, \theta) = \prod_{i=1}^{n} \theta^{x_i}(1-\theta)^{1-x_i} = \theta^t(1-\theta)^{n-t}$, where $t = \sum_{i=1}^{n} x_i$.

The likelihood equation is $\dfrac{\partial \log L}{\partial \phi} = \dfrac{t - n\theta}{\theta(1-\theta)} = 0$ and gives the solution $\hat{\theta} = \dfrac{t}{n} = \bar{x}$. However for $t = 0$ or $t = n$ the likelihood of the sample is given by

$$L(0, \theta) = (1-\theta)^n$$

$$L(1, \theta) = \theta^n$$

and $\dfrac{\partial \log L}{\partial \theta} = 0$ has no solution in Ω. We however take

$$\hat{\theta} = 0 \quad \text{if } \Sigma\, x_i = 0$$
$$= 1 \quad \text{if } \Sigma\, x_i = n$$

which are values on the boundary of Ω.

We observe that the boundary points $\theta = 0$ and $\theta = 1$ correspond to a degenerate random variable.

Note that the pmf of X belongs to one parameter exponential family with $K(x) = x$ so that the method of moments has also the same problem since in this case the moment equation and likelihood equation are identical. We refer to discussion in Example 6.4.2 and Example 7.1.3 where similar problem had arisen for method of moments.

The distinction between whether Ω is open or closed set is mathematically important but in practice may not be very crucial. However asymptotic distribution of the estimator at the boundary point may be non-normal. We illustrate this by way of an example.

EXAMPLE 7.2.2 Let $\{X_i\}_1^n$ be the i.i.d. $N(\mu, 1)$ where apriori it is known that $\mu \geq 0$ so that $\Omega = [0, \infty)$. The log likelihood of the sample is

$$\log L(x, \mu) = c - \frac{\Sigma\,(x_i - \mu)^2}{2}$$

$\dfrac{\partial \log L}{\partial \mu} = n(\bar{x} - \mu)$ and $\dfrac{\partial^2 \log L}{\partial \mu^2} = -n$, at any $\mu > 0$ and at $\mu = 0$ we consider differentiation w.r.t. μ from right. For observed $\bar{x} \in [0, \infty)$, $\hat{\mu} = \bar{x}$ but if $\bar{x} < 0$ then $\dfrac{\partial \log L}{\partial \mu} < 0$ for any $\mu \in [0, \infty)$ and $\log L$ is decreasing. Therefore maximum of $\log L$ occurs at $\mu = 0$. Hence, we define the MLE to be

$$\hat{\mu}_1 = \bar{x} \quad \text{if } \bar{x} \geq 0$$
$$= 0 \quad \text{if } \bar{x} < 0$$

Now consider

$$\sqrt{n}(\hat{\mu}_1 - \mu) = \sqrt{n}(\bar{x} - \mu) + \sqrt{n}(\hat{\mu}_1 - \bar{x})$$

Now $\sqrt{n}(\bar{x} - \mu) \xrightarrow{d} N(0, 1)$ for any $\mu \in [0, \infty)$ and

$$P[\sqrt{n}(\hat{\mu}_1 - \bar{x}) \mid < \varepsilon] \geq P[\hat{\mu}_1 = \bar{x}] = P(\bar{x} \geq 0) = 1 - \Phi[-\sqrt{n}\mu]$$

For any $\mu > 0$, $1 - \Phi(-\sqrt{n}\mu) \to 1$ as $n \to \infty$ and therefore $\hat{\mu}_1 \sim AN\left(\mu, \frac{1}{n}\right)$ for any $\mu > 0$. But at $\mu = 0$, $P[\bar{x} \geq 0] = 1/2$ and the d.f. of $\sqrt{n}(\hat{\mu}_1 - 0)$ at $\mu = 0$ is given by

$$\begin{aligned} G_n(u) &= 0 \quad \text{if } u < 0 \\ &= \Phi[\sqrt{n}u] \quad \text{if } u \geq 0 \end{aligned}$$

which is not normal. Thus, Theorem 7.1.1 holds for all values of $\mu \in \Omega = [0, \infty)$ except $\mu = 0$.

We also note that MSE $(\hat{\mu}_1) <$ MSE $(\bar{X})$ for any $\mu \in [0, \infty)$. This follows for the fact that

$$\text{MSE}(\hat{\mu}_1) = \int_0^\infty (\bar{x} - \mu)^2 g(\bar{x}, \mu)\, d\bar{x} + \int_{-\infty}^0 \mu^2 g(\bar{x}, \mu)\, d\bar{x}$$

$$\text{MSE}(\bar{X}) = \int_0^\infty (\bar{x} - \mu)^2 g(\bar{x}, \mu) + \int_{-\infty}^0 (\bar{x} - \mu)^2 g(\bar{x}, \mu)\, d\bar{x}$$

Therefore

$$\text{MSE}(\bar{X}) - \text{MSE}(\hat{\mu}_1) = \int_{-\infty}^0 \bar{x}(\bar{x} - 2\mu)\, g(\bar{x}, \mu)\, d\bar{x} \qquad (7.2.1)$$

Now as $\bar{x} < 0$ over the range of integration and as $\mu \geq 0$ we have $(\bar{x} - 2\mu) < 0$ over the range of integration and therefore the RHS of (7.2.1) is non-negative and

$$\text{MSE}(\bar{X}) \geq \text{MSE}(\hat{\mu}_1) \text{ for any } \mu \in [0, \infty)$$

On the other hand if we delete point $\mu = 0$ and consider the parameter space $\Omega = (0, \infty)$, Theorem 7.1.1 holds and we can take the MLE of μ to be $\bar{X}$ as the asymptotic distribution of $\bar{X}$ and $\hat{\mu}_1$ is same for $\mu \in (0, \infty)$. The inclusion of $\mu = 0$ in Ω or otherwise is a problem of model specification and we will not go in to further details here.

Exercise 7.2.1 (1) Let $\{X_i\}_1^n$ be i.i.d. $b(1, \theta)$ where a priori it is known that $\theta \in \left[\frac{1}{4}, \frac{3}{4}\right]$.

Show that technically MLE of θ is

$$\hat{\theta}_1 = \frac{1}{4} \quad \text{if } \bar{x} \le \frac{1}{4}$$
$$= \bar{x} \quad \text{if } \bar{x} \in \left(\frac{1}{4}, \frac{3}{4}\right)$$
$$= \frac{3}{4} \quad \text{if } \bar{x} \ge \frac{3}{4}$$

Show that $\hat{\theta}_1 \sim AN\left(\theta, \frac{\theta(1-\theta)}{n}\right)$ for any $\theta \in \left(\frac{1}{4}, \frac{3}{4}\right)$ but at $\theta = \frac{1}{4}$ and $\theta = \frac{3}{4}$, the MLE $\hat{\theta}_1$ is not asymptotically normal. Show that MSE($\hat{\theta}_1$) < MSE($\bar{X}$) for any $\theta \in \left[\frac{1}{4}, \frac{3}{4}\right]$.

(2) Let $(X_1, \ldots, X_n)$ be i.i.d. Poisson with mean λ. Show that $\hat{\lambda} = \bar{x}$ is MLE when $\bar{x} > 0$. Discuss the case when $\bar{x} = 0$ and show that

$$\hat{\lambda}_1 = \bar{x} \quad \text{if } \bar{x} > 0$$
$$= \varepsilon \quad \text{if } \bar{x} = 0$$

is such that $\sqrt{n}(\hat{\lambda}_1 - \lambda) \xrightarrow{d} N(0, \lambda)$.

(3) For the sample of size n from bivariate pmf $f(x, y, \lambda, p) = e^{-\lambda} \frac{\lambda^x}{x!} \binom{x}{y} p^y (1-p)^{x-y}$, $y = 0, 1, 2, \ldots, x$, $x = 0, 1, 2, \ldots$, $0 < p < 1$, $\lambda > 0$, obtain MLE $(\hat{\lambda}, \hat{p})$ with due emphasis on critical cases (i) $\sum x_i = 0$, $\sum y_i = 0$ and (ii) $\sum x_i > 0$ but $\sum y_i = 0$.

7.3 Cramèr Family

Section 7.2 showed that the method of maximum likelihood when applied to one or m-parameter exponential family leads to CAN estimator of θ with asymptotic variance equal to CRLB. Cramèr (1946) extended this result to a larger family of distributions, or models under consideration, by specifying regularity conditions on the family in a suitable way. The family of distributions satisfying Cramèr regularity conditions will be called a Cramèr family and we will show by the example of Cauchy location model that Cramèr family is larger than the exponential family. We will first consider the case where θ is a real parameter.

We thus have a random sample of size n or $\{X_i\}_1^n$ are i.i.d. with pdf belonging to the family $\{f(x, \theta), \theta \in \Omega \subset R_1\}$. Cramèr required that certain regularity conditions are satisfied in an open interval $N_\rho(\theta_0) \subset \Omega$ where θ_0 is the "true value" of the parameter i.e. the random sample is drawn from the distribution indexed by θ_0. Since $N_\rho(\theta_0)$ is properly contained in Ω it follows that θ_0 is an interior point of Ω and we are not sampling from a distribution where the indexing parameter θ is on the boundary of Ω. We thus can as well assume that Ω is an open set of R_1. Other Cramèr conditions were to ensure that Fisher information $I(\theta)$ is strictly positive and finite at each $\theta \in \Omega$ and an additional condition on the third order derivative of $\log f(x, \theta)$ w.r.t. θ.

Cramèr regularity conditions on the family $\{f(x, \theta), \theta \in \Omega \subset R_1\}$ are then as follows:

C–1 : Support $S_\theta = \{x \mid f(x, \theta) > 0\}$ does not depend on θ so that $S_\theta = S$.

C–2 : The parameter space Ω is an open interval of R_1.

C–3 : $\log f(x, \theta)$ is such that $\dfrac{\partial \log f}{\partial \theta}, \dfrac{\partial^2 \log f}{\partial \theta^2}$ and $\dfrac{\partial^3 \log f}{\partial \theta^3}$ exists for almost all values of $x \in S$ for every $\theta \in N_\rho(\theta_0)$ for some $\rho > 0$.

C–4 : $\int_S f(x, \theta)\, dx = 1$ can be differentiated twice under integral sign w.r.t. θ so that $E\left(\dfrac{\partial \log f}{\partial \theta}\right) = 0$ and $E\left(\dfrac{\partial \log f}{\partial \theta}\right)^2 = E\left(-\dfrac{\partial^2 \log f}{\partial \theta^2}\right) = I(\theta)$, the Fisher information is positive.

C–5 : $\left|\dfrac{\partial^3 \log f}{\partial \theta^3}\right| \le M(x)$ where $E_\theta(M(x) < \infty$. Here $M(x)$ may depend on θ_0 and ρ.

It is easy to verify that the above regularity conditions hold for any one parameter exponential family when the parameter space Ω is restricted to an open interval. It may be pointed out here that the family $\{N(\mu, 1), \mu \ge 0\}$ considered in Example 7.2.2 will not be a Cramèr family, but its sub-family obtained by deleting the point $\mu = 0$ would be a Cramèr family.

As an example of a family which is not an exponential family but is a Cramèr family consider the Cauchy pdf given by

$$f(x, \mu) = \frac{1}{\pi} \frac{1}{1 + (x - \mu)^2}, \; x \in R_1, \mu \in R_1$$

Here conditions C–1 and C–2 hold and

$$\log f(x, \mu) = -\log \pi - \log [1 + (x - \mu)^2]$$

Now for fixed x, $1 + (x - \mu)^2$ being a polynomial of degree two is an analytic function of μ and therefore $\log [1 + (x - \mu)^2]$ is also analytic function of μ and C–3 is satisfied. The checking of validity of differentiation under integral sign is a little complicated. But we observe that $\dfrac{\partial f}{\partial \mu} = \dfrac{1}{\pi} \dfrac{2(x - \mu)}{[1 + (x - \mu)^2]^2}$ is itself continuous function of μ and is an integrable function over any finite interval (a, b) and this integral is uniformly convergent as $a \to -\infty$ and $b \to +\infty$, since the integrand behaves like $\dfrac{1}{x^3}$ near $\pm\infty$. Therefore $\int_{R_1} f(x, \mu)\, dx = 1$ can be differentiated under integral sign w.r.t. μ once to obtain $\int_{R_1} \dfrac{\partial f}{\partial \mu} dx = 0$. Consider second order differentiation. Now

$$\frac{\partial^2 f}{\partial \mu^2} = \frac{1}{\pi} \frac{-2}{[1+(x-\mu)^2]^2} + \frac{1}{\pi} \frac{8(x-\mu)^2}{[1+(x-\mu)^2]^3}$$

which is continuous and integrable over any interval (a, b) and this integral is uniformly convergent as $a \to -\infty$ and $b \to +\infty$ since the two terms on RHS behave like $\frac{1}{x^4}$ near $\pm\infty$.

It now easily follows that $E\left(\frac{\partial \log f}{\partial \mu}\right) = 0$ and

$$I(\mu) = E\left(\frac{\partial \log f}{\partial \mu}\right)^2 = \int_{R_1} \left(\frac{\partial f}{\partial \mu}\right)^2 \frac{1}{f}\, dx$$

$$= \frac{4}{\pi} \int_{R_1} \frac{(x-\mu)^2}{[1+(x-\mu)^2]^3}\, dx = \frac{8}{\pi} \int_0^{\infty} \frac{t^2}{(1+t^2)^3}\, dt$$

Putting $t^2 = w$ we have

$$I(\mu) = \frac{4}{\pi} \int_0^{\infty} \frac{w^{1/2}}{(1+w)^4}\, dw = \frac{4}{\pi}\, \beta(3/2, 3/2) = \frac{1}{2}$$

Now by straightforward calculations we can show that

$$\frac{\partial^3 \log f}{\partial \mu^3} = \frac{-12(x-\mu)}{[1+(x-\mu)^2]^2} + \frac{16(x-\mu)^3}{[1+(x-\mu)^2]^2}$$

Again $\int_{R_1} \left| \frac{\partial^3 \log f}{\partial \mu^3} \right| f(x, \mu)\, dx$ exists as the integrand behaves like $\frac{1}{x^3}$ near $\pm\infty$ and in fact here $E\left(\frac{\partial^3 \log f}{\partial \mu^3}\right) = 0$, for any $\mu \in N_\rho(\mu_0)$.

To emphasize the fact that the conditions C–3 and C–5 need hold only in $N_\rho(\theta_0)$, consider $(X_1, \ldots, X_n)$ as i.i.d exponential with mean θ so that $\log f(x, \theta) = -\log \theta - \frac{x}{\theta}$. Then

$$\frac{\partial \log f}{\partial \theta} = -\frac{1}{\theta} + \frac{x}{\theta^2}$$

$$\frac{\partial^2 \log f}{\partial \theta^2} = +\frac{1}{\theta^2} - \frac{2x}{\theta^3}$$

$$\frac{\partial^3 \log f}{\partial \theta^3} = -\frac{2}{\theta^3} + \frac{6x}{\theta^4}$$

Now for $\theta \in (\theta_0 - \rho, \theta_0 + \rho)$ where $\theta_0 - \rho > 0$

$$\left| \frac{\partial^3 \log f}{\partial \theta^3} \right| \le \frac{2}{\theta^3} + \frac{6x}{\theta^4} \le \frac{2}{(\theta_0 - \rho)^3} + \frac{6x}{(\theta_0 - \rho)^4} = M(x)$$

and $E_\theta(M(x)) = \dfrac{2}{(\theta_0 - \rho)^3} + \dfrac{6\theta}{(\theta_0 - \rho)^4}$ exists for any θ.

Note that the function $\dfrac{2}{\theta^3} + \dfrac{6x}{\theta^4}$ for $\theta \in \Omega = (0, \infty)$ is not bounded as $\dfrac{2}{\theta^3}$ can be made arbitrarily large by selecting θ sufficiently small.

As an example of a family of distributions which is not a Cramèr family, we consider the double exponential or Laplace distribution with

$$f(x, \theta) = \frac{1}{2} \exp \{- | x - \theta | \}, x \in R_1, \theta \in R_1$$

Here although C_1 and C_2, hold $\log f(x, \theta) = - \log 2 - | x - \theta |$ and as $- | x - \theta |$ is not differentiable function of θ for fixed x at $\theta = x$ we have a problem. Note that as θ varies over $N_\rho(\theta_0)$ the set of exceptional values of $x \in R_1$ where $\dfrac{\partial \log f}{\partial \theta}$ does not exist, would be $(\theta_0 - \rho, \theta_0 + \rho)$ and has positive probability under each $\theta \in R_1$. Again any class of distributions such as $\{U(0, \theta), \theta > 0\}$ or exponential distribution with location θ with pdf $f(x, \theta) = \exp \{- (x - \theta)\}, x > \theta, \theta \in R_1$ will not be Cramèr family since the range of the r.v. or support of the pdf depends on parameter θ and C–1 is violated.

One can in a similar way define m-parameter Cramèr family. Let $\{f(x, \theta), \theta \in \Omega \subset R_m\}$ be a family of pdfs indxed $\theta = (\theta_1, \ldots, \theta_m)'$. This will be Cramèr family if following regularity conditions hold in $N_\rho(\theta_0)$.

C′–1 Support $S_\theta = \{x \mid f(x, \theta) > 0\}$ does not depend on θ or $S_\theta = S$.

C′–2 Partial derivatives of $\log f(x, \theta)$ w.r.t. components of θ exist up to third order for almost all values of $x \in S$ for every $\theta \in N_\rho(\theta_0)$ where θ_0 is the true value of the parameter.

C′–3 $\int_s f(x, \theta)\, dx = 1$ can be differentiated twice under the integral sign w.r.t. components of θ. So that $E\left(\dfrac{\partial \log f}{\partial \theta_r}\right) = 0, r = 1, 2, \ldots, n$ and

$$J_{rs} = E\left(\frac{\partial \log f}{\partial \theta_r} \frac{\partial \log f}{\partial \theta_s}\right) = E\left(-\frac{\partial^2 \log f}{\partial \theta_r \partial \theta_s}\right)$$ is such that the Fisher information matrix $J = ((J_{rs}))$ is positive definite.

C′–4 For any $(\theta_r, \theta_s, \theta_t)$ $r = 1, 2, \ldots, m$, $s = 1, 2, \ldots, m$, $t = 1, 2, \ldots, m$ and

$$\left| \frac{\partial^3 \log f}{\partial \theta_r \partial \theta_s \partial \theta_t} \right| \le M_{rst}(x) \text{ for any } \theta \in N_\rho(\theta_0) \text{ where } E_\theta(M_{rst}(x)) < \infty.$$

One can verify that an m-parameter exponential family is also a Cramèr family and as an illustration of a non-exponential two parameter family we consider Cauchy distribution with pdf

$$f(x, \mu, \sigma) = \frac{1}{\pi\sigma} \frac{1}{1 + \left(\frac{x - \mu}{\sigma}\right)^2}, \; x \in R_1, \mu \in R_1, \sigma > 0$$

As an example of a family of distribution which is not a Cramèr family one can consider two parameter Laplace or double exponential distribution with pdf

$$f(x, \mu, \sigma) = \frac{1}{2\sigma} \exp\left\{-\frac{|x - \mu|}{\sigma}\right\}, \; x \in R_1, \mu \in R_1, \sigma > 0$$

Here it is interesting to observe that for each μ fixed (known) and $\sigma > 0$ we have one-parameter exponential family and therefore also a Cramèr family but for each σ fixed (known) and $\mu \in R_1$ we have neither an exponential nor a Cramèr family. For Laplace distribution C′-1 and C′-2 are satisfied, but not the other conditions. When the support depends on θ, as in the case of $U(\mu - \sigma, \mu + \sigma)$, C′-1 is violated and we do not have a Cramèr family. Cramèr (1946), under the regularity conditions C-1 through C-5, proved that with probability approaching one as $n \to \infty$, the likelihood equation admits a solution $\hat{\theta}$ which is CAN with $AV(\hat{\theta}) = \frac{1}{nI(\theta)}$. However, the questions of uniqueness of solution and whether $\hat{\theta}$ provides the maximum of the likelihood were not addressed. Huzurbazar (1948) proved that for large samples with very high probability there is at least a relative maximum at $\hat{\theta}$ and the likelihood equation admits a unique consistent solution. Thus Cramèr and Huzurbazar results showed that for sufficiently large samples, with probability close to one, the likelihood equation has unique consistent solution which is CAN with asymptotic variance equal to the CRLB $= \frac{1}{nI(\theta)}$. In the next section we will state and prove these results as a theorem and label it as Cramèr–Huzurbazar Theorem.

7.4 Cramèr-Huzurbazar Theorem

Let $\{f(x, \theta), \theta \in \Omega \subset R_1\}$ be a one parameter Cramèr family satisfying regularity conditions C-1 through C-5. Then we have the following results.

R–1: With probability approaching one as $n \to \infty$, the likelihood equation admits a consistent solution $\hat{\theta}$. [Cramèr (1946)].

R–2: $\sqrt{n}(\hat{\theta} - \theta) \xrightarrow{d} N\left(0, \frac{1}{I(\theta)}\right)$ or $\hat{\theta} \sim AN\left(\theta, \frac{1}{nI(\theta)}\right)$ [Cramèr (1946)].

R–3: With probability approaching one as $n \to \infty$, at $\hat{\theta}$ there is a relative maximum of the likelihood or $P\left[\left(\frac{\partial^2 \log L}{\partial \theta^2}\right)_{\hat{\theta}} < 0\right] \to 1$ as $n \to \infty$ (Huzurbazar, 1948).

R–4 : Consistent solution $\hat{\theta}$ of the likelihood equation is essentially unique or the probability that the likelihood equation admits two consistent solutions $\hat{\theta}_1$ and $\hat{\theta}_2$ tends to zero as $n \to \infty$ (Huzurbazar, 1948).

To prove the result we consider the behaviour of $\log L(x, \theta)$ in the closed interval $[\theta_0 - \delta, \theta_0 + \delta] \in N_\rho(\theta_0)$ where $0 < \delta < \rho$. Let $y = \frac{f(x, \theta)}{f(x, \theta_0)}$ be the ratio of pdfs at θ and θ_0. Then $E_{\theta_0}(Y) = \int_s \frac{f(x, \theta)}{f(x, \theta_0)} f(x, \theta_0)\, dx = 1$. Now consider $-\log Y$. Then as $\varphi(u) = -\log u$ is a strictly convex function with $\varphi''(u) = \frac{1}{u^2} > 0$ we have by Jensen's inequality $E(-\log Y) > -\log E(Y) = 0$ and therefore

$$E(-\log Y) = \int -\log \frac{f(x, \theta)}{f(x, \theta_0)} f(x, \theta_0)\, dx > 0$$

or $$E_{\theta_0}(\log Y) = E_{\theta_0}\left(\log \frac{f(x, \theta)}{f(x, \theta_0)}\right) = -I(\theta, \theta_0) < 0 \quad (7.4.1)$$

The quantity $I(\theta, \theta_0)$ defined in (7.4.1) is known as Kullback Leibler information per unit of observation. We also note that by Taylor expansion of $\log f(x, \theta)$ around θ_0

$$\log f(x, \theta) = \log f(x, \theta_0) + \left(\frac{\partial \log f}{\partial \theta}\right)_{\theta_0} (\theta - \theta_0)$$

$$+ \frac{(\theta - \theta_0)^2}{2}\left(\frac{\partial^2 \log f}{\partial \theta^2}\right)_{\theta = \theta_0} + 0(\theta - \theta_0)^2 \quad (7.4.2)$$

Then

$$E_{\theta_0}\left[-\log \frac{f(x, \theta)}{f(x, \theta_0)}\right] = \frac{(\theta - \theta_0)^2}{2} I(\theta_0) + 0(\theta - \theta_0)^2$$

or $$I(\theta, \theta_0) = \frac{(\theta - \theta_0)^2}{2} I(\theta_0) + 0(\theta - \theta_0)^2$$

Now Let $Y_i = \log \frac{f(x_i, \theta)}{f(x_i, \theta_0)}$, then $\{Y_i\}_1^n$ are i.i.d.r.v.s with $E_{\theta_0}(Y_i) = -I(\theta, \theta_0)$.

Therefore by WLLN, under θ_0,

$$\frac{1}{n}\Sigma Y_i = \frac{1}{n}\log \frac{L(x, \theta)}{L(x, \theta_0)} \xrightarrow{p} -I(\theta, \theta_0) < 0 \quad (7.4.3)$$

Consider the sets

$$E_n = \{x \mid \log L(x, \theta_0 - \delta) < \log L(x, \theta_0)\}$$

$$F_n = \{x \mid \log L(x, \theta_0 + \delta) < \log L(x, \theta_0)\}$$

Then by (7.4.3) under θ_0, $P_{\theta_0}\left\{\log \dfrac{L(x, \theta)}{L(x, \theta_0)} < 0\right\} \to 1$ as $n \to \infty$ and thus $P(E_n) \to 1$. Similarly we have $P(F_n) \to 1$ and therefore $P(E_n \cap F_n) \to 1$ as $n \to \infty$. Thus for any $x \in E_n \cap F_n$, $\log L(x, \theta_0 - \delta) < \log L(x, \theta_0)$ and $\log L(x, \theta_0 + \delta) < \log L(x, \theta)$. Since the function $\log L(x, \theta)$ is continuous it must then first increase and then decrease as θ varies over the interval $[\theta_0 - \delta, \theta_0 + \delta]$. Therefore, $\dfrac{\partial \log L}{\partial \theta}$ is initially positive and then negative. In view of the regularity condition C–3, $\dfrac{\partial \log L}{\partial \theta}$ is continuous in θ and it now follows that there exists a $\hat{\theta} \in [\theta_0 - \delta, \theta_0 + \delta]$ such that $\dfrac{\partial \log L}{\partial \theta} = 0$ at $\hat{\theta}$. This gives $P[\,|\hat{\theta} - \theta_0| \le \delta] \ge P(E_n \cap F_n) \to 1$ as $n \to \infty$, and therefore $\hat{\theta}$ is consistent. Thís show that with probability approaching one as $n \to \infty$, the likelihood equation admits a solution $\hat{\theta}$ which is consistent. This completes the proof of R–1.

We next consider R–3. Expanding $\left(\dfrac{\partial^2 \log L}{\partial \theta^2}\right)_{\hat{\theta}}$ around θ_0 we have

$$\frac{1}{n}\left(\frac{\partial^2 \log L}{\partial \theta^2}\right)_{\hat{\theta}} = \frac{1}{n}\left(\frac{\partial^2 \log L}{\partial \theta^2}\right)_{\theta_0} + (\hat{\theta} - \theta_0)\,\frac{1}{n}\left(\frac{\partial^3 \log L}{\partial \theta^3}\right)_{\theta^*} \tag{7.4.4}$$

where $\theta^* \in (\theta_0 - \delta, \theta_0 + \delta)$.

Now

$$\frac{1}{n}\left(\frac{\partial^2 \log L}{\partial \theta^2}\right)_{\theta_0} = \frac{1}{n}\Sigma\left(\frac{\partial^2 \log f(x_i\theta)}{\partial \theta^2}\right)_{\theta_0} = \frac{1}{n}\sum_{i=1}^{n} U_i$$

where U_i are i.i.d.r.v.s with $E(U_i) = -I(\theta_0)$ and by WLLN, the first term on RHS of (7.4.4) converges in probability to $-I(\theta_0)$. Now $(\hat{\theta} - \theta_0) \xrightarrow{p} 0$ and by C–5 $\left|\dfrac{1}{n}\left(\dfrac{\partial^3 \log L}{\partial \theta^3}\right)_{\theta^*}\right| \le \dfrac{1}{n}\sum_{i=1}^{n} M(x_i)$. Since $E(M(x_i))$ exists, by WLLN $\dfrac{1}{n}\Sigma M(x_i) \xrightarrow{p} A(\theta_0)$ and therefore $\dfrac{1}{n}\left(\dfrac{\partial^3 \log L}{\partial \theta^3}\right)_{\theta^*} \xrightarrow{p} B(\theta_0)$. Therefore the second term on RHS of (7.4.4) converges in probability to zero and

$$\frac{1}{n}\left(\frac{\partial^2 \log L}{\partial \theta^2}\right)_{\hat{\theta}} \xrightarrow{p} -I(\theta_0) \tag{7.4.5}$$

It now follows that $P\left[\left(\frac{\partial^2 \log L}{\partial \theta^2}\right)_{\hat{\theta}} < 0\right] \to 1$ as $n \to \infty$ as $\left\{x \mid \left(\frac{\partial^2 \log L}{\partial \theta^2}\right)_{\hat{\theta}} < 0\right\} \supseteq (E_n \cap F_n)$.

This shows that with probability approaching one as $n \to \infty$ there is a relative maximum of the likelihood at $\hat{\theta}$. This completes the proof of R–3.

We now prove R–4 by the method of contradiction. Let $x \in E_n \cap F_n$ and if possible let $\hat{\theta}_1$ and $\hat{\theta}_2$ be two distinct consistent roots of the likelihood equation. Then by Rolle's theorem there must exist $\hat{\theta}_3 = \alpha\hat{\theta}_1 + (1-\alpha)\,\hat{\theta}_2$ $\in N_\delta(\theta_0)$ such that $\frac{1}{n}\left(\frac{\partial^2 \log L}{\partial \theta^2}\right)_{\hat{\theta}_3} = 0$. However expanding $\frac{1}{n}\left(\frac{\partial^2 \log L}{\partial \theta^2}\right)_{\hat{\theta}_3}$ around θ_0, we show as in the proof of R–2, that $\frac{1}{n}\left(\frac{\partial^2 \log L}{\partial \theta^2}\right)_{\hat{\theta}_3} \xrightarrow{p} -I(\theta_0)$.

Thus we have a contradiction and the likelihood equation can admit two roots in $N_\delta(\theta_0)$ only on the compliment of $(E_n \cap F_n)$ the probability of which tends to zero as $n \to \infty$. This completes the proof of R–4.

We now consider the proof of R–2. Again we expand $\frac{1}{n}\left(\frac{\partial \log L}{\partial \theta}\right)_{\hat{\theta}}$ around θ_0 by Taylor series. Then for $x \in E_n \cap F_n$,

$$0 = \frac{1}{n}\left(\frac{\partial \log L}{\partial \theta}\right)_{\hat{\theta}} = \frac{1}{n}\left(\frac{\partial \log L}{\partial \theta}\right)_{\theta_0} + (\hat{\theta}-\theta_0)\frac{1}{n}\left(\frac{\partial^2 \log L}{\partial \theta^2}\right)_{\theta_0}$$

$$+\frac{(\hat{\theta}-\theta_0)^2}{2}\cdot\frac{1}{n}\left(\frac{\partial^3 \log L}{\partial \theta^3}\right)_{\theta'}$$

where $\theta' \in N_\delta(\theta_0)$.

After some routine algebra, we obtain

$$\sqrt{n}(\hat{\theta}-\theta_0) = \frac{1}{\sqrt{n}}\left(\frac{\partial \log L}{\partial \theta}\right)_{\theta_0} \Big/ D \tag{7.4.6}$$

where $D = -\frac{1}{n}\left(\frac{\partial^2 \log L}{\partial \theta^2}\right)_{\theta_0} - \frac{(\hat{\theta}-\theta_0)^2}{2}\cdot\frac{1}{n}\left(\frac{\partial^3 \log L}{\partial \theta^3}\right)_{\theta'}$

Now $\frac{1}{\sqrt{n}}\left(\frac{\partial \log L}{\partial \theta}\right)_{\theta_0} = \frac{1}{\sqrt{n}}\sum_{i=1}^{n} W_i$, where $E(W_i) = 0$ and Var $(W_i) = I(\theta_0)$.

Therefore, by CLT, $\frac{1}{\sqrt{n}} \sum_{i=1}^{n} W_i \xrightarrow{d} N(0, I(\theta_0))$. Similarly, by WLLN $-\frac{1}{n}\left(\frac{\partial^2 \log L}{\partial \theta^2}\right)_{\theta_0} = \frac{1}{n} \Sigma(-U_i) \xrightarrow{p} I(\theta_0)$ as U_i are i.i.d.r.v.s with $E(U_i) = -I(\theta_0)$. As shown in R-2, the second term in D would converge in probability to 0 as $n \to \infty$. Therefore $D \xrightarrow{p} I(\theta_0)$. Therefore RHS of (7.4.6) converges in distribution to $N\left(0, \frac{1}{I(\theta_0)}\right)$. This shows that $\sqrt{n}(\hat{\theta} - \theta) \xrightarrow{d} N\left(0, \frac{1}{I(\theta)}\right)$ which completes the proof of R-2.

Remark 7.4.1 The general method of proof used above can be summarized as follows. Suppose we want to prove that a certain mathematical property Q holds under conditions say (A), (B) and (C). Then for $(x_1, \ldots, x_n)$ in the support of the joint pdf of $(X_1, \ldots, X_n)$ we determine set H_n where conditions (A), (B) and (C) hold. If $P(H_n) \to 1$ or equivalently $P(H_n^c) \to 0$ as $n \to \infty$ then we say that the property Q holds with probability approaching one as $n \to \infty$. From the proofs of R–1 and R–3 it is clear that for $x \in E_n \cap F_n$, the log likelihood function $\log L(x, \theta)$ is convex in $N_\delta(\theta_0)$ and therefore has a relative maximum at some point $\hat{\theta} \in N_\delta(\theta_0)$. From the regularity conditions it follows that $\left(\frac{\partial \log L}{\partial \theta}\right)_{\hat{\theta}} = 0$ and $\left(\frac{\partial^2 \log L}{\partial \theta^2}\right)_{\hat{\theta}} < 0$ and as $P(E_n \cap F_n) \to 1$ as $n \to \infty$ these properties hold with probability approaching one as $n \to \infty$.

An extension of Cramèr-Huzurbazar theorem for m-parameter Cramèr family was first given by Chanda (1954). Under the regularity conditions C′–1, through C′–5, it was shown that

R′-1 : With probability approaching one as $n \to \infty$, the likelihood equations $\frac{\partial \log L}{\partial \theta_r} = 0$, $r = 1, 2, \ldots, m$ admit a consistent solution $\hat{\theta}$.

R′-2 : $\sqrt{n}(\hat{\theta} - \theta) \xrightarrow{d} N^{(m)}(0, J^{-1}(\theta))$.

R′-3 : With probability approaching one as $n \to \infty$, the matrix $\left[-\left(\frac{\partial^2 \log L}{\partial \theta_r \partial \theta_s}\right)\right]$ is positive definite at $\hat{\theta}$.

R′-4 : $\hat{\theta}$ is essentially unique or the probability that the likelihood equation admits two consistent solutions $\hat{\theta}_1$, $\hat{\theta}_2$ tends to zero as $n \to \infty$.

Although the extended Cramèr-Huzurbazar theorem was extensively used, there was a feeling in the statistics community that Chanda's proof was not completely correct. The first specific counter example to some arguments used in Chanda's proof was given by Tarone and Gruenhage (1975). The problem was pinpointed by observing that Rolle's theorem is not valid in higher dimensions. The situation was rectified by Foutz (1977) by proving

Cramèr-Huzurbazar Theorem in m-dimensions $m \geq 1$ by using the implicit function theorem.

As this is a first course we will only present an outline of the proof and for more details refer to Foutz (1977). For a thorough treatment of the implicit function theorem we refer to Apostole (1994) and Rudin (1964).

Let $h_r = \dfrac{\partial \log L}{\partial \theta_r}$, $r = 1, 2, \ldots, m$ and $h = (h_1, \ldots, h_m)'$ then h is mapping from $S^n \times \Omega_m$ to R_m. The question then is under what conditions $h(x, \theta) = 0$ has a unique solution, for the fixed $x = x_0$, given by $\hat{\theta}(x_0)$. The implicit function theorem shows that under certain regularity conditions there exists an open set $U \subset S^n$ and an open neighbourhood of θ_0, $N_\delta(\theta_0)$ for which $h(x, \theta)$ is one to one function and h^{-1} exists. Then for each $x \in U$, $h^{-1}(0)$ exists and determines a function $\hat{\theta}(x)$ such that $h(x, \hat{\theta}(x)) = 0$. Apart from the condition that $\dfrac{\partial h_r}{\partial \theta_s}$, $r = 1, 2, \ldots, m$, $s = 1, 2, \ldots, m$ exist and are continuous the following two conditions are required for the validity of the implicit function theorem.

(A) For each fixed x, the vector $0 \in R_m$, is in the range space of h.

(B) The Jacobian $H = \dfrac{\partial(h_1, \ldots, h_m)}{\partial(\theta_1, \ldots, \theta_m)}$ is non-singular at $\theta = \theta_0$.

Now identify h with the vector of score functions and note that C′-3 implies existence of continuous partial derivatives $\dfrac{\partial h_r}{\partial \theta_s}$. We now show that conditions (A) and (B) hold with probability tending to one as $n \to \infty$. Therefore the implicit function theorem holds with probability approaching one as $n \to \infty$.

Let θ_0 be the "true value" of θ and consider the random vector $h = \left(\dfrac{\partial \log L}{\partial \theta_r}, \ldots, \dfrac{\partial \log L}{\partial \theta_m}\right)'_{\theta_0}$. Then by WLLN, for each $r = 1, 2, \ldots, m$, $\dfrac{1}{n}\left(\dfrac{\partial \log L}{\partial \theta_r}\right)_{\theta_0} \xrightarrow{p} 0$ as $n \to \infty$. Therefore with probability approaching one as $n \to \infty$ the zero vector belongs to the range space of h as θ varies over $N_\delta(\theta_0)$.

Similarly consider $h_{rs} = \dfrac{\partial^2 \log L}{\partial \theta_r \partial \theta_s}$ then as $\dfrac{1}{n}\dfrac{\partial^2 \log L}{\partial \theta_r \partial \theta_s} \xrightarrow{p} -J_{rs}(\theta)$, the matrix $H = \left(-\dfrac{\partial^2 \log L}{\partial \theta_r \partial \theta_s}\right)_{\theta_0}$ is positive definite and is therefore non-singular with probability approaching one as $n \to \infty$. Therefore both conditions (A) and (B) hold with probability approaching one as $n \to \infty$ and the implicit function theorem holds with probability approaching one as $n \to \infty$.

We remark here that even in exponential family, in order that $\hat{\theta}$ is well defined we need the condition that for the likelihood equations or equivalently the moment equations based on sufficient statistic T given by

$$\frac{1}{n}\sum_{i=1}^{n} T_r(x_i) = E(T_r(x)) = \eta_r(\theta),\ r = 1, 2, \ldots, m,$$

the observed vector $\left(\frac{1}{n}\Sigma T_1(x_i), \ldots, \frac{1}{n}\Sigma T_m(x_i)\right)'$ must belong to range space of $(\eta_1(\theta), \ldots, \eta_m(\theta))'$. This condition may hold only with probability approaching one as $n \to \infty$ as was observed in Example 7.1.3 for $m = 1$. For $m = 2$ one can refer to discussion in Example 6.4.2. For multinomial distribution in k-cells such a problem would arise if some of the observed cell frequencies are zero.

7.5 Multinomial with Cell Probabilities Depending on a Parameter

Now consider a multinomial distribution in k-cells with cell probabilities depending on a real or vector valued parameter θ. These models arise in genetics quite often. A Classical example is given by Rao (1973) on frequencies of blood groups O, A, B and AB. Thus we have here $k = 4$ and the cell probabilities are given by

$$P(O) = r^2,\ P(A) = p^2 + 2pr,\ P(B) = q^2 + 2qr\ P(AB) = 2pq$$

Here $p + q + r = 1$ so that cell probabilities depend on a two dimensional parameter. The above probabilities can be obtained based on assumption of random mating and that O is recessive to A and B and the relative frequencies of O, A and B are r, p, q respectively. Another example is where the relative frequencies of G and g are θ and $1 - \theta$ respectively. The progeny would then be classified as GG, Gg, gG and gg and under random mating the probabilities of four cells corresponding to four genotypes are

$$P(GG) = \theta^2,\ P(Gg) = \theta(1 - \theta) = P(gG) \text{ and } P(gg) = (1 - \theta)^2$$

Note that the well known Mendelian hypotheses 9:3:3:1 corresponds to $\theta = 3/4$. If the physical attributes of Gg and gG are same then the distribution is multinomial distribution in 3 cells with cell probabilities θ^2, $2\theta(1 - \theta)$ and $(1 - \theta)^2$ respectively.

Another context in which such a multinomial distribution arises is when we have a grouped data. For example consider an income distribution modelled by Pareto distribution with pdf $f(x, \lambda, \sigma) = \frac{\lambda\sigma^\lambda}{x^{\lambda+1}}$, $x > \sigma > 0$, $\lambda > 0$. Many a time exact income data is not available and we get a grouped data by income being reported as belonging to one of the intervals say $[\sigma_0, 2\sigma_0)$, $[2\sigma_0, 3\sigma_0)$, …, $[k\sigma_0, \infty)$ when $\sigma = \sigma_0$ is known. Otherwise the data is reported as frequencies of the income intervals given by $[a_0, a_1), [a_1, a_2), \ldots, [a_{k-1}, \infty)$, where

$$0 = a_0 < a_1 < a_2 \ldots < a_{k-1} < a_k = \infty$$

In the second case we have a multinomial distribution in k-cells with cell probabilities given by

$$p_1 = P(0, a_1] = \int_{\sigma}^{a_1} f(x, \lambda, \sigma)\, dx = 1 - (\sigma/a_1)^{\lambda}$$

$$p_r = \int_{a_{r-1}}^{a_r} f(x, \lambda, \sigma) = \left(\frac{\sigma}{a_{r-1}}\right)^{\lambda} - \left(\frac{\sigma}{a_r}\right)^{\lambda}, r = 2, \ldots, k-2$$

$$p_k = \left(\frac{\sigma}{a_{k-1}}\right)^{\lambda}$$

In general the multinomial distribution in k-cells with cell probabilities depending on a parameter θ has pmf

$$f(x, \theta) = \prod_{r=1}^{k} [p_r(\theta)]^{x_r} \tag{7.5.1}$$

with $x_r = 0$ or 1 and $\sum_{r=1}^{k} x_r = 1$ and $\theta \in \Omega$ is such that $p_r(\theta) > 0$, $\sum_{r=1}^{k} p_r(\theta) = 1$.

This gives $\log f(x, \theta) = \sum_{r=1}^{k} x_r \log p_r(\theta)$.

Suppose θ is real and Ω is an open interval. As S the support of f does not depend on θ the conditions C-1 and C-2 of the previous section hold. For C-3 we now assume that $\log p_r(\theta)$, $r = 1, 2, \ldots, k$ have derivatives up to 3rd order. Then

$$\frac{\partial \log f(x, \theta)}{\partial \theta} = \sum_{r=1}^{k} x_r \frac{\partial \log p_r}{\Sigma\, \partial \theta}$$

and as $E(x_r) = p_r(\theta)$ and

$$E\left(\frac{\partial \log f}{\partial \theta}\right) = \sum_{r=1}^{k} p_r \frac{\partial \log p_r}{\Sigma\, \partial \theta} = \sum_{r=1}^{k} \frac{\partial p_r}{\partial \theta}$$

But as $\sum_{r=1}^{k} p_r(\theta) = 1$ for any $\theta \in \Omega$, $\sum_{r=1}^{k} \frac{\partial p_r}{\partial \theta} = 0$. Hence $E\left(\frac{\partial \log f}{\partial \theta}\right) = 0$. Next consider

$$E\left(\frac{\partial \log f}{\partial \theta}\right)^2 = \sum_{r=1}^{k} x_r^2 \left(\frac{\partial \log p_r}{\partial \theta}\right)^2 + \sum_{r \neq s}\sum x_r x_s \frac{\partial \log p_r}{\partial \theta} \frac{\partial \log p_s}{\partial \theta}$$

But $E(x_r^2) = p_r(\theta)$ and $x_r x_s = 0$ when $r \neq s$.

Therefore $$E\left(\frac{\partial \log f}{\partial \theta}\right)^2 = \Sigma p_r \left(\frac{\partial \log p_r}{\partial \theta}\right)^2$$

Similarly $$\frac{\partial^2 \log f}{\partial \theta^2} = \sum_{r=1}^{k} x_r \left(\frac{\partial^2 \log p_r}{\partial \theta^2}\right)$$

and $$E\left(\frac{\partial^2 \log f}{\partial \theta^2}\right) = \Sigma p_r \left(\frac{\partial^2 \log p_r}{\partial \theta^2}\right)$$

From $\sum_{r=1}^{k} p_r \frac{\partial \log p_r}{\partial \theta} = 0$ it follows that

$$\sum_{r=1}^{k} p_r \frac{\partial^2 \log p_r}{\partial \theta} + \sum_{r=1}^{k} p_r \left(\frac{\partial \log p_r}{\partial \theta}\right)^2 = 0$$

Therefore, Fisher information

$$I(\theta) = \Sigma p_r \left(\frac{\partial \log p_r}{\Sigma\, \partial \theta}\right)^2 = -\Sigma p_r \frac{\partial^2 \log p_r}{\partial \theta^2}$$

Thus C-4 holds. Next consider

$$\left(\frac{\partial^3 \log f}{\partial \theta^3}\right) = \sum_{r=1}^{k} x_r \left(\frac{\partial^3 \log p_r}{\partial \theta^3}\right)$$

Now as $x_r = 0$ or 1, we have

$$\left|\left(\frac{\partial^3 \log f}{\partial \theta^3}\right)\right| < \sum_{r=1}^{k} \left|\left(\frac{\partial^3 \log p_r}{\partial \theta^3}\right)\right| = C(\theta)$$

In order that C-5 holds we now require that $C(\theta)$ for $\theta \in N_\rho(\theta_0)$ must be bounded so that $\sup_{\theta \in N_\rho(\theta_0)} C(\theta)$ can be taken as $M(x)$. With this condition the multinomial distribution in k-cells with cell probabilities depending on a real parameter θ will lead us to Cramèr family and the results of the pervious section hold and the MLE $\hat{\theta}$ can be obtained as a solution of the likelihood equation which is essentially unique and is CAN with $AV(\hat{\theta}) = \frac{1}{nI(\theta)}$. One can prove these results under much weaker assumptions and also for $\theta = (\theta_1, \ldots, \theta_m)'$. We refer to Rao (1973) for further details. We next consider some illustrative examples.

EXAMPLE 7.5.1 Consider the model given by $P(GG) = \theta^2$, $P(gG$ or $Gg) = 2\theta(1 - \theta)$ and $P(gg) = (1 - \theta)^2$ when Gg and gG are not distinguishable.

Then $\log f(x, \theta) = x_2 \log 2 + (2x_1 + x_2) \log \theta + (2x_3 + x_2) \log (1 - \theta)$

where $0 < \theta < 1$, $x_1 + x_2 + x_3 = 1$ and $x_i = 0$ or 1, $i = 1, 2, 3$.

Now observing that $2x_1 + 2x_2 + 2x_3 = 2$ we have

$$(2x_1 + x_2) + (2x_3 + x_2) = 2 \text{ or } 2x_3 + x_2 = 2 - (2x_1 + x_2)$$

Therefore we have

$$\log f(x_1, x_2, \theta) = x_2 \log 2 + (2x_1 + x_2)\,[\log \theta - \log (1 - \theta)] + 2 \log (1 - \theta)$$

which can be shown to be a one-parameter exponential family with $u(\theta) = \log \theta - \log (1 - \theta)$, $k(x) = 2x_1 + x_2$, $v(\theta) = 2 \log (1 - \theta)$ and $w(x) = x_2 \log 2$. Since $x_i = 0$ or 1 with $x_1 + x_2 \le 1$ the support of S does not depend on θ and $\Omega = (0, 1)$ is open in R_1. Further $u'(\theta) = \dfrac{1}{\theta(1-\theta)} > 0$ over Ω. Further $(1, 2x_1 + x_2)$ are linearly independent as can be seen by following argument. Let $a + b(2x_1 + x_2) = 0$ for all $(x_1, x_2) \in S$. Then taking $x_1 = 0$, $x_2 = 0$ we have $a = 0$ and taking $x_1 = 0$ and $x_2 = 1$ we have $b = 0$. Thus the pdf belongs to one parameter exponential family with $T(x) = \sum_{i=1}^{n} (2x_{1i} + x_{2i})$ as complete sufficient statistic. The likelihood equation is same as moment equation given by

$$\sum_{i=1}^{n} (2x_{1i} + x_{2i}) = 2n\theta^2 + 2n\theta(1 - \theta) = 2n\theta$$

which leads to

$$\hat{\theta} = \frac{1}{2n} \sum_{i=1}^{n} (2x_{1i} + x_{2i}) = \frac{2n_1 + n_2}{2n}$$

where (n_1, n_2) are observed cell frequencies. The Fisher information $I(\theta)$ can be obtained in a straightforward way as

$$\frac{\partial \log f}{\partial \theta} = \frac{2x_1 + x_2}{\theta} - \frac{2x_3 + x_2}{1 - \theta}$$

$$-\frac{\partial^2 \log f}{\partial \theta^2} = -\left[-\frac{2x_1 + x_2}{\theta^2} - \frac{2x_3 + x_2}{(1 - \theta)^2}(-1)(-1)\right]$$

Taking expectations we have

$$I(\theta) = \frac{2\theta^2 + 2\theta(1 - \theta)}{\theta^2} + \frac{2(1 - \theta)^2 + 2\theta(1 - \theta)}{(1 - \theta^2)}$$

$$= \frac{2}{\theta} + \frac{2}{1 - \theta} = \frac{2}{\theta(1 - \theta)}$$

Thus $\hat{\theta} \sim AN\left(\theta, \dfrac{\theta(1 - \theta)}{2n}\right)$.

We leave it as an exercise to the reader to verify that Var $(\hat{\theta}) = \frac{\theta(1-\theta)}{2n}$ for each n, a result that can be proved by using variance covariance structure of $(n_1, n_2)'$ in the multinomial setup. Note that $E(\hat{\theta}) = \theta$ and as $\hat{\theta}$ is a function of complete sufficient statistic it is MVUE of θ attaining CRLB to its variance.

Suppose gG and Gg are distinguishable then we have multinomial distribution in 4 cells with cell probabilities $p_1(\theta) = \theta^2$, $p_2(\theta) = \theta(1-\theta) = p_3(\theta)$ and $p_4(\theta) = (1-\theta)^2$ and data would consist of cell frequencies $(n_1, n_2, n_3, n_4)'$ with $\sum n_i = n$. One can show that in this case

$$\log f(x, \theta) = (2x_1 + x_2 + x_3) \log \theta + (2x_4 + x_2 + x_3) \log (1-\theta)$$

$$= k(x) \log \frac{\theta}{1-\theta} + 2 \log (1-\theta) \text{ where } k(x) = 2x_1 + x_2 + x_3$$

since $2x_1 + x_2 + x_3 = 2 - (2x_4 + x_3 + x_2)$.

Again we have one parameter exponential family with $u(\theta) = \log \frac{\theta}{1-\theta}$ and $T(x) = \sum_{i=1}^{n} (2x_{1i} + x_{2i} + x_{3i})$ as complete sufficient statistic. Routine calculations will show that $\hat{\theta} = \frac{2n_1 + n_2 + n_3}{2n}$ and $I(\theta) = \frac{2}{\theta(1-\theta)}$ so that $AV(\hat{\theta}) = \frac{\theta(1-\theta)}{2n}$ which is also the exact variance of $\hat{\theta}$.

EXAMPLE 7.5.2 Fisher (1954) had used a multinomial distribution in 4 cells to analyse Carver's data on two varieties of maize classified as starchy vs sugary and further cross classified with the colour Green and White. The cell probabilities depend on a parameter $\theta \in (0, 1)$ which is known as a linkage factor. The probabilities are given below along with the observed frequencies as obtained in an experiment.

Starchy		Sugary	
Green	White	Green	White
$\frac{2+\theta}{4}$	$\frac{1+\theta}{4}$	$\frac{1+\theta}{4}$	$\frac{\theta}{4}$
1977	906	904	32

Note that the log likelihood of (n_1, n_2, n_3, n_4) for a sample of size n is given by

$$\log L = \text{Constant} + n_1 \log (2+\theta) + (n_2 + n_3) \log (1-\theta) + n_4 \log \theta$$

Further $\log p_i(\theta)$, $i = 1, 2, 3, 4$ are analytic functions of $\theta \in (0, 1)$ and

$$\frac{\partial \log L}{\partial \theta} = \frac{n_1}{2+\theta} - \frac{(n_3 + n_2)}{1-\theta} + \frac{n_4}{\theta}$$

$$\frac{\partial^2 \log L}{\partial \theta^2} = \frac{-n_1}{(2+\theta)^2} - \frac{(n_3+n_2)}{(1-\theta)^2}(-1)(-1) - \frac{n_4}{\theta^2}$$

$$\frac{\partial^3 \log L}{\partial \theta^3} = -\left[\frac{-2n_1}{(2+\theta)^3} - \frac{2(n_2+n_3)}{(1-\theta)^3} - \frac{2n_4}{\theta^3}\right]$$

Now the likelihood equation is given by $\frac{\partial \log L}{\partial \theta} = 0$ or

$$Q(\theta) = n\theta^2 - \theta\{n_1 - 2(n_2+n_3) - n_4\} - 2n_4 = 0$$

and routine calculation will show that the Fisher information for the sample is given by

$$nI(\theta) = \frac{n(1+2\theta)}{2\theta(1-\theta)(2-\theta)}$$

The family of distributions is a Cramèr family but not an exponential family. Observe that

$$\left|\frac{\partial^3 \log f}{\partial \theta^3}\right| \le \frac{2x_1}{(2+\theta)^3} + \frac{2(x_2+x_3)}{(1-\theta)^3} + \frac{2x_4}{\theta^3}$$

and for any $\theta \in (\theta_0 - \delta, \theta_0 + \delta)$

$$\left|\frac{\partial^3 \log f}{\partial \theta^3}\right| \le \frac{2x_1}{(2+\theta_0-\delta)^3} + \frac{2(x_2+x_3)}{(1-\theta_0-\delta)^3} + \frac{2x_4}{(\theta_0-\delta)^3} = M(x)$$

and $E_\theta(M(x)) < \infty$ and C–5 also holds.

To obtain MLE of θ, we observe that the likelihood equation is quadratic in θ. The roots of the quadratic equations are

$\frac{-b \pm \sqrt{b^2 - 4ac}}{2a}$ where $a = n$, $b = -n_1 + 2(n_2+n_3) + n_4$ and $c = -2n_4$.

Now there is a possibility of both roots being real or both complex. However observe that $\frac{\partial^2 \log L}{\partial \theta^2} < 0$ for any set of frequencies $(n_1, n_2, n_3, n_4)'$, i.e. $\log L$ is concave and further $Q(0) = -2n_4 < 0$ and $Q(1) = 3(n_2+n_3) > 0$ for most of the vectors $(n_1, n_2, n_3, n_4)'$ and therefore there is exactly one root in (0, 1) the other root being negative as the product of the two roots is $-2n_4 < 0$. We therefore take the positive root as the MLE. Note that this argument fails only if $(n_1, n_2, n_3, n_4)'$ is such that $n_4 = 0$ and $n_2 = n_3 = 0$ and therefore $n_1 = n$. Now the probability of obtaining such a sample is $\left(\frac{2+\theta}{4}\right)^n$ which goes to zero as $n \to \infty$. For the Carver's data the MLE $\hat{\theta} = 0.035712$.

Exercise 7.5.1 (i) For the classical model on the frequencies of blood groups O, A, B and AB obtain likelihood equations and the information matrix $J(p, q)$. For the data 0 : 374 A : 436 B : 132 AB : 58. Obtain MLE $(\hat{p}, \hat{q})'$ and estimate of its asymptotic variance covariance matrix (Rao, 1973).

Exercise 7.5.2 According to a genetic model the sex and color blindness are associated and multinomial distribution has cell probabilities.

Male		Female	
Normal	Color blind	Normal	Color blind
$p/2$	$q/2$	$\frac{p^2}{2} + pq$	$q^2/2$

where $p + q = 1$. Obtain MLE of p and the Fisher information $I(p)$. Obtain MLE of $q^2/2$ and its asymptotic variance. Another CAN estimator for $q^2/2$ is given by n_4/n. Compare the values of two estimators for the data $(442, 38, 514, 6)'$ and their asymptotic variances.

Exercise 7.5.6 For the Pareto distribution with parameters (λ, σ) the data is grouped in $\sigma < a_1 < a_2 \ldots < a_{k-1} < \infty$. Suppose σ is known which without loss of generality can be taken to be unity. Set up likelihood equations for the data $(n_1, n_2, \ldots, n_k)'$. Converting to the log scale i.e. using $y = \log x$ we have $0 < \log a_1 < \ldots < \log a_k < \infty$ where y has exponential distribution with pdf $g(y, \lambda) = \lambda \cdot e^{-\lambda y}$, $\lambda > 0$, $y > 0$. Set up likelihood equations. For a distribution grouped in 3 cells with $a_1 = 2$, $a_2 = 3$ set up likelihood equations in both systems and try to solve them for observed frequencies $(60\ 30\ 10)'$ and determine MLE of λ using exponential model.

7.6 Solution of Likelihood Equations

First consider the case of a real parameter θ so that the MLE is determined by solving the equation $\frac{\partial \log L}{\partial \theta} = 0$. We have seen that in case of a single parameter exponential family, the likelihood equation is equivalent to moment equation based on sufficient statistic $T(x) = \frac{1}{n} \Sigma\, k(x_i)$ and using results of 7.2 we have

$$T(x) = \frac{1}{n} \Sigma\, k(x_i) = E(k\,(x)) = \frac{-v'(\theta)}{u'(\theta)} = \eta(\theta) \qquad (7.6.1)$$

In case $\eta(\theta) = \theta$ we have $\hat{\theta} = T(x)$ itself. For example such a situation occurs in $N(\theta, 1)$, $b(1, \theta)$, Poisson (θ) and exponential distribution with mean θ. On the other hand in the general case, as $\frac{\partial \eta}{\partial \theta} \neq 0$ and $\frac{\partial \eta}{\partial \theta}$ is continuous, we have η^{-1} is well defined and $\hat{\theta} = \eta^{-1}(T(x))$. For example, such a situation occurs in the Pareto distribution with $f(x, \lambda) = \frac{\lambda}{x^{\lambda+1}}$, $x \geq 1$, $\lambda > 0$ where $T(x) = \frac{1}{n} \sum_{i=1}^{n} \log x_i$ and $\eta(\lambda) = \frac{1}{\lambda}$ so that $\hat{\lambda} = \frac{n}{\Sigma \log x_i}$. Here η^{-1} is fairly easy to obtain explicitly. On the other hand there are many situations in which η^{-1} is not easy to obtain. We consider a couple of examples.

EXAMPLE 7.6.1 Consider a Gamma pdf given by

$$f(x, \lambda) = \frac{1}{\Gamma(\lambda)} x^{\lambda-1} e^{-x}, x > 0, \lambda > 0$$

Here $k(x) = \log x$ and $\eta(\lambda) = \frac{d \log \Gamma(\lambda)}{d\lambda}$. The moment equation based on x itself is $\bar{x} = \lambda$ and $\bar{X}$ is the moment estimator of λ which is CAN with $AV(\bar{X}) = \frac{\lambda}{n}$. On the other hand the likelihood equation or the moment equation based on $\log x$ is

$$\frac{1}{n} \Sigma \log x_i = E(\log X) = \frac{d}{d\lambda} \log \Gamma(\lambda) = \psi(\lambda) \qquad (7.6.2)$$

As $I(\lambda) = \frac{d^2}{d\lambda^2} \log \Gamma(\lambda) = \frac{d\psi}{d\lambda} > 0$, we have ψ is monotone increasing and therefore has unique inverse ψ^{-1} and $\hat{\lambda} = \psi^{-1}\left(\frac{1}{n} \Sigma \log x_i\right)$. The MLE $\hat{\lambda}$ is CAN with

$$AV(\hat{\lambda}) = \frac{1}{nI(\lambda)} = \frac{1}{n \frac{d^2}{d\lambda^2} (\log \Gamma(\lambda))}$$

The function $\psi(\lambda) = \frac{d}{d\lambda} \log \Gamma(\lambda)$ is known as digamma function and $\frac{d^2}{d\lambda^2} \log \Gamma(\lambda) = \psi'(\lambda)$ as trigamma function which occurs in Applied Mathematics quite often. These functions have been studied extensively, not only for real $\lambda > 0$ but also for complex values of λ as well.

The function $\frac{d \log \Gamma(\lambda)}{d\lambda} = \psi(\lambda)$ has been tabulated extensively. The first such tables were compiled by E. Pairman (1919) edited by *K*. Pearson. However they define digamma function as $\frac{d}{d\lambda} \log \Gamma(\lambda + 1)$. Note that as $\Gamma(\lambda + 1) = \lambda\Gamma(\lambda)$ for $\lambda > 0$, we have $\frac{d}{d\lambda} \log \Gamma(\lambda + 1) = \frac{1}{\lambda} + \psi(\lambda)$ and values of $\psi(\lambda)$ can be obtained from Pairman (1919). More recently Abromowitz and Stegun (1964) give values of $\psi(\lambda)$ for $\lambda = 1.00\ (0.01)\ 2.00$ from which values of $\psi(\lambda)$ for $\lambda \in (0, 1)$ can be obtained by the recurrence relation $\psi(z + 1) = \frac{1}{z} + \psi(z)$. In general $\psi(z + p) = \sum_{r=0}^{p-1} \frac{1}{z + r} + \psi(z)$ for any $z \in (0, 1)$ and p an integer and thus values of $\psi(z + p)$ for any integer p can be obtained.

To solve the equation $\frac{1}{n} \Sigma \log x_i = \psi(\lambda)$, we use repeated bisection method or use inverse interpolation for the observed value of $\frac{1}{n} \Sigma \log x_i$. Since $\bar{X}$ is consistent and $\hat{\lambda}$ is also consistent it is recommended that we first determine an interval $\bar{x} \pm h$ which contains the given observed value of $\frac{1}{n} \Sigma \log x_i$ and then use repeated bisection method. If digamma function is available in tabulated form we can use methods of inverse interpolation.

EXAMPLE 7.6.2 Consider a truncated Poison distribution with mean λ but where zero class is missing. Here the pmf is given by

$$f(x, \lambda) = \frac{e^{-\lambda}}{1 - e^{-\lambda}} \frac{\lambda^x}{x!}, \quad x = 1, 2, \ldots, \lambda > 0$$

$$= \frac{1}{e^{\lambda} - 1} \frac{\lambda^x}{x!}, \quad x = 1, 2, \ldots, \lambda > 0$$

The pmf belongs to one parameter exponential family with $k(x) = x$. Now $\frac{\partial \log f}{\partial \lambda} = \frac{x}{\lambda} - \frac{e^{\lambda}}{e^{\lambda} - 1}$ so that $E(X) = \frac{\lambda e^{\lambda}}{e^{\lambda} - 1}$ and then the likelihood equation is given by

$$\bar{x} = \frac{\lambda e^{\lambda}}{e^{\lambda} - 1} = \eta(\lambda)$$

Again we have MLE $\hat{\lambda}$ is given by $\eta^{-1}(\bar{x})$. The function η^{-1} is not explicitly available and we must again use numerical methods to obtain the MLE $\hat{\lambda}$. We determine λ_1, λ_2 such that $\bar{x} \in (\eta(\lambda_1), \eta(\lambda_2))$ and then use repeated bisection method.

The likelihood equation in case of one parameter exponential family has a special structure, namely the variables $(x_1, \ldots, x_n)$ and θ are here separated. On the other hand if the likelihood equation is such that variables are separated i.e. it is of the type $T(x) = \eta(\theta)$ then $\frac{\partial \log L}{\partial \theta} = \Lambda(\theta)[T(x) - \eta(\theta)]$ where $\Lambda(\theta) \neq 0$ and under usual regularity conditions $L(x, \theta)$, the joint pdf of $(X_1, \ldots, X_n)$ belongs to one parameter exponential family with $T(x)$ as complete sufficient statistic which is MVUE for $\eta(\theta)$.

Now in case of m-parameter exponential family we have a similar situation. If $T = (T_1, \ldots, T_m)'$ then likelihood equations are $T = \eta(\theta)$, where $\eta_r(\theta_1, \ldots, \theta_m)$ is $E(T_r)$, $T_r = \frac{1}{n} \sum_{i=1}^{n} K_r(x_i)$. The MLE $\hat{\theta}$ is given by $\eta^{-1}(T)$. Note that $\frac{\partial(\eta_1 \ldots \eta_m)}{\partial(\theta_1 \ldots \theta_m)}$ being non-singular, η^{-1} is well defined. In multiparameter case the method of repeated bi-section can not be easily generalized and multivariate inverse interpolation formula are also not easy to apply.

In case of a single parameter Cramèr family such as Cauchy, the variables $(x_1, \ldots, x_n)'$ and θ are not seperable and the likelihood equation is given by

$$\frac{\partial \log L}{\partial \theta} = -\sum_{i=1}^{n} \frac{2(x_i - \theta)}{[1 + (x_i - \theta)^2]} = 0$$

This is an algebraic equation of degree $(2n - 1)$ in θ and explicit solution is not available. We then use classical iteration procedures to obtain a numerical solution for observed values $(x_1, \ldots, x_n)'$. In Newton-Raphson method we start the iterative procedure with T_1 as a trial value and obtain successive iterations by

$$T_{r+1} = T_r - \left\{ \left(\frac{\partial \log L}{\partial \theta} \right) \Big/ \left(\frac{\partial^2 \log L}{\partial \theta^2} \right) \right\}_{\theta = T_r} \tag{7.6.2}$$

Fisher (1925) proposed a modification of the Newton-Raphson method

$$T_{r+1} = T_r + \left\{ \left(\frac{\partial \log L}{\partial \theta} \right) \Big/ nI(\theta) \right\}_{\theta = T_r} \tag{7.6.3}$$

Note that Fisher modification of Newton-Raphson method consists in using $E\left(\frac{-\partial^2 \log L}{\partial \theta^2} \right) = nI(\theta)$ in the denominator, and the iterative procedure given by (7.6.3) is also known as Fisher's method of Scoring or Fisher-Newton-Raphson method. The difference $T_{r+1} - T_r$ is known as the correction to T_r. Since the solution $\hat{\theta}$ is not known, it is recommended that the iterative procedure be stopped when $| T_{r-1} - T_r | < \delta$, where δ is prespecified level of accuracy.

Kale (1961) showed that if T_1 the initial or trial solution is chosen as a consistent estimator then under Cramèr-conditions the iterative procedures are valid in large samples. Essentially the result shows that the double limit of $T_{rn} - \hat{\theta}_n$ converges in probability to zero as $r \to \infty$ and $n \to \infty$. The method of proof is similar to the one used in Cramèr-Huzurbazar Theorem, and the proof is obtained by showing that the conditions which imply the convergence of the iterative procedures are satisfied with probability approaching one as $n \to \infty$. Kale (1962) showed that similar results are valid for the multiparameter Newton-Raphson method and Fisher Newton Raphson method also known as method of Scoring. Iterative procedures defined respectively are

$$T_{r+1} = T_r - \left[\left(\left(\frac{\partial^2 \log L}{\partial \theta_r \partial \theta_s} \right) \right)^{-1} \left(\frac{\partial \log L}{\partial \theta_1} \cdots \frac{\partial \log L}{\partial \theta_m} \right)' \right]_{\theta = T_r} \tag{7.6.4}$$

$$T_{r+1} = T_r + \left[\frac{1}{n} J^{-1}(\theta) \left(\frac{\partial \log L}{\partial \theta_1} \cdots \frac{\partial \log L}{\partial \theta_m} \right)' \right]_{\theta = T_r} \tag{7.6.5}$$

Kale (1961) considered the application of Newton-Raphson method, Method of scoring and some other methods to Carver's data discussed in the previous section. Starting with trial solution as the CAN estimator $T_1 = \frac{1}{n}\{n_1 - n_2 - n_3 + n_4\} = 0.057046$. It is shown that for both methods $T_5 = 0.035713$ and the correction $| T_5 - T_4 | = 6 \times 10^{-6}$ for the Newton Raphson process and that for method of scoring the correction is 5×10^{-6}. For Carver's data we have $\hat{\theta} = 0.035712$ so the error in 5th iteration for both the processes is 10^{-6}. Kale (1962) illustrates the use of method of scoring and Newton Raphson process for the two parameter Gamma distribution using simulated random sample of size 15 only.

Exercise 7.6.1 (1) Draw a random sample of size 25 from Cauchy with $\mu = 0$ from $f(x, \mu) = \frac{1}{\pi} \frac{1}{1 + (x - \mu)^2}$, $x \in R_1$, $\mu \in R_1$ set up the likelihood equation and use method of scoring for parameters with $T_1 = x_{(13)}$, the sample median. Note that here $I(\theta) = \frac{1}{2}$ and method of scoring is much easy to apply than Newton-Raphson method.

2. For Carver's data use the method of repeated bisection in the interval (0, 1) to determine $\hat{\theta}$ correct to five places of decimals.

3. For Carver's data as a trial solution use

(i) $T_1' = 1 - \frac{4n_2}{n}$

(ii) $T_1'' = \frac{4n_4}{n}$

and then in each case use the method of scoring and compare the iterations required to achieve accuracy to the fifth place of decimal for $\hat{\theta}$. Compare these iterations with those given by initial value $T_1 = \frac{1}{n}\{n_1 - (n_2 + n_3) + n_4\}$. Compare the variances of T_1, T_1' and T_1''.

4. A biologist obtained a sample of number of eggs in the unopened flower heads of the black Knapweed by the Knapweed fly. The flower heads in which no eggs were laid were not included in the sample. A truncated Poisson distribution with zero class missing was used as a model since the event of a flower head without any egg could not be ascribed to the event $X = 0$ under the complete Poisson model. Obtain MLE of λ.

No. of eggs per flower head	1	2	3	4	5	$\geq$ 6
No. of flower heads	96	32	9	7	1	0

For more details refer to Cohen (1960).

5. Consider $(X_1, \ldots, X_n)$ i.i.d. $N(\mu, 1)$. Then the likelihood equation is $\frac{\partial \log L}{\partial \mu} = n(\bar{x} - \mu) = 0$. Starting with trial value $T_1 = \tilde{\mu}$ sample median, show that the method of scoring as well as Newton-Rphson process gives $T_2 = \bar{x}$ for any data set $(x_1, \ldots, x_n)'$.

7.7 Asymptotically Most Efficient Estimator

We have seen earlier that in Cramèr family which includes exponential family under certain regularity conditions the MLE $\hat{\theta}$ is $AN\left(\theta, \frac{1}{nI(\theta)}\right)$.

Fisher had conjectured that if $T_1 \sim AN\left(\theta, \frac{v_1(\theta)}{n}\right)$ then

$$AV(T_1) = v_1(\theta)/n \geq \frac{1}{nI(\theta)} \tag{7.7.1}$$

and defined asymptotic relative efficiency of T_1. ARE $(T_1, \hat{\theta}) = AV(\hat{\theta})/AV(T_1) = \frac{1}{v_1(\theta)\, I(\theta)}$ and observed that the MLE $\hat{\theta}$ is asymptotically most efficient as it attains the lower bound $\frac{1}{nI(\theta)}$ which will be referred to as the Fisher Lower Bound (FLB). A hotly debated controversy arose between K. Pearson and Fisher on the issue of ARE of moment estimator and the maximum likelihood estimator. We have already seen that in the common models such as $N(\theta, 1)$, Poisson (θ), $b(1, \theta)$ or exponential with mean θ, the moment estimator based on X is same as $\hat{\theta}$. However in the Gamma distribution with shape parameter λ and pdf

$$f(x, \lambda) = \frac{1}{\Gamma(\lambda)}\, e^{-x}\, x^{\lambda-1}, x > 0, \lambda > 0$$

the moment estimator $\bar{X} \sim AN(\lambda, \lambda^2/n)$ whereas $\hat{\lambda} \sim AN\left(\lambda, \frac{1}{\frac{n\partial^2 \log \Gamma(\lambda)}{\partial\lambda^2}}\right)$.

Since explicit calculation of $\hat{\lambda}$ and $AV(\hat{\lambda})$ for any λ was difficult, an Indian Statistician named Koshal working in Indian Council of Agricultural Research (ICAR) observed that the experimental data set indicated that the estimate of the $AV(\bar{X})$ is much larger than that of $AV(\hat{\lambda})$.

Today in the light of the knowledge of trigamma function and because of CRLB result applied to $\bar{X}$ an unbiased stimator for λ, we know that

$$\text{Var}\ (\bar{X}) = \frac{\lambda^2}{n} \geq \frac{1}{nI(\lambda)} = \frac{1}{\frac{n\partial^2 \log \Gamma(\lambda)}{\partial\lambda^2}}$$

and the moment estimator of λ would be less efficient than the MLE $\hat{\lambda}$.

Koshal (1933, 1935) had considered fitting a Gamma density by method of moments and the method maximum likelihood and had shown that the method of MLE gives a much better fit than that by the method of moments. The evidence was numerical and sample based which Pearson refused to accept and wrote a strong rebuttal and critique of Koshal in a paper in Biometrika (1936). Fisher (1937) defended Koshal vigourously.

In selected papers of Fisher edited by Shewhart (1950) which contains

Fisher's own comments on each paper, Fisher's comments on this paper indicate that he found Pearson's attack on Koshal as 'flagrantly unfair'. Fisher remarks that Koshal was "right on the matter under dispute" and "in grave danger of injury to his position and prospects through the prestige that his attacker still enjoyed". For more details of this controversy between Pearson and Fisher we refer to Box (1978).

We note that the Fisher Lower Bound (FLB) of the asymptotic variance of a CAN estimator of θ is same as CRLB to the variance of an unbiased estimator of θ. Indeed Rao (1992) mentions that CRLB was derived by him in 1943 in response to the following question asked in his class by a M.A. student V.M. Dandekar. "Whether there is an analogous lower bound to variance of an unbiased estimator in finite samples?" Later V.M. Dandekar became an eminent economist and member of Planning Commission of India and headed National Sample Survey for several years. He was a distinguished intellectual of India, always raising basic fundamental questions in many fields.

As the turn of the events later showed Hodges and LeCam (1953) produced a counter example to show that Fisher lower bound given in (7.7.1) is not valid even for the simplest case of samples from $\{N(\theta, 1), \theta \in R_1\}$. Recall that for a sample of size n from $N(\theta, 1)$, $\overline{X} \sim N(\theta, 1/n)$ is MLE as well as MVUE and its asymptotic variance and finite sample variance attains FLB and CRLB respectively. We now construct a class of CAN estimators of θ, T_α for $0 < \alpha < 1$ such that

$$\begin{aligned} AV(T_\alpha) &= \frac{1}{n}, \ \forall\ \theta \neq 0 \\ &= \alpha^2/n \text{ at } \theta = 0 \end{aligned}$$

Now for $N(\theta, 1)$ case FLB = CRLB = $1/n$, and therefore we have a class of CAN estimators of θ for which FLB result does not hold.

Let $(X_1, \ldots, X_n)'$ be i.i.d. $N(\theta, 1)$ and let $\{a_n\}_1^\infty$ be a sequence of real positive numbers such that $a_n \to 0$ and $\sqrt{n}\, a_n \to \infty$ as $n \to \infty$. For example we can take $a_n = n^{-1/4}$ or $a_n = \dfrac{1}{\log n}$ etc. Let $0 < \alpha < 1$. Define

$$\begin{aligned} T_\alpha &= \bar{x} \quad \text{if } |\bar{x}| \geq a_n \\ &= \alpha\bar{x} \quad \text{if } |\bar{x}| < a_n \end{aligned}$$

Then we show that $\sqrt{n}(T_\alpha - \theta) \xrightarrow{d} N(0, v(\theta))$ where $v(\theta) = 1$ if $\theta \neq 0$ and $v(\theta) = \alpha^2$ if $\theta = 0$. To prove this consider the identity

$$\sqrt{n}(T_\alpha - \theta) = \sqrt{n}(\bar{x} - \theta) + \sqrt{n}(T_\alpha - \bar{x}) \tag{7.7.2}$$

Note that (7.7.2) holds, $\forall\ x \in R_n$ and $\forall\ \theta \in R_1$.

Now $\sqrt{n}(\bar{x} - \theta) \sim N(0, 1)$, $\forall\ n \geq 1$ and thus $\sqrt{n}(\bar{x} - \theta) \xrightarrow{d} N(0, 1)$. Further let $Y_n = \sqrt{n}(T_\alpha - \bar{x})$ and let $\theta \neq 0$. Then

$$P[\,|\,Y_n\,|<\varepsilon] \geq P[T_\alpha = \bar{x}\,] = P[\,|\,\bar{x}\,|\geq a_n]$$
$$= 1 - \Phi[\sqrt{n}(a_n - \theta) + [\sqrt{n}(-a_n - \theta)]$$

For $\theta > 0$, $\sqrt{n}\,(a_n - \theta) \to -\infty$ and $\sqrt{n}(-a_n - \theta) \to -\infty$. Thus $P[\,|\,Y_n\,|<\varepsilon] \to 1$ as $n \to \infty$ or $Y_n \xrightarrow{p} 0$. Therefore $\sqrt{n}(T_\alpha - \theta) \xrightarrow{d} N(0, 1)$, for any $\theta > 0$. Similarly for $\theta < 0$, we can show that $\sqrt{n}\,(T_\alpha - \theta) \xrightarrow{d} N(0, 1)$.

For $\theta = 0$ consider the identity given by

$$\sqrt{n}(T_\alpha) = \sqrt{n}(\alpha\bar{x}\,) + \sqrt{n}(T_\alpha - \alpha\bar{x}\,) \tag{7.7.3}$$

Now $\sqrt{n}(\alpha\,\bar{x}\,) \sim N(0, \alpha^2)$ under $\theta = 0$ and thus $\sqrt{n}(\alpha\,\bar{x}\,) \xrightarrow{d} N(0, \alpha^2)$. Further if $Z_n = \sqrt{n}(T_\alpha - \alpha\bar{x}\,)$ then

$$P[\,|\,Z_n\,|<\varepsilon] \geq P[T_\alpha = \alpha\,\bar{x}\,] = P[\,|\,\bar{x}\,|< a_n]$$
$$= \Phi\;[\sqrt{n}\,a_n] - \Phi[-\sqrt{n}\,a_n]$$

As $\sqrt{n}a_n \to \infty$, $P[\,|\,Z_n\,|<\varepsilon] \to 1$ as $n \to \infty$ and therefore under $\theta = 0$, $\sqrt{n}(T_\alpha - \theta) \xrightarrow{d} N(0, \alpha^2)$. Therefore $T_\alpha \sim AN(\theta, v(\theta)/n)$ where $v(\theta) = 1$ if $\theta \neq 0$ and $v(\theta) = \alpha^2$ or $\theta = 0$. Thus T_α is CAN such that $AV(T_\alpha) = \dfrac{v(\theta)}{n} \leq \dfrac{1}{nI(\theta)}$ with strict inequality at $\theta = 0$.

This shows that the FLB is not valid and T_α is more efficient than $\bar{x}$. T_α constructed by Hodges and reported in LeCam (1953) was called as 'supeficient'. The statistical community was surprised at this counter example and reacted in two different ways. One was to restrict the class of CAN estimators so that FLB result holds. Wolfowitz (1965), Rao [(1962), (1963)] followed this line of attack. The other approach was to show that the problem of 'superefficiency' or improvement of $\bar{x}$ using T_α in the above way can be done at only a single point or finite number of points only. LeCam (1953) showed that such an improvement of $\bar{x}$ in $N(\theta, 1)$ case cannot be obtained for $\theta \in (a, b)$ an interval of R_1. Later Ibragimov and Has'minski (1981) showed that for $a_n = n^{-1/4}$ and for the sequence of parameter points $\theta = c/\sqrt{n}$, we have

$$\lim_{n\to\infty} n\{E(\bar{x} - \theta_n)^2 \mid \theta_n\} = 1$$

$$\lim_{n\to\infty} n\{(E(T_\alpha - \theta_n)^2 \mid \theta_n\} > 1$$

Kale (1985) obtained the exact distribution of T_α. Using the fact that $\bar{x} \sim N(\theta, 1/n)$ for each n, one can easily show that $G_n(t, \theta) = P[T_\alpha \leq t \mid \theta]$ the d.f. of T_α is given by

$$\begin{aligned} G_n(t, \theta) &= \Phi[\sqrt{n}(\mathrm{t} - \theta)],\ t < -a_n \\ &= \Phi[\sqrt{n}(-a_n - \theta)],\ -a_n \leq t < -\alpha a_n \\ &= \Phi[\sqrt{n}(t/\alpha - \theta)],\ -\alpha a_n \leq t < \alpha a_n \end{aligned} \tag{7.7.3}$$

$$= \Phi[\sqrt{n}(a_n - \theta)],\ \alpha a_n \le t < a_n$$

$$= \Phi[\sqrt{n}(a_n - \theta)],\ a_n \le t$$

$G_n(t, \theta)$ is everywhere continuous and is differentiable at all t except $t = -a_n, -\alpha a_n, \alpha a_n, a_n$. The pdf of T_α is given by

$$g_n(t, \theta) = \varphi[\sqrt{n}, (t - \theta)]\sqrt{n}\ t \notin (-a_n, a_n)$$

$$= \frac{\sqrt{n}}{\alpha}\varphi\left[\sqrt{n}\left(\frac{t}{\alpha} - \theta\right)\right], t \in (-\alpha a_n, \alpha a_n)$$

$$= 0 \quad t \in (\alpha a_n, a_n) \cup (-a_n, -\alpha a_n)$$

Observe that probability of any subset of $(\alpha a_n, a_n) \cup (-a_n, -\alpha a_n)$ is zero under any θ. It now follows that

$$\frac{\partial^2 \log g_n(t, \theta)}{\partial \theta^2} = -n \quad \text{if } t \notin (-a_n, a_n) \text{ or } t \in (-\alpha a_n, \alpha a_n)$$

$$= 0 \text{ otherwise}$$

and that Fisher information $I_{T_\alpha}(\theta) = n$ and T_α preserves all the information in the sample. Thus T_α is sufficient. Indeed

$$\frac{\partial T_\alpha}{\partial \bar{x}} = \pm 1 \quad \text{if } t \notin (-a_n, a_n)$$

$$= \alpha \quad \text{if } \bar{x} \in (-a_n, a_n)$$

and $\frac{\partial T_\alpha}{\partial \bar{x}}$ does not exist at $\bar{x} = a_n$ or $-a_n$. Since $P\left[\left|\frac{\partial T}{\partial \bar{x}}\right| \ne 0\right] = 1$ under each $\theta \in R_1$, T_α to $\bar{x}$ is one-one transformation and T_α is minimal sufficient and complete. Thus superefficiency of T_α based on asymptotic variance is somewhat illusory.

We can show that for $\theta_n = \pm a_n(1 + \alpha)/2$, as $n \to \infty$, MSE$(T_\alpha \mid \theta_n) \to \infty$ and the reduction in MSE by T_α at $\theta = 0$ is achieved at the cost of increase in MSE of T_α at infinitely many points in the neighbourhood of $\theta = 0$. The anamolous behaviour of T_α as compared to $\bar{x}$ may be attributed to the fact that whereas the convergence in distribution of $\sqrt{n}(\bar{x} - \theta)$ to $N(0, 1)$ is uniform in θ the convergence in distribution of $\sqrt{n}(T_\alpha - \theta)$ to $N(0, 1)$ is not uniform in θ.

Beside the method of maximum likelihood and moment and percentile method there are many other methods to generate CAN estimators. For discrete or grouped data this includes minimum chi-square where we minimize

Pearson $\chi^2 = \sum_{i=1}^{k} \frac{(O_i - E_i(\theta))^2}{E_i(\theta)}$ where O_i = observed cell frequency of the ith cell and $E_i(\theta)$ = expected cell frequency in the ith cell when θ is the true value. The minimization of Neyman modification of χ^2, namely

$$\chi_N^2 = \sum_{i=1}^{k} \frac{(O_i - E_i(\theta))^2}{O_i}$$

also leads to CAN estimator of θ under suitable regularity conditions. We also mention L, R and M estimators and for the details refer to Serfling (1980). However, we will not go into details here.

The method of maximum likelihood is perhaps most widely used method to generate CAN estimators when the model is fully specified except for a finite dimensional labelling parameter θ. Even here there are situations such as Neyman-Scott problem, truncated parameter space and/or truncated sample space where the method leads to estimators which are not CAN and do not perform well. We refer to Rao (1962) for such situations. It is emphasized that the best course of action is to obtain MLE $\hat{\theta}$ and study its asymptotic distribution by various methods including those based on computer simulations as well as bootstrap/subsampling etc.

8

Tests of Hypotheses-I

8.1 Historical Perspective

In several situations rather than estimating an indexing parameter θ in the model $\{f(x, \theta), \theta \in \Omega\}$, we are interested in checking whether θ has a specified value. For example if we observe the sex of a new born child such that θ denotes the probability of the event that a male child is born then we may be interested in testing whether $\theta = 1/2$ i.e. parity between the two sexes which has infact been a problem of interest since 18th Century. For example Laplace studied the ratio of births of boys to that of girls observed in different cities in Europe and obtained an estimate of $\hat{\theta} = \frac{22}{43} = 0.5116279$. The observed difference from the hypothetical value $\theta = 1/2$ is 0.0116279. Does this deviation from parity value $\theta = 1/2$ provide enough evidence against the hypothesis of parity? For more details we refer to E.S. Pearson (1978).

Similarly, Arbuthnot (1710) (Kendall and Plackett, 1977) observed that in the city of London during the period 1629 to 1710, each year the number of male births exceeded the number of females. Implicitly using a model of Bernoulli series of trials Arbuthnot calculated the probability of such an event as $(1/2)^{82}$ and using logarithms evaluated the same as $\frac{1}{4836} \times 10^{-21}$ $\simeq (2.0678246)10^{-25}$. Arbuthnot argued that this probability is extremely small and therefore the regularity with which the number of male births exceed the number of female births may be due to 'Divine Providence'. On the other hand Laplace argued that the observed deviation is significant in that the probability of such a deviation from hypothetical value is very small and the deviation may be due to 'regular causes'. In another context Laplace studied further the phenomenon of "excess of births of boys over those of girls" and claimed a specific cause namely the practice of abondoning newly born girls to Foundling Hospitals in many cities in Europe.

The above thinking was first used by K. Pearson in proposing the well-known chi-squared test of goodness of fit, and later by Fisher in proposing various tests of significance. The logic here according to Fisher (1959) was that of a simple disjunction namely "Either an exceptionally rare chance has occurred or else the null hypothesis is not true."

The question arises how to define "an exceptionally rare chance" or an event of small probability. One could do it in a normative way i.e. by prescribing a level such as 5%, 1% etc., and then determine a cutoff point in the tails of the distribution of a statistic T which is used as a test statistic. For example Laplace used $\hat{\theta} = \frac{\Sigma x_i}{n}$ and Arbuthnot T = number of years in which the number of male births is larger than female births. Under the hypothesis of parity i.e. $\theta = 1/2$, Laplace model uses $T = n\hat{\theta} \sim B(n, 1/2)$ and Arbudhnot $T \sim B(82, 1/2)$, and the tails of the distribution of the test statistic $n\hat{\theta}$ and T can be defined in a natural way. If we plot the pdf of the test statistic tail areas get defined in a natural way. On the other hand why one should restrict to tail areas only is not very clear. If exceptionally rare chance event is defined normatively by 5% or 1% level, then any event with probability less then 5% or 1% level would be eligible to become 'an exceptionally rare chance event'.

For example in the Bernoulli series model used by Arbuthnot every sequence of 82 trials has same probability $1/2^{82}$ but one would not suspect a sequence in which 41 times male births exceeded female births whereas in remaining 41 years female births exceeded the male births. The total number of such sequences is $\binom{82}{41}$. Among such sequences should one suspect the sequence in which we have the difference between male and female briths is alternately positive and negative? In a similar manner suppose that in the first 41 years male birhs exceeded female births and in the next 41 years reverse situation is observed then should this be regarded as a rare chance event? Similarly in a coin tossing situation a captain that wins all the toses in a series of games is regarded as lucky but as per the Bernoulli series model every sequence of outcomes has the same probability say $1/2^n$ where n is the total number of games. Savage (1976) reports about an informal discussion which he had with Fisher on this matter. Fisher regarded the issue as an academic one and felt that which events are to be regarded as rare events is clear from the scientific context on which the model is based. Thus the values in the tail areas of the distribution of the test statistic are regarded as significant and not those in the central part of the distribution. In the problem of testing of goodness fit, the large values of χ^2 belonging to right tail are regarded as significant not the small values of χ^2 although one could argue that "the fit is too good to be true".

This being a first course in parametric inference, assume that the given model $\{L(x, \theta), \theta \in \Omega\}$ is true and consider the problem of testing of hypotheses in this restricted context only. Then for an event E we can compare $P_\theta(E)$ under different values of θ and define E to be event of small probability in terms of $P_\theta(E)$, for fixed E and variations in θ. This naturally leads to comparison of the likelihood $L(x, \theta)$ for fixed data x and variations in θ. This approach will be considered. Typically then the problem is posed as follows.

We have a sample of size n from $\{f(x, \theta), \theta \in \Omega\}$ and to test whether the data x supports the hypotheses $\theta \in \Omega_1$ a subset of Ω or $\theta \in \Omega_2 = \Omega - \Omega_1$, the compliment of Ω_1 relative to Ω. Note that we could view the above problem as that of estimating $\psi(\theta)$ by taking $\psi(\theta)$ to be indicator function of Ω_1 i.e. $\psi(\theta) = 1$ if $\theta \in \Omega_1$ and zero otherwise. However the theory of MVU estimation does not apply in most cases of interest as U_ψ is usually empty if Ω_1 is large as the following example shows.

EXAMPLE 8.1.1. Consider a problem commonly occurring in SQC. Out of a large batch of production a sample of size 20 is drawn and the item is inspected and classified as defective or nondefective. We assume that the model is such that $(X_1, \ldots, X_{20})$ are i.i.d. $b(1, \theta)$ where $\theta = P[X_i = 1] = P[i\text{-th item is defective}]$. Suppose $\Omega_1 = (0, \theta_0]$ where θ_0 is a specified tolerence level of percentage defective such as 10%. Then $\Omega_2 = (\theta_0, 1)$. Now $\bar{X}_{20}$ is MVUE, and CAN for θ as well as it is minimal complete sufficient statistic for subfamilies $\{L(x, \theta), \theta \in \Omega_1\}$ and $\{L(x, \theta), \theta \in \Omega_2\}$. If possible let $\varphi(\bar{x}_{20})$ be unbiased for $\psi(\theta) = 1$ if $\theta \le \theta_0$ and zero otherwise. Then using completeness of subfamilies we must have $\varphi(\bar{x}_{20}) = 1$ as well as $\varphi(\bar{x}_{20}) = 0$ for all values of $\bar{x}_{20}$ which is a contradiction. This shows that U_ψ is empty and MVU estimation approach is not possible. Further $\psi(\theta)$ has zero derivate at all points $\theta \ne \theta_0$ and at θ_0, $\psi(\theta)$ is not differentiable. Therefore we can not obtain a CAN estimator of $\psi(\theta)$ based on $\bar{X}_n$ using the techniques given in earlier chapters, although $\bar{X}_n$ is itself CAN for θ.

On the other hand if Ω is too small say consisting of two competing models given by distribution functions $F_1(x)$ and $F_2(x)$ only, then U_ψ is too large and then also MVUE approach fails as the following example shows

EXAMPLE 8.1.2 (Rao, 1973). Suppose we have a sample of size one on $f(x, \theta) = \theta x^{\theta-1}, 0 < x < 1, \theta > 0$ and $\Omega = \{1/2, 1\}$ i.e. it consists of two points only namely $\theta = 1/2$ and $\theta = 1$. Thus we have two competing models given by $f(x, 1)$ and $f(x, 1/2)$. Let $\Omega_1 = \{1/2\}$ and $\Omega_2 = \{1\}$. Consider $\varphi_m(x) = ax^m + b$, then

$$E[\varphi_m(x) \mid \theta = 1] = \frac{a}{m+1} + b \quad \text{if } m > -1$$

and
$$E[\varphi_m(x) \mid \theta = 1/2] = \frac{a}{2m+1} + b \quad \text{if } m > -1/2.$$

Therefore for $m > -1/2$, $\varphi_m(x)$ is unbiased for $\psi(\theta)$ the indicator function of Ω_1 if $\dfrac{a}{m+1} + b = 0$ and $\dfrac{a}{(2m+1)} + b = 1$. Solving these two equations for a and b in terms of m, we have for $m \ne -1$ or $m \ne -1/2$ or $m \ne 0$

$$\varphi_m(x) = -\frac{(m+1)(2m+1)}{m} x^m + \frac{2m+1}{m} \tag{8.1.1}$$

is unbiased for $\psi(\theta)$.

Now under $\theta = 1$, we have

$$\text{Var}\,[\varphi_m(x)] = E[\varphi_m^2(x)]$$
$$= \frac{(2m+1)^2}{m^2}\int_0^1 [1 - 2(m+1)x^m + (m+1)^2 x^{2m}]\,dx.$$

As $m > -1/2$ the integral above exists and

$$\text{Var}\,[\varphi_m(x)] = (2m+1). \qquad (8.1.2)$$

Now consider $m \to -1/2$ then (8.1.2) shows that inf Var $[\varphi_m(x)] = 0$ and thus zero is the lower bound for any $T \in U_\psi$. However we cannot have any $T \in U_\psi$ which attains the lower bound zero as $E(T \mid \theta = 1) = 0$ and $E[T \mid \theta = 1/2] = 1$. Therefore there does not exist any MVUE of $\psi(\theta)$ although U_ψ is not empty.

In the parametric set up that we have assumed in this course it is more natural to pose the problem as that of testing hypotheses $H_1 : \theta \in \Omega_1$ against $H_2 : \theta \in \Omega_2 = \Omega - \Omega_1$ on the basis of $X = x$ assuming that the distribution of X is specified by the model $\{L(x, \theta);\ \theta \in \Omega\}$. In this formulation the important characteristic of the procedure is the two types of possible errors namely (i) false rejection of H_1, equivalent to false acceptance of H_2 and (ii) false acceptance of H_1 equivalent to false rejection of H_2. Thus the emphasis would not be on defining events of small probability but study the probabilities of two types of error.

In the next section we will illustrate this approach and consider the problem from the above view point.

8.2. Critical Regions and Test Functions

Let $(X_1, X_2, \ldots, X_n)$ be a random sample of size n from $\{f(x, \theta),\ \theta \in \Omega\}$. On the basis of the data $x = (x_1, \ldots, x_n)$ we have to decide whether $H_1 : \theta \in \Omega_1$ holds or $H_2 : \theta \in \Omega_2 = \Omega_1^c$ holds. One of the simplest way to define such a procedure is to divide the sample space into two mutually exclusive sets say E_1 and $E_2 = E_1^c$ such that if $x \in E_1$ then we accept H_1 (or equivalently reject H_2) and if $x \in E_2 = E_1^c$ then we accept H_2 (reject H_1). Then the two error probabilities are

(i) $P_\theta(E_1)$ for $\theta \in \Omega_2$: false acceptance of H_1
(ii) $P_\theta(E_2)$ for $\theta \in \Omega_1$: false rejection of H_1.

Note that $P_\theta(E_2) = 1 - P_\theta(E_1)$ as (E_1, E_2) is a disjoint partition of the sample space. Thus it is enough to study either $P_\theta(E_1)$ or $P_\theta(E_2)$. Note that $P_\theta(E_2)$ is the probability of rejecting H_1 when θ is the true value of the parameter. Thus if we define a subset C_{H_1} of the sample space as a critical region for H_1 i.e. if $x \in C_{H_1}$ then we reject H_1 otherwise we accept it. Then the test procedure is defined by the critical region C_{H_1} or equivalently its indicator function $\varphi(x) = 1$ if $x \in C_{H_1}$ and zero otherwise. Now $P_\theta(C_{H_1}) = E_\theta(\varphi(x))$

and the function $P_\theta(C_{H_1})$ for $\theta \in \Omega_1$ represents the error of false rejection of H_1 and $1 - P_\theta(C_{H_1})$ for $\theta \in \Omega_2$ represents the error of false acceptance of H_1 since when $x \notin C_{H_1}$ we accept H_1. One could also formulate the problem in terms of a critical region C_{H_2} which rejects H_2 when $x \in C_{H_2}$. However $C_{H_2} = C^c_{H_1}$ and thus it is adequate to formulate the problems in terms of either C_{H_1} or C_{H_2}. So far there is a perfect duality between H_1, H_2 and C_{H_1}, C_{H_2}. We will therefore formulate the problem in terms of H_1 and C_{H_1} only. We first consider an example.

EXAMPLE 8.2.1. Consider the situation occurring commonly in SQC as given in Example 8.1.1. Let H_1 be the hypotheses that $\theta \in \Omega_1 = (0, .1]$ and H_2 be the hypotheses that $\theta \in \Omega_2 = (.1, 1)$. A critical region for hypotheses H_1 (or equivalently an acceptance region for H_2) can be defined in terms of $T = \sum_{i=1}^{n} X_i$. For example let $C_3 = \{3, 4, \ldots, 20\}$. Thus if observe T in C_3 we reject H_1 and say that the batch is not of good quality i.e. $\theta > .1$. This would imply rejecting the batch and returning the same to the manufacturer.

Now the critical region for H_1 given by C_3 can be described as $[T \geq 3]$. Error probabilities are given by $P_\theta(C_3)$ for $\theta \in \Omega_1 = (0, .1]$ and $1 - P_\theta(C_3)$ for $\theta \in \Omega_2 = (.1, 1)$. Consider now some other possible critical regions $C_4 = [T \geq 4]$ so that

$$C_3 = [T \geq 3] = C_4 \cup [T = 3] \text{ or } C_2 = [T \geq 2] \text{ i.e. } C_3 \cup [T = 2] = C_2.$$

Note that $C_4 \subset C_3 \subset C_2$. Then the following table which can be obtained from Tables of Binomial Distributions gives the error probabilities of false rejection of H_1 and false acceptance of H_1 for values of $\theta = .01$, .05 and .10 and $\theta = .20$, .30 and .40.

	Error probabilities of false rejection of H_1			Error probabilities of false acceptance of H_1		
	$\theta = .01$	$\theta = .05$	$\theta = .10$	$\theta = .20$	$\theta = .30$	$\theta = .40$
C_4	.000	.016	.133	.441	.107	.016
C_3	.001	.075	.323	.206	.036	.004
C_2	.017	.264	.608	.069	.008	.001

For any C_i, $i = 2, 3, 4$, $P_\theta(C_i)$ for $\theta \leq .10$ is an error of false rejection and $1 - P_\theta(C_i)$ for $\theta > .1$ is an error of false acceptance.

Note that if we compare the critical region C_3 with C_2 the probability of false rejection of H_1 is lower than that of C_2 but the probability of error of other type, namely probability of false acceptance of H_1 is higher. On the other hand if we compare C_3 and C_4 then probability of false rejection of H_1 is smaller for C_4 than that for C_3 but the probability of false acceptance of

H_1 behaves in the opposite manner. The reader will instantly recognize it to be a general phenomenon. Let C and D be two critical regions such that $C \subset D$. Then for $\theta \in \Omega_1$, $P_\theta(C) \le P_\theta(D)$ and for $\theta \in \Omega_2$, $1 - P_\theta(C) \ge 1 - P_\theta(D)$ or probabilities of errors of two types behave in such a way that if one decreases then the other increases. Thus by enlarging or shrinking a given critical region we can decrease one type of error probability at the cost of increase in the error probability of other type.

In fact consider now the general problem where we have a random sample of size n from $\{f(x, \theta), \theta \in \Omega\}$ and we want to test H_1: $\theta \in \Omega_1$ against $H_2 : \theta \in \Omega_2 = \Omega_1^c$. Let A be a critical region for H_1 and A^C its acceptance region. Define $\beta(A, \theta) = P(x \in A \mid \theta)$ then $\beta(A, \theta)$ for $\theta \in \Omega_1$ is the error of false rejection of H_1 and $1 - \beta(A, \theta)$ for $\theta \in \Omega_2$ is the error of false acceptance of H_1. An ideal test procedure would be one for which $\beta(A, \theta)$, $\theta \in \Omega_1$ and $1 - \beta(A, \theta)$; $\theta \in \Omega_2$ are minimized simultaneously by selecting A in a proper way. However, even in the simplest problem where $\Omega_1 = \{\theta_1\}$ and $\Omega_2 = \{\theta_2\}$, we cannot select A such that both $\beta(A, \theta_1)$ and $1 - \beta(A, \theta_2)$ are minimized simultaneously. This follows from the fact that $\beta(A, \theta_1)$ can be reduced only by removing sample points from A i.e. shrinking A, but this results in reducing $\beta(A, \theta_2)$ which increases $1 - \beta(A, \theta_2)$. On the other hand $1 - \beta(A, \theta_2)$ can be reduced only by increasing $\beta(A, \theta_2)$ or by enlarging the set A which increases $\beta(A, \theta_1)$.

To get out of this dilemma we weigh the importance of the two types of errors based on other considerations, and try to control that error which is more important and then subject to this constraint, minimize the error of the other type. Thus if we regard false rejection of H_1 as more serious error than false acceptance of H_1 then we control $\beta(A, \theta)$ for $\theta \in \Omega_1$ by specifying a level α i.e. $\beta(A, \theta) \le \alpha$, $\forall\ \theta \in \Omega_1$ and subject to this condition, minimize $1 - \beta(A, \theta)$ for each $\theta \in \Omega_2$ by varying A. On the other hand if we consider the error in false acceptance of H_1 is more serious then we must have $1 - \beta(A, \theta) \le \alpha$, $\forall\ \theta \in \Omega_2$ and subject to this constraint, minimize $\beta(A, \theta)$ for each $\theta_1 \in \Omega_1$ by varying A. This is the classical formulation of the problem due to Neyman and Pearson (1933). Their fundamental contribution was to show that there is a paradigm by which the best critical region can be determined such that one of the errors is controlled at a preassigned level α and subject to this restriction the other error is minimized.

However in such an approach the perfect duality between H_1 and H_2 is lost. To emphasize this asymmetry between H_1 and H_2 that we have introduced, we denote by H_0 that hypotheses among H_1 and H_2 the false rejection of which is regarded as the more serious error and call it the null hypotheses. The other hypotheses will be called as the alternative hypotheses and will be denoted by H_A.

The problem of choosing H_0 the null hypothesis out of H_1 and H_2 may involve variety of considerations. For example consider the illustrative Example 8.2.2 where

H_1 : Batch is of good quality which is equivalent to $\theta \le .1$

H_2 : Batch is not of good quality which is equivalent to $\theta > 1$.

From the consumer's view point rejecting H_1 when H_1 is true i.e. reject a Batch when in fact it is of good quality is less serious an error when compared to accepting H_1 (rejecting H_2) when H_1 is false (H_2 is true). Thus here H_0 should be taken as H_2. From the manufacturer's view point rejecting a batch when in fact it is of good quality may result in a considerable financial loss. On the other hand rejecting H_2 when H_2 is true, i.e. marketing a batch of bad quality may result in higher profits. Thus from the view point of manufacturer it is tempting to select H_0 as H_1. However marketing bad batches too often may lead to loss of reputation and eventually reduce his share of the market. Thus if the manufacturer has short term interest in making quick money and then leaving the market, he may select H_1 as H_0. But if he has long term interest then he would take H_2 as H_0.

We will now assume that such deliberations have been made and the problem is posed as testing $H_0 : \theta \in \Omega_0$ vs $H_A : \theta \in \Omega_A$, on the basis of data $X = x$ arising out of the model $\{L(x, \theta), \theta \in \Omega_0 \cup \Omega_A\}$. Thus if we have a critical region C for H_0, then we reject H_0 if $x \in C$ and $P_\theta(C)$ for $\theta \in \Omega_0$ is the error probability of false rejection and is called as Type I error. The error probability of false acceptance is called as Type II error and is measured by $1 - P_\theta(C)$ for $\theta \in \Omega_A$. Let $0 < \alpha < 1$ be a fixed level which is selected as the upper bound for Type I error i.e. C is such that $P_\theta(C) \le \alpha, \forall\ \theta \in \Omega_0$ and let $\mathbb{D}_\alpha$ denote the collection of all critical regions so that

$$\mathbb{D}_\alpha = \{C \mid P_\theta(C) \le \alpha, \forall\ \theta \in \Omega_0\} \tag{8.2.1}$$

Among $\mathbb{D}_\alpha$ we want to select C^* such that $1 - P_\theta(C^*) \le 1 - P_\theta(C)$ for all $\theta \in \Omega_A$ which is same as determining C^* such that

$$P_\theta(C^*) \ge P_\theta(C),\ \forall\ C \in \mathbb{D}_\alpha \text{ and } \forall\ \theta \in \Omega_A \tag{8.2.2}$$

The prescribed number α is called the level of significance of the test procedure and $\sup_{\theta \in \Omega_0} P_\theta(C)$ its size. $P_\theta(C)$, $\theta \in \Omega_0$ is called the size function and is denoted by $\beta(C, \theta)$. Similarly $P_{\theta_1}(C)$ for a given $\theta_1 \in \Omega_A$ is called the power of the critical region C at $\theta = \theta_1$ and $P_\theta(C) = \beta(C, \theta)$ for $\theta \in \Omega_A$ the power function of C. Indeed we can pose the problem of determining Best Critical Region in terms of the function $\beta(C, \theta)$ for $\theta \in \Omega_0 \cup \Omega_A$. Among all critical regions C of level α such that

$$\beta(C, \theta) \le \alpha,\ \forall\ \theta \in \Omega_0$$

determine C^* such that

$$\beta(C^*, \theta) \ge \beta(C, \theta),\ \forall\ \theta \in \Omega_A,\ \forall\ C \in \mathbb{D}_\alpha$$

i.e. the power of C^* is maximum at each point $\theta \in \Omega_A$.

We can slightly generalize the idea of a critical region by defining a test function $\varphi(x)$ such that (1) $0 \le \varphi(x) \le 1$ and (2) $\varphi(x)$ is the conditional probability of rejecting H_0 when x is the data. Note that if $\varphi(x) = 1$ if $x \in C$ and zero otherwise then $\varphi(x)$ corresponds to a test defined by the critical region C. The test function $\varphi(x)$ allows us an extra freedom, namely rejecting H_0 with a probability $\varphi(x)$ when x is observed rather than either rejecting H_0 or accepting H_0 with no option in between. Now $E_\theta(\varphi(x)) = \beta_\varphi(\theta)$ is called as the power function of the test defined by φ. $\beta_\varphi(\theta_1)$ for $\theta_1 \in \Omega_A$ is the power of the test at θ_1 and $\beta_\varphi(\theta_0)$ for $\theta_0 \in \Omega_0$ is the size of the test at $\theta_0 \in \Omega_0$. The test φ is called as the level α test if

$$\sup_{\theta \in \Omega_0} \beta_\varphi(\theta) \le \alpha \Leftrightarrow \beta_\varphi(\theta) \le \alpha \,\forall, \theta \in \Omega_0 \tag{8.2.3}$$

Let $\mathbb{D}_\alpha$ denote the class of all level α tests. Then φ^* is Uniformly Most Powerful (UMP) level α test if $\varphi^* \in \mathbb{D}_\alpha$ and

$$\beta_{\varphi^*}(\theta) \ge \beta_\varphi(\theta),\ \forall\ \varphi \in \mathbb{D}_\alpha,\ \forall\ \theta \in \Omega_A \tag{8.2.4}$$

Our object is to determine the UMP test and in the later sections we will present a constructive method of obtaining UMP tests when such tests exist. We will start with the problem where $\Omega_0 = \{\theta_0\}$ and $\Omega_1 = \{\theta_1\}$ i.e. discriminating between two completely specified distributions with pdf of X given by $L(x, \theta_0)$ under H_0 or $L(x, \theta_1)$ under H_A. This problem is known as testing a simle null hypothesis against a simple alternative. This is the simplest problem and as we will show later it always has a solution. Since Ω_A consists of a single point only, φ^* the best test for such a problem is called as the Most Powerful (MP) test rather than UMP since for uniformity we need several distinct values $\theta \in \Omega_A$. We note that it is also common to denote the alternative hypotheses by H_1: $\theta \in \Omega_1$ and we will use H_A and H_1 and Ω_A and Ω_1 interchangeably.

Exercise 8.2. (i) Let n chickens be selected at random and let $(X_1, \ldots, X_n)$ denote their initial weights. Let the animals be provided a special diet D for a period of 4 weeks and let $(Y_1, \ldots, Y_n)$ be their weights at the end of treatment period. Denote by $Z_i = Y_i - X_i$ which represents gain in weight due to special diet D. Suppose that $(Z_1, \ldots, Z_n)$ can be treated as i.i.d. $N(\theta, 2)$ and suppose we can recommend special diet D if the gain is at least 2. For $n = 18$, $\overline{Z} \sim N(\theta, 1/9)$ consider the critical regions $C_1 = \{z \mid \bar{z} > 2 + 1/3\}$, $C_2 = \{z \mid \bar{z} < 2 + 2/3\}$ for $H_1 : \theta \ge 2$ and $H_2 : \theta < 2$. Calculate error probabilities of C_1 and C_2 for $\theta = 1, 1.5, 2, 2.5, 3$. Discuss how you will select H_0 and H_A.

(ii) Suppose that the life of an electronic item is exponential with mean θ and let $\psi(t_0, \theta) = P[X > t_0] = e^{-t_0/\theta}$. For issuing warranties up to time t_0, we want to test whether $e^{-t_0/\theta} \ge 1 - \delta$ where δ is sufficiently small. Then $H_1 : e^{-t_0/\theta} \ge 1 - \delta$ and $H_2 : e^{-t_0/\theta} < 1 - \delta$. Transferring these hypotheses in terms of θ by taking natural logs we have H_1 : $\theta \ge t_0/[(-\log(1 - \delta)] = \theta_0$ and $H_2 : \theta < t_0/[1 - \log(1 - \delta)]$. Let $t_0 = 30$ and $\delta = .1$. Propose a suitable critical region based on $T = \sum_{i=1}^{20} X_i \sim G(20, \theta)$ and obtain its error probabilities at $\theta = \frac{1}{2}\theta_0, \frac{3}{4}\theta_0, \theta_0, \frac{4}{3}\theta_0, 2\theta_0$. Discuss how you will choose H_0 and H_A.

8.3 Neyman-Pearson Lemma and the MP Test

Let $(X_1, \ldots, X_n)$ be a random sample of size n on X and let $H_0 : X$ has pdf $f_0(x)$ and $H_1 : X$ has pdf $f_1(x)$. Note that the joint pdf of the sample under H_0 is $L_0(x) = \prod_{i=1}^{n} f_0(x_i)$ for $x \in S_0$ and under H_1 it would be $L_1(x) = \prod_{i=1}^{n} f_1(x_i)$ for $x \in S_1$ where S_0 and S_1 are supports of $L_0(x)$ and $L_1(x)$ respectively. Let φ be a test function then

$$\left.\begin{aligned} \text{size of } \varphi = E[\varphi \mid H_0] &= \int_{S_0} L_0(x)\,dx \\ \text{power of } \varphi = E[\varphi \mid H_1] &= \int_{S_1} L_1(x)\,dx \end{aligned}\right\} \tag{8.3.1}$$

Let $\mathbb{D}_\alpha = \{\varphi \mid E[\varphi \mid H_0] \leq \alpha\}$ denote the class of all level α tests. Then to obtain the MP test of level α we want to maximize $E[\varphi \mid H_1]$ by varying φ over $\mathbb{D}_\alpha$. The famous Neyman–Pearson Lemma shows constructively that the MP test φ^* can be determined for any $\alpha \in [0, 1]$ and φ^* is essentially unique.

Now consider $x \in S_0^c \cap S_1^c$ i.e. where x does not belong to support of $L_0(x)$ and $L_1(x)$. Such points do not contribute to the size or the power and therefore can be excluded from consideration. Let $x \in S_0^c \cap S_1$ i.e. for which $L_0(x) = 0$ and $L_1(x) > 0$ then we must define $\varphi^*(x) = 1$ as this increases the power to the highest possible extent without increasing the size of the test. Similarly if $x \in S_0 \cap S_1^c$ i.e. for which $L_0(x) > 0$ but $L_1(x) = 0$ we must define $\varphi^*(x) = 0$ as this does not affect the power and at the same time decreases the size to the maximum possible extent. The points $x \in S_0 \cap S_1$ i.e. where $L_0(x) > 0$ and $L_1(x) > 0$ need a careful consideration. The fundamental lemma of Neyman and Pearson shows us a way for defining $\varphi^*(x)$.

THEOREM 8.3.1 [N-P Lemma (Part A)]

(i) Any test $\varphi_k(x)$ of the form

$$\begin{aligned} \varphi_k(x) &= 1 && \text{if } L_1(x) > kL_0(x) \\ &= \gamma(x) && \text{if } L_1(x) = kL_0(x) \\ &= 0 && \text{if } L_1(x) < kL_0(x) \end{aligned} \tag{8.3.2}$$

for some $0 \leq k < \infty$ and $0 \leq \gamma(x) \leq 1$ is a MP test of level $E[\varphi_k(x) \mid H_0]$ for testing $H_0 : X$ has joint pdf $L_0(x)$ against $H_1 : X$ has joint pdf $L_1(x)$.

(ii) The test $\varphi_\infty(x)$ corresponding to $k = \infty$ is given by

$$\begin{aligned} \varphi_\infty(x) &= 0 && \text{if } L_0(x) > 0 \\ &= 1 && \text{if } L_0(x) = 0 \end{aligned} \tag{8.3.3}$$

and is a MP test of level zero.

Proof: Note that the size of $\varphi_k = E[\varphi_k \mid H_0]$ in this case is same as the level of φ_k. Let φ' be any test with level $E[\varphi_k \mid H_0]$ i.e. $E\{\varphi' \mid H_0] \leq E[\varphi_k \mid H_0]$, then we show that $E[\varphi_k \mid H_1] \geq E[\varphi' \mid H_1]$ to claim the result (i). Define

$$Q(x) = [\varphi_k(x) - \varphi'(x)]\,[L_1(x) - kL_0(x)].$$

For $x \in E_1$, where $E_1 = \{x \mid L_1(x) > k\, L_0(x)\}$,

since $0 \leq \varphi'(x) \leq 1$, we have $Q(x) \geq 0$.

For $x \in E_2 = \{x \mid L_1(x) = k\, L_0(x)\}$, $Q(x) = 0$.

For $x \in E_3 = \{x \mid L_1(x) < k\, L_0(x)\}$, as $\varphi_k(x) = 0$ and $0 \leq \varphi'(x) \leq 1$

we have $Q_k(x) \geq 0$. Therefore $\int Q(x)\, dx \geq 0$ which implies that

$$E[\varphi_k \mid H_1] - E[\varphi' \mid H_1] \geq k\{E[\varphi_k \mid H_0] - E[\varphi' \mid H_0]\} \qquad (8.3.4)$$

As φ' is level $E[\varphi_k \mid H_0]$ test, RHS of (8.3.4) is non-negative. Therefore $E[\varphi' \mid H_1] \leq E[\varphi_k \mid H_1]$ and φ_k is MP test of level $E[\varphi_k \mid H_0]$.

To prove (ii) observe that $E[\varphi_\infty \mid H_0] = 0$ and any level zero test φ' must have its size also zero or $E[\varphi' \mid H_0] = 0$. But $E[\varphi' \mid H_0] = \int_{S_0} \varphi'(x)\, L_0(x)\, dx$ and $0 \leq \varphi' \leq 1$ and $L_0(x) > 0$ on S_o. Therefore we must have $\varphi'(x) = 0$ on S_0. Now

$$E[\varphi_\infty \mid H_1] - E[\varphi' \mid H_1] = \int_{S_1} (\varphi_\infty - \varphi')\, L_1(x)\, dx \qquad (8.3.5)$$

Now $\varphi_\infty - \varphi = 0$ on S_0. Further $S_1 = (S_1 \cap S_0^c) \cup (S_1 \cap S_0)$ therefore RHS of (8.3.5)

$$= \int_{S_1 \cap S_0^c} (1 - \varphi')\, L_1(x)\, dx$$

$$\geq 0$$

as $0 \leq \varphi' \leq 1$ and on $S_1 \cap S_0^c$, $L_1(x) > 0$. This completes the proof of (ii) of Part *A* or *N-P* lemma.

We next show that for any $\alpha \in [0, 1]$ there exists a test of the form (8.3.2) with $\gamma(x) = \gamma$ a constant such that $E[\varphi_k(x) \mid H_0] = \alpha$ i.e. For given α we can determine a corresponding k and γ so that the test $\varphi_k(x)$, by Theorem 8.3.1, is MP test of level α.

THEOREM 8.3.2 [N-P lemma (Part B)]
For every α, $0 \leq \alpha \leq 1$ there exists a test $\varphi_k(x)$ of the form (8.3.2) with $\gamma(x) = \gamma$ such that $E[\varphi_k(x) \mid H_0] = \alpha$.

Proof: Let $\alpha = 0$ then we can take $k = \infty$ and $\varphi_\infty(x)$ is MP test of level zero as seen earlier. We next show that for $0 < \alpha \le 1$ we can select constants k and γ such that there exists a MP test of the form (8.3.2) with $\gamma(x) = \gamma$ of size α.

$$E[\varphi_k(x) \mid H_0] = \int_{S_0} \varphi_k(x)\, L_0(x)\, dx = \int_A \varphi_k(x)\, L_0(x)\, dx + \gamma \int_B L_0(x)\, dx$$

where $A = S_0 \cap E_1$ and $B = S_0 \cap E_2$. Note that on $S_0 \cap E_3$ we have $\varphi_k(x) = 0$. On A and B, $L_0(x) > 0$ and if we define a r.v. $Y = L_1(X)/L_0(X)$ on $(S_0 \cap A) \cup (S_0 \cap B)$ then the above integrals are $P[Y > k \mid H_0]$ and $\gamma P[Y = k \mid H_0]$, respectively and thus

$$E[\varphi_k(x) \mid H_0] = 1 - P[Y \le k \mid H_0] + \gamma P[Y = k \mid H_0].$$

We must thus determine k and γ such that

$$P[Y \le k \mid H_0] - \gamma P[Y = k \mid H_0] = 1 - \alpha.$$

If there exists *a* k_0 such that $P[Y \le k \mid H_0] = 1 - \alpha$ then we take $\gamma = 0$ and $k = k_0$. If such a k_0 does not exist then there exists a k_1 such that

$$P[Y < k_1 \mid H_0] \le 1 - \alpha < P[Y \le k_1 \mid H_0].$$

Then we take $k = k_1$ and $\gamma = \dfrac{P[Y \le k_1 \mid H_0] - (1-\alpha)}{P(Y = k_1 \mid H_0)}$. For this pair (k, γ) we have $E[\varphi_k(x) \mid H_0] = \alpha$.

Thus for any $\alpha \in [0, 1]$ we can determine a test of the form (8.3.2) which by part A of N-P lemma is MP test of its size which is its level also. The parts (A) and (B) of N-P lemma show that for any $\alpha \in [0, 1]$ there exists a MP level α test which is given by (8.3.2). Later in Part C we show that any other MP level α test can differ from the one given by (8.3.2) on the set $E_2 = \{x \mid L_1(x) = kL_0(x)\}$ only and therefore MP test is essentially unique. But before this we consider a few examples.

EXAMPLE 8.3.1. Let $(X_1, \ldots, X_n)$ be a random sample of size n from exponential distribution with the mean θ. Suppose that we want to test $H_0 : \theta = \theta_0$ vs $H_1 : \theta = \theta_1$ where $\theta_1 < \theta_0$. Then

$$\frac{L_1(x)}{L_0(x)} = \left(\frac{\theta_0}{\theta_1}\right)^n \exp\left\{\frac{T}{\theta_0} - \frac{T}{\theta_1}\right\} \text{ where } T = \Sigma\, x_i.$$

Now $\dfrac{L_1(x)}{L_0(x)} \gtreqless k$ according as $T \lesseqgtr C$ since $\left(\dfrac{\theta_0}{\theta_1}\right)^n$ is a positive constant and $\left(\dfrac{1}{\theta_0} - \dfrac{1}{\theta_1}\right) < 0$. Further $P_\theta\left[\dfrac{L_1(x)}{L_0(x)} = k\right] = 0$ for any θ as this probability is equal to $P_\theta[T = C]$ where T is $G(n, \theta)$, a continuous r.v. Thus we have to define k or equivalently C such that $P_\theta[T < C] = \alpha$ or $\displaystyle\int_0^C \frac{t^{n-1}}{\theta_0^n \Gamma_{(n)}} e^{-t/\theta_0}\, dt = \alpha$. Thus $C = \theta_0 \gamma_{n,\alpha}$ where $\gamma_{n,\alpha}$ is the $100\alpha\%$ point of the Gamma distribution with shape parameter n.

If the alternative hypothesis is $H_1 : \theta = \theta_1$ where $\theta_1 > \theta_0$ then we would reject H_0 if $T > C'$ i.e. for large values of T where $C' = \theta_0 \gamma_{n,1-\alpha}$. We observe that the MP test depends on the observations $(X_1, \ldots, X_n)$ only through the minimal sufficient statistics $T = \Sigma X_i$. Further as T is a continuous r.v. there is no need for randomization on the set

$$E_2 = \{x \mid L_1(x) = kL_0(x)\} = \{x \mid T(x) = C\},$$

as this set has probability zero under $\theta = \theta_0$. Note that the constant $C = \theta_0 \gamma_{n,\alpha}$ or $C' = \theta_0 \gamma_{n,1-\alpha}$ depends only on θ_0 and whether the alternative is $\theta_1 < \theta_0$ or $\theta_1 > \theta_0$ i.e. the relative position of θ_1 w.r.t. θ_0 and not on the exact value of θ_1.

EXAMPLE 8.3.1. We consider now an example in which X is a discrete so that $\lambda(x) = \dfrac{L_1(x)}{L_0(x)}$ is also discrete and we need a randomization on the boundary $E_2 = \{x \mid \lambda(x) = k\}$. Consider the case when we have a sample of size one and pmf under H_0 and H_1 are given by

$$H_0 : P[X = x] = \frac{1}{2^{x+1}},\ x = 0, 1, 2, \ldots$$

and $$H_A : P[X = x] = \frac{1}{4}\left(\frac{3}{4}\right)^x,\ x = 0, 1, 2, \ldots.$$

Thus under H_0 we have geometric distribution with $\theta = 1/2$ and under H_A it is geometric with $\theta = \frac{1}{4}$. Now

$$\lambda(x) = \frac{L_1(x)}{L_0(x)} = \frac{1}{4^{x+1}} 3^x \Big/ \frac{1}{2^{x+1}} = \frac{3^x}{2^{x+1}} \gtreqless k$$

which is equivalent to $x \gtreqless c$. The M.P. level $\alpha = .05$ say test is given by selecting (c, γ) such that

$$P[X > c \mid H_0] + \gamma P[X = c \mid H_0] = \alpha = .05.$$

Now for H_0: $P[X > t] = \sum\limits_{r=t}^{\infty} \dfrac{1}{2^{r+1}} = \dfrac{1}{2^{t+1}}$. Thus if there exists an integer t_0 such that $\dfrac{1}{2^{t_0+1}} = .05$ then we select $c = t_0$ and $\gamma = 0$. However for $\alpha = .05 = \dfrac{1}{20}$ we have $\dfrac{1}{32} < \dfrac{1}{20} < \dfrac{1}{16}$ and $P[X > 3] > \dfrac{1}{20}$, but $P[X > 4] < \dfrac{1}{20}$. Thus we select $t_0 = 4$ and find γ such that $P[X > 4] + \gamma P[X = 4] = .05$, i.e.

$\frac{1}{32}+\frac{\gamma}{32}=\frac{1}{20}$ or $\gamma=\frac{12}{20}$. Therefore MP test of level $\alpha = .05$ is given by

$$\begin{aligned}\varphi(x) &= 1 \quad \text{if } x = 5, 6, 7, \ldots\\ &= \frac{3}{5} \quad \text{if } x = 4\\ &= 0 \quad \text{if } x = 0, 1, 2, 3.\end{aligned}$$

The power of this MP test is

$$\begin{aligned}E[\varphi(x) \mid H_1] &= 1 \sum_{x=5}^{\infty}\left(\frac{1}{4}\right)\left(\frac{3}{4}\right)^x + \frac{3}{5}\left(\frac{1}{4}\right)\left(\frac{3}{4}\right)^4\\ &= \left(\frac{3}{4}\right)^5 + \frac{3}{5}\left(\frac{3}{4}\right)^4 \frac{1}{4}.\end{aligned}$$

EXAMPLE 8.3.3. Let X be Cauchy with location θ and known scale $\sigma = 1$ and $H_0 : \theta = 0$ and $H_1 : \theta = 1$. We show that the test $\varphi(x) = 1$ if $1 < x < 3$ and zero otherwise is MP test of its size. Note that

$$\lambda(x) = \frac{L_1(x)}{L_0(x)} = \frac{1 + x^2}{2 + x^2 - 2x}.$$

From NP Lemma (Part A), if we show that the given test $\varphi(x)$ corresponds to a value k then the result is proved. Since $\lambda(x)$ is a continuous function, the region $\lambda(x) > k$ is defined by $x^2(1 - k) + 2kx + (1 - 2k) > 0$. The set $E_2 = \{x \mid \lambda(x) = k\}$ consists of two roots of the quadratic equation $x^2(1 - k) + 2kx + (1 - 2k) = 0$ and $P(E_2) = 0$. Thus we need to show that $E_1 = \{x \mid 1 < x < 3]$ corresponds to a value of k such that $\{x \mid x^2(1 - k) + 2kx + (1 - 2k) > 0\} = E_1$. Now another way to describe $E_1 = \{x \mid (3 - x)(x - 1) > 0\}$ or $E_1 = \{x \mid -x^2 + 4x - 3 > 0\}$. Comparing two quadratic equations if we take $k = 2$ then $\varphi_2(x)$ corresponds to the critical region $(1, 3)$. The level of this test is same as size given by

$$P[1 < X < 3 \mid H_0] = \frac{1}{\pi}\int_1^3 \frac{dx}{(1 + x^2)} = \frac{1}{\pi}[\tan^{-1}(3) - \tan^{-1}(1)].$$

The power of $\varphi_2(x) = E[\varphi_2(x) \mid H_1]$ $= \frac{1}{\pi}\int_1^3 \frac{dx}{1 + (x - 1)^2}$

$$= \frac{1}{\pi}\int_0^2 \frac{du}{1 + u^2} = \frac{1}{\pi}\tan^{-1}(2).$$

In general the level α, the maximum tolerable type I error is prescribed in the context of the problem and we determine the MP level test by determining corresponding (k, γ) which we will denote by $(k_\alpha, \gamma_\alpha)$. As seen from the

Theorem 8.3.2, N-P lemma Part (B), the determination of $(k_\alpha, \gamma_\alpha)$ depends on the distribution of $Y = L_1(X)/L_0(X)$ under H_0. We now give below a few examples to show that the distribution of Y under H_0 can be complicated and consequently the determination of $(k_\alpha, \gamma_\alpha)$ may be not so easy.

EXAMPLE 8.3.4. Consider a random sample of size one and we want to obtain MP test for $H_0 : X \sim N(0, 1)$ vs $H_1 : X \sim C(0, 1)$. Then $L_1(x) = \frac{1}{\pi}\frac{1}{1+x^2}$, $x \in R_1$ and $L_0(x) = \frac{1}{\sqrt{2\pi}} e^{-x^2/2}$, $x \in R_1$. Note that here $S_0 = S_1 = R_1$ and $\lambda(x) = \frac{\exp(x^2/2)}{1+x^2}\sqrt{2/\pi}$. As $\lambda(x)$ depends on x only through x^2, we have $\lambda(x) = \lambda(-x)$, $\forall\, x \in R_1$ and if $x \in E_1 = \{x \mid \lambda(x) > k\}$ then $-x \in E_1$ and E_1 is symmetric about origin. Again $P(\lambda(x) = k \mid H_0) = 0$; since for any fixed k, $\exp\left(\frac{x^2}{2}\right) = k\sqrt{\pi/2}(1 + x^2)$ holds for only a finite number of points and can not hold for x in an interval (a, b). Thus the MP level α test is given by $\varphi_k(x) = 1$ if $\frac{\exp\{x^2/2\}}{1+x^2} > c$, where c is determined by $E[\varphi_k(x) \mid H_0] = \alpha$.

To obtain the critical region corresponding to $\varphi_k(x)$ we study the function $\exp\{x^2/2\}/(1 + x^2)$ in detail. We can restrict attention to $x > 0$ and consider equivalent variable $y = x^2$. Let $u(y) = \exp(y/2)/(1 + y)$, $y > 0$. Then

$$u'(y) = -\exp(y/2)/(1+y)^2 + \frac{1}{2}\exp(y/2)/(1+y) = \frac{\exp(y/2)}{2(1+y)^2}(y-1)$$

$u'(y) = 0$ at $y = 1$ and $u'(y) < 0$ for $0 < y < 1$ and $u'(y) > 0$ for $y > 1$. Hence $u(y)$ for $y \geq 0$ attains the minimum value at $y = 1$ a local maximum at $y = 0$ and goes to ∞ as $y \to \infty$. Returning back to x we have $\lambda(0) = .7979$, $\lambda(1) = .6527$ and $\lambda(x)$ is decreasing in $(0, 1)$ and increasing in $(1, \infty)$. The behaviour of $\lambda(x)$ for $x < 0$ is exactly the mirror image having minimum $\lambda(-1) = .6527$ and $\lambda(x) \to \infty$ as $x \to -\infty$.

The horizontal line $\lambda(x) = k$ intersects the graph of $\lambda(x)$,

(i) in two points if $k > .7979$
(ii) in three points if $k = .7979$
(iii) in four points if $k \in (.6527\ .7979)$
(iv) in two points if $k = .6527$
(v) in no point if $k < .6527$

Thus the MP test $\varphi_k(x)$ has size given by

$$E[\varphi_k(x) \mid H_0] = 1 \text{ if } 0 \leq k \leq .6527.$$

If $k \in (.6527, .7979)$ then

$$E[\varphi_k(x) \mid H_0] = P[-c_1 < X < c_1 \mid H_0] + P[X < -c_2 \mid H_0] + P[X > c_2 \mid H_0]$$

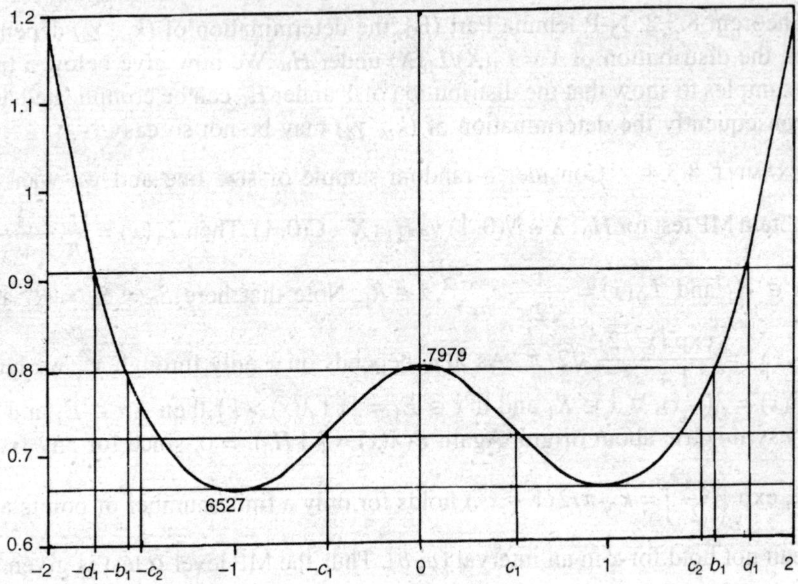

Fig. 8.1 Graph of $\lambda(x)$ $\sqrt{\frac{2}{\pi}}$ exp $(x^2/2)/(1 + x^2)$.

and

$$E[\varphi_k(x) \mid H_0] = P[X < -d_1] + P[X > d_1 \mid H_0] \text{ if } k > .7979.$$

For the standard level $\alpha = .05$ the MP test will have critical region $| x | > d_1$ provided $\lambda(d_1) > .7979$. Now from normal tables for $\alpha = .05$, $d_1 = 1.96$ with $\lambda(d_1) = 1.2124 > .7979$. The power of the MP test is $1 - \frac{2}{\pi} \tan^{-1}(1.96)$.

EXAMPLE 8.3.5. We now give an example where the r.v. X is continuous under H_0 and H_1 but $\lambda(x)$ is discrete. Let X be $U(0, 1)$ under H_0 and its pdf under H_1 be given

$$f(x) = \frac{4}{3}, \quad 0 < x < 1/4 \text{ or } \frac{3}{4} < x < 1$$

$$= \frac{2}{3}, \quad \frac{1}{4} \le x \le \frac{3}{4}$$

Here

$$\lambda(x) = \frac{2}{3}, \quad \frac{1}{4} \le x \le 3/4$$

$$= \frac{4}{3}, \quad 0 < x < \frac{1}{4} \text{ or } \frac{3}{4} < x < 1.$$

Then $y = \lambda(x)$ has d.f. under H_0 given by

$$G_0(y) = 0 \text{ if } y < 2/3$$

$$= \frac{1}{2} \text{ if } \frac{2}{3} \le y < \frac{4}{3}$$

$$= 1 \text{ if } y \ge \frac{4}{3}.$$

The pmfs of Y under H_0 and H_1 are given by

Y	$\frac{2}{3}$	$\frac{4}{3}$
H_0	$\frac{1}{2}$	$\frac{1}{2}$
H_1	$\frac{1}{3}$	$\frac{2}{3}$

The MP level α test given by N-P lemma can be obtained by determining $(k_\alpha, \gamma_\alpha)$ such that

$$1 - P[Y \le k_\alpha] + \gamma P[y = k_\alpha] = \alpha$$

or

$$1 - G(k_\alpha) + \gamma P[Y = k_\alpha] = \alpha.$$

For example if $\alpha = .05$ we must have $(k_\alpha, \gamma_\alpha)$ such that under H_0: $P[Y < k_\alpha] \le .95 < P[Y \le k_\alpha]$. Using $G_0(y)$ defined above we must have $k_\alpha = 4/3$ and therefore $\gamma \cdot \frac{1}{2} = .05$ or $\gamma = .10$. Thus, the MP test rejects H_0 when $x \in \left(0, \frac{1}{4}\right) \cup \left(\frac{3}{4}, 1\right)$ with probability .10 and otherwise accepts H_0. The power of this test is $.10\left(\frac{2}{3}\right) = .0667$.

EXAMPLE 8.3.6. In all the examples considered above we had $S_0 = S_1$. We now consider a situation in which $S_0 \ne S_1$. Let $(X_1, \ldots, X_n)$ be i.i.d. $U(0, \theta)$ and let $H_0 : \theta = \theta_0$ and $H_1 : \theta = \theta_1 > \theta_0$. Here

$$L(x, \theta_0) = \frac{1}{\theta_0^n} \Pi\, I_{(0,\theta_0)}(x_i) = L_0(x) \text{ and}$$

$$L(x, \theta_1) = \frac{1}{\theta_1^n} \Pi\, I_{(0,\theta_1)}(x_i) = L_1(x).$$

Note that $S_0 \cap S_1 = (0, \theta_0)$ as $\theta_1 > \theta_0$ and $S_0^c \cap S_1 = (\theta_0, \theta_1)$. Thus if any of the observed x_i's exceed θ_0 then the MP test must reject H_0. This is equivalent to defining $\varphi(x) = 1$ if $x_{(n)} > \theta_0$. Recall that $T = x_{(n)}$ is minimal sufficient for $U(0, \theta)$ and as such, $L(x, \theta) = g(t, \theta)\, h(x)$, where $g(t, \theta) = \frac{nt^{n-1}}{\theta^n} I_{(0,\theta)}(t)$. Hence

$$\frac{L_1(x)}{L_0(x)} = \frac{L(x, \theta_1)}{L(x, \theta_0)} = \frac{g(t, \theta_1)}{g(t, \theta_0)}.$$

The MP test thus can be defined in terms of t only, and we consider $\lambda(t) = \dfrac{g(t, \theta_1)}{g(t, \theta_0)}$ over $0 < t < \theta_0$. Then $\lambda(t) = \left(\dfrac{\theta_0}{\theta_1}\right)^n$ for $0 < t < \theta_0$ and is constant. Here $\lambda(t)$ takes only one value under H_0, $\left(\dfrac{\theta_0}{\theta_1}\right)^n$ with probability one and under H_A over $(0, \theta_0)$ it takes the same value with probability $\left(\dfrac{\theta_0}{\theta_1}\right)^n$. Note that under H_A over $[\theta_0, \theta_1)$, $\lambda(t)$ is not defined but could be taken as $+\infty$ (an improper value) which is consistent with our rule that the MP test rejects H_0 i.e. $\varphi(t) = 1$, if $t > \theta_0$. To determine MP level α test we must determine $(k_\alpha, \gamma_\alpha)$ such that

$$1 - P[\lambda(t) \le k_\alpha \mid H_0] + \gamma P[\lambda(t) = k_\alpha] = \alpha$$

The d.f. of $\lambda(t)$ is given by

$$G_0(y) = 0 \quad \text{if } y < \left(\frac{\theta_0}{\theta_1}\right)^n$$

$$= 1 \quad \text{if } y \ge \left(\frac{\theta_0}{\theta_1}\right)^n$$

We must take $k_\alpha = \left(\dfrac{\theta_0}{\theta_1}\right)^n$ and then $\gamma = \alpha$.

Thus the MP test $\varphi^*(t) = 1$ if $t > \theta_0$

$= \alpha$ if $0 < t < \theta_0$.

$$\text{The power of } \varphi^*(t) = \alpha \int_0^{\theta_0} \frac{nt^{n-1}}{\theta_1^n}\, dt + \int_{\theta_0}^{\theta_1} \frac{nt^{n-1}}{\theta_1^n}\, dt$$

$$= \alpha \left(\frac{\theta_0}{\theta_1}\right)^n + 1 - \left(\frac{\theta_0}{\theta_1}\right)^n$$

$$= 1 - \left(\frac{\theta_0}{\theta_1}\right)^n (1 - \alpha). \qquad \text{(A)}$$

Now consider a test φ_1 defined as:

$$\varphi_1(t) = 1 \text{ if } t > \theta_0$$

$$= \gamma(t) \text{ if } 0 < t < \theta_0$$

where $0 \le \gamma(t) < 1$ is such that

$$\int_0^{\theta_0} \gamma(t) \frac{nt^{n-1}}{\theta_1^n}\, dt = \alpha \qquad \text{(B)}$$

$$\text{The power of } \varphi_1(t) = \int_{\theta_0}^{\theta_1} \frac{nt^{n-1}}{\theta_1^n}\,dt + \int_0^{\theta_0} \gamma(t)\,\frac{nt^{n-1}}{\theta_1^n}\,dt$$

$$= 1 - \left(\frac{\theta_0}{\theta_1}\right)^n + \left(\frac{\theta_0}{\theta_1}\right)^n \int_0^{\theta_0} \gamma(t)\,\frac{nt^{n-1}}{\theta_0^n}\,dt$$

$$= 1 - \left(\frac{\theta_0}{\theta_1}\right)^n (1-\alpha) \qquad \text{in view of (B).}$$

This example shows that there exists several level α tests given by (B) whose power is same as that of MP test. For example consider a specific choice of $\gamma(t)$ given by

$$\gamma(t) = 0 \text{ if } 0 < t < a$$

$$= 1 \text{ if } a \le t < \theta_0$$

Then a is given by the condition that $\int_a^{\theta_0} \frac{nt^{n-1}}{\theta_1^n}\,dt = \alpha$ or $1 - \left(\frac{a}{\theta_0}\right)^n = \alpha$ or $a = \theta_0(1-\alpha)^{1/n}$ and

$$\varphi_\alpha(t) = 1 \text{ if } t > a = \theta_0(1-\alpha)^{1/n}$$

$$= 0 \quad \text{otherwise}$$

is a level α test and has same power as the MP test given by N-P lemma. Similarly if we take

$$\gamma(t) = 1 \text{ if } 0 < t < b = \theta_0\,\alpha^{1/n}$$

and zero otherwise this will lead to a level α test having the same power as the MP test given by N-P lemma.

This example shows that in general MP level α test is not unique.

EXAMPLE 8.3.7. Here we consider the same distributional set up as in the last example but consider $H_0 : \theta = \theta_0$ and $H_A : \theta = \theta_1 < \theta_0$. Here $S_0 = (0, \theta_0)$ and $S_1 = (0, \theta_1) \subset S_0$. Hence for any $x_i \in S_0 \cap S_1^c$ i.e. $x_i \in (\theta_1, \theta_0)$ we accept H_0 or $\varphi(x) = 0$ if any $x_i > \theta_1$ which is equivalent to $t = \max_i (x_i) = x_{(n)} > \theta_1$. Thus $\varphi(t) = 0$ if $t \in (\theta_1, \theta_0)$. Now in this case

$$\lambda(t) = \left(\frac{\theta_0}{\theta_1}\right)^n \qquad \text{for } 0 < t < \theta_1$$

$$= 0 \qquad \text{for } \theta_1 < t < \theta_0$$

We leave it as an exercise to the reader (using the technique outlined in the above example) to show that any test $\varphi(t)$ such that

$$\begin{aligned}\varphi(t) &= \gamma(t) && 0 < t < \theta_1 \\ &= 0 && \theta_1 \le t < \theta_0\end{aligned}$$

is MP level α test where $0 \le \gamma(t) \le 1$ is chosen such that $\int_0^{\theta_1} \gamma(t) \frac{nt^{n-1}}{\theta_0^n} dt = \alpha$.

The power of this test is $\left(\frac{\theta_1}{\theta_0}\right)^n \alpha$ and one possible choice of $\gamma(t)$ is

$$\begin{aligned}\gamma(t) &= 1 && \text{if } 0 < t < \theta_0(\alpha)^{1/n} \\ &= 0 && \text{otherwise.}\end{aligned}$$

Note that if $\theta_1 < \theta_0(\alpha)^{1/n}$, the power of this test is one.

The above examples show that MP test given by N-P lemma parts (A) and (B) is not unique. However, we show in the next theorem that it is essentially unique.

THEOREM 8.3.3 [N-P lemma (Part C)]

For every $\alpha \in (0, 1]$, the MP test given by N-P lemma part (B) is essentially unique in that if $\varphi_{k_\alpha}(x)$ and $\varphi'(x)$ are both MP level α tests given by N-P lemma then $\varphi_{k_\alpha}(x) \ne \varphi'(x)$ only on set $E_2 = \{x \mid L_1(x) = k_\alpha L_0(x)\}$. Note that $E[\varphi_{k_\alpha}(x) \mid H_0] = E[\varphi'(x) \mid H_0]$ and $E[\varphi_{k_\alpha}(x) \mid H_1] = E[\varphi'(x) \mid H_1]$ and therefore if

$$Q(x) = [\varphi_{k_\alpha}(x) - \varphi'(x)]\,[L_1(x) - k_\alpha L_0(x)]$$

then $\int Q(x) = 0$. However as seen before $Q(x) \ge 0$ and therefore $Q(x) = 0$. Thus $\varphi_{k_\alpha}(x) = \varphi'(x_1)$ on E_1 and E_3. However, since on E_2, $L_1(x) = k_\alpha L_0(x)$ we can define $\varphi_{k_\alpha}(x) \ne \varphi'(x)$ for which $Q(x) = 0$ and MP level α tests can differ only on E_2.

Remark 8.3.1: Using N-P lemma and Neyman factorizability criterion we can show that if under H_0 the joint pdf is $L(x, \theta_0)$ and under H_1 it is $L(x, \theta_1)$. Then $L(x, \theta_1) \gtreqless k\, L(x, \theta_0)$ according as $g(t, \theta_1) \gtreqless k\, g(t, \theta_0)$ where T is minimal sufficient statistic for $\{L(x, \theta), \theta \in \Omega\}$. Therefore the regions E_1, E_2, E_3 depend on x only through T and if $\varphi(x)$ is MP test then $\varphi_k(t)$ the corresponding test defined in terms of minimal sufficient statistic is MP test of same level (size) and same power. Another way to look at this situation is if $\varphi(x)$ is any test and we define $\psi_\theta(t) = E_\theta[\varphi(x) \mid T = t]$. Then as $0 \le \varphi(x) \le 1$, $0 \le \psi_\theta(t) \le 1$. Further as T is minimal sufficient $\psi_\theta(t)$ does not depend on θ and equals $\psi(t)$ which is a test function depending on observations only through minimal sufficient statistic T and $E[\varphi(x) \mid \theta] = E_\theta\{E[\varphi(x) \mid T = t]\}$. Thus for any test function $\varphi(x)$ there is an equivalent test function $\psi(t)$ which depends on x only through minimal sufficient statistic T.

In the next section we show that for one parameter exponential family the MP test given by N-P lemma also provides a solution to more complex problems such as testing $H_0 : \theta = \theta_0$ against $H_1 : \theta > \theta_0$ or $H_0 : \theta \leq \theta_0$ against $H_1 : \theta > \theta_0$

Exercise 8.3. (i) Show that for a MP level α test, power $\geq$ level [Hint : take $\varphi'(x) \equiv \alpha$, $\forall\, x \in S_0 \cup S_1$].

(ii) Let $H_0 \sim U(0, 1)$ and $H_1 : X \sim f_1(x)$ where

$$f_1(x) = 4x, \quad 0 < x < 1/2$$

$$= 4 - 4x, \quad \frac{1}{2} \leq x < 1.$$

Obtain MP level α test and its power.

(iii) Let $(X_1, \ldots, X_n)$ be i.i.d. exponential with location θ i.e. $f(x, \theta) = \exp\{-(x - \theta)\}$, $x \geq \theta$. Obtain MP level α test to test $H_0 : \theta = \theta_0$ against $H_1 : \theta = \theta_1$. Discuss the two cases $\theta_1 > \theta_0$ and $\theta_1 < \theta_0$ separately.

(iv) Let $(X_1, \ldots, X_n)$ be i.i.d. Pareto with shape parameter λ i.e. $f(x, \lambda) = \dfrac{\lambda}{x^{\lambda+1}}$, $x \geq 1$. Obtain MP level α test for testing $H_0 : \lambda = \lambda_0$ against $H_1 : \lambda = \lambda_1$. Discuss the two cases $\lambda_1 > \lambda_0$ and $\lambda_1 < \lambda_0$ separately.

(v) On the basis of a single observation, using NP lemma obtain MP level α test for testing $H_0 : X \sim N(0, 1)$ against $H_1 : X \sim DE(0, 1)$. Obtain power of this test and verify that power > level.

(vi) Let $X \sim N(0, 1)$ under H_0 and $N(0, 2)$ under H_1. Suppose $\varphi(x) = 1$ if $|x| \geq 1$ and zero otherwise. Is $\varphi(x)$ MP test of level equal to its size?

(vii) Let X be a discrete r.v. with pmf under H_0 and H_1 given by

$$p_0(x) = .05 \text{ for } x = 1, 2, \ldots, 20$$

$$p_1(x) = .60 \text{ for } x = 1,\ p_1(x) = .15 \text{ for } x = 2, 3$$

and $$p_1(x) = \frac{.10}{17} \text{ for } x = 4, 5, \ldots, 20.$$

Define $\varphi_1(x) = 1$ for $x = 1, 2$ and zero otherwise and $\varphi_2(x) = 1$ if $x = 1$, and $\varphi_2(x) = 1/2$ if $x = 2, 3$, and zero otherwise. Show that both φ_1, φ_2 are MP test of level $\alpha = .10$ and power .75. Do φ_1, φ_2 satisfy NP-lemma?

(viii) Let X be a non-negative r.v. with pdf under H_0 given be Weibull density $f_0(x) = xe^{-x^2/2}$, $x > 0$ and under H_1 given by $f_1(x) = \sqrt{\dfrac{2}{\pi}}\, e^{-x^2/2}$, $x > 0$, (folded normal). Based on a single observation obtain MP level α test and its power.

(ix) Let $f(x, \theta) = \theta(2x) + (1 - \theta)\, 2(1 - x)$, $0 < x < 1$, $0 \leq \theta \leq 1$. On the basis of a single observation obtain MP level $\alpha = .05$ test to test (a) $H_0 : \theta = 1/4$, $H_1 : \theta = 3/4$, (b) $H_0 : \theta = 3/4$, $H_1 : \theta = 1/4$, (c) $H_0 : \theta = 1$, $H_1 : \theta = 1/2$.

8.4 Uniformly Most Powerful (UMP) Tests

Suppose we want to test $H_0 : \theta = \theta_0$ against $H_A : \theta > \theta_0$. Here null hypothesis is simple and alternative is composite. Let $\mathbb{D}_\alpha = \{\varphi \mid E[\varphi(x) \mid \theta_0] \leq \alpha\}$ be the class of all level α tests. A test $\varphi^* \in \mathbb{D}_\alpha$ is called UMP if

$$E[\varphi^* \mid \theta] \geq E[\varphi \mid \theta],\ \forall\, \theta > \theta_0,\ \forall\, \varphi \in \mathbb{D}_\alpha \qquad (8.4.1)$$

Let $\beta_\varphi(\theta)$ denote $E[\varphi \mid \theta]$ which we call as the power function of the test φ. Note that $\beta_\varphi(\theta_0)$ is the type I error and $\beta_\varphi(\theta_1)$ for $\theta_1 > \theta_0$ is the power at θ_1 or $1 - \beta_\varphi(\theta_1)$ is the type II error at θ_1. We now show how N-P lemma can be used to obtain UMP test by way of an example.

EXAMPLE 8.4.1. Let $(X_1, \ldots, X_n)$ be a random sample of the size n from $\{N(\theta, 1), \theta \in R_1\}$. Suppose we want to test $H_0 : \theta = \theta_0$ against $H_A : \theta > \theta_0$. Consider a sub-problem of testing $H_0 : \theta = \theta_0$ vs $H'_A : \theta = \theta_1 > \theta_0$ a specific alternative. Then using NP lemma we have MP level α test for the subproblem is given by

$$\varphi_1(\bar{x}) = 1 \quad \text{if } \bar{x} > \theta_0 + \frac{\xi_{1-\alpha}}{\sqrt{n}}$$

$$= 0 \quad \text{otherwise}$$

and its power function is

$$\beta_\varphi(\theta) = 1 - \Phi[\sqrt{n}(\theta_0 - \theta) + \xi_{1-\alpha})]$$

where $\xi_{1-\alpha}$ is $100(1-\alpha)\%$ point of $N(0, 1)$. Observe that the MP test φ_1 does not depend on the particular alternative θ_1 chosen for the sub-problem. Therefore φ_1 is MP level α test for $H_0 : \theta = \theta_0$ vs $H'_A : \theta = \theta_1$ for each $\theta_1 > \theta_0$ and therefore we conclude that φ_1 is UMP level α test for the original problem $H_0 : \theta = \theta_0$ vs $H_A : \theta > \theta_0$.

We also note that $\beta_\varphi(\theta)$ is monotone increasing function of θ as $\Phi[\sqrt{n}(\theta_0 - \theta) + \xi_{1-\alpha}]$ is decreasing function of θ. Further $\lim_{\theta\to\infty} \beta_\varphi(\theta) = 1$ and $\beta_\varphi(\theta_0) = \alpha$.

EXAMPLE 8.4.2. Consider Example 8.3.6 of testing $H_0 : \theta = \theta_0$ against $H'_A : \theta = \theta_1 > \theta_0$ in $\{U(0, \theta)\ \theta > 0\}$. Here we showed that the MP level α test is given by

$$\varphi_1(t) = 1 \text{ if } t > \theta_0$$

$$= \gamma(t) \text{ if } 0 < t < \theta_0$$

where $\gamma(t)$ satisfies the condition $E[\gamma(t) \mid \theta_0] = \alpha$. We have also seen that the above MP level α test is not unique.

Since φ_1 does not depend on the particular value of θ_1 but its relative position w.r.t. θ_0, namely $\theta_1 > \theta_0$ we have φ_1 is UMP level α test for testing $H_0 : \theta = \theta_0$ against $H_A : \theta > \theta_0$. This example also shows that the UMP level α test is not necessarily unique.

We now show that the results of Example 8.4.1 for the $N(\theta, 1)$ hold in general for one parameter exponential family.

Let $(X_1, \ldots, X_n)$ be random sample of size n from pdf belonging one parameter exponential family so that

$$\log L(x, \theta) = u(\theta) \sum K(x_i) + nv(\theta) + \sum w(x_i).$$

Suppose we are interested in testing $H_0 : \theta = \theta_0$ against $H_A : \theta > \theta_0$ then consider the subproblem $H_0 : \theta = \theta_0$ vs $H'_A : \theta = \theta_1 > \theta_0$. Note that here $S_0 = S_1 = S$ and therefore

$$\log \lambda (x) = \log \frac{L_1(x)}{L_0(x)} = \sum K(x_i)[u(\theta_1) - u(\theta_0)] + n[v(\theta_1) - v(\theta_0)]$$

and by N-P lemma the MP level α test for H_0 against H'_A, assuming that $u(\theta_1) > u(\theta_0)$, is given by

$$\begin{aligned} \varphi_1(x) &= 1 \quad \text{if } \sum K(x_i) = T > k_\alpha \\ &= \gamma_\alpha \quad \text{if } \sum K(x_i) = T = k_\alpha \\ &= 0 \quad \text{if } \sum K(x_i) = T < k_\alpha. \end{aligned} \tag{8.4.2}$$

If $u(\theta_1) < u(\theta_0)$ then the MP level α test is given by

$$\begin{aligned} \varphi_2(x) &= 1 \quad \text{if } \sum K(x_i) = T < k'_\alpha \\ &= \gamma'_\alpha \quad \text{if } \sum K(x_i) = T = k'_\alpha \\ &= 0 \quad \text{if } \sum K(x_i) = T > k'_\alpha. \end{aligned} \tag{8.4.3}$$

Here $T = \sum K(x_i)$ is minimal sufficient statistic for $\{L(x, \theta), \theta \in \Omega\}$. Further for the test given by (8.4.2) k_α and γ_α are determined by the condition

$$1 - P_{\theta_0}[T \le k_\alpha] + \gamma_\alpha P_{\theta_0}[T = k_\alpha] = \alpha \tag{8.4.4}$$

If $u(\theta_1) < u(\theta_0)$ then the MP test is given by (8.4.3) and k'_α and γ'_α are determined by

$$P_{\theta_0}[T \le k_\alpha] - P_{\theta_0}[T = k_\alpha] + \gamma'_\alpha P_{\theta_0}[T = k_\alpha] = \alpha \tag{8.4.5}$$

In both cases the MP test depends on the distribution of T under H_0 and whether $u(\theta_1) > u(\theta_0)$ or $u(\theta_1) < u(\theta_0)$ and not on the specific choice of θ_1. Now as $\frac{du}{d\theta} \neq 0$, $u(\theta)$ is increasing or decreasing. In case $u(\theta)$ is increasing $u(\theta_1) > u(\theta_0)$ and MP level α test for H_0 against H'_A is given by φ_1 i.e. (8.4.2). Since φ_1 does not depend on the specific value of θ_1 as for any $\theta_1 \in H_A$, $u(\theta_1) - u(\theta_0) > 0$, φ_1 is UMP level α test for testing H_0 against H_A. On the other hand if $u(\theta)$ is decreasing then $u(\theta_1) < u(\theta_0)$ for every $\theta_1 \in H_A$ and φ_2 given by (8.4.3) is UMP level α test for testing H_0 against H_A.

Similarly, we can show that for testing $H_0 : \theta = \theta_0$ against $H_A : \theta < \theta_0$ if $u(\theta)$ is increasing i.e. $u(\theta_1) < u(\theta_0)$, φ_2 is UMP level α test whereas if $u(\theta)$ is decreasing i.e. $u(\theta_1) > u(\theta_0)$ φ_1 will be UMP level α test for the problem.

We next show that the power function of these UMP tests are monotone. We consider the case where $u(\theta)$ is increasing and we are testing $H_0 : \theta = \theta_0$

against $H_A : \theta > \theta_0$. As seen earlier in this section, the UMP test in this case is given by φ_1 as defined in (8.4.2). The power function of this test is given by

$$\beta_{\varphi_1}(\theta) = 1 - P[T \le k_\alpha \mid \theta] + \gamma_\alpha P[T = k_\alpha \mid \theta] \qquad (8.4.6)$$

We will show that $\beta_{\varphi_1}(\theta_2) \ge \beta_{\varphi_1}(\theta_1)$ if $\theta_2 > \theta_1$ so that $\beta_{\varphi_1}(\theta_1)$ is increasing. Consider testing $H_0 : \theta = \theta_1$ against $H_A : \theta = \theta_2$ where $\theta_1 < \theta_2$. Then as $u(\theta_2) > u(\theta_1)$ the MP level $\alpha_1 = [\beta_{\varphi_1}(\theta_1)]$ test is given by

$$\begin{aligned} \varphi_1'(x) &= 1 \quad \text{if } T > c_\alpha \\ &= \delta_\alpha \quad \text{if } T = c_\alpha \\ &= 0 \quad \text{if } T < c_\alpha \end{aligned}$$

where c_α, δ_α are determined by

$$1 - P[T \le c_\alpha \mid \theta_1] + \delta_\alpha P[T = c_\alpha \mid \theta_1] = \beta_{\varphi_1}(\theta_1).$$

This implies that $k_\alpha = c_\alpha$ and $\gamma_\alpha = \delta_\alpha$ and φ_1' is same test as φ_1 or $\varphi_1' \equiv \varphi_1$. But $\beta_{\varphi_1'}(\theta_2) \ge \beta_{\varphi_1'}(\theta_1)$ as the power of a MP test is at least as large as its level as seen in Exercise 8.3(i). But as $\varphi_1' \equiv \varphi_1$ we have $\beta_{\varphi_1}(\theta_1) \le \beta_{\varphi_1}(\theta_2)$ and the power function of $\beta_{\varphi_1}(\theta)$ is increasing. Similarly if $u(\theta)$ is decreasing then one can show that the UMP test φ_2 for this problem has power function increasing in θ. The remaining cases of $H_0 : \theta = \theta_0$ vs $H_A : \theta < \theta_0$ and $u(\theta)$ increasing (decreasing) can be handled in a similar way. For example consider H_0: $\theta = \theta_0$ vs $H_A : \theta < \theta_0$ and $u(\theta)$ increasing. Then UMP level α test is given by φ_2 as given in (8.4.2) with constant (k_1', γ_α') given by (8.4.5). Now consider testing $H_0 : \theta = \theta_1$ against $H_A : \theta = \theta_2 < \theta_1$ at level $\beta_{\varphi_2}(\theta_1)$. Then MP test of level $\beta_{\varphi_2}(\theta_1)$ is given by

$$\begin{aligned} \varphi_2'(x) &= 1 \quad \text{if } T < c_\alpha \\ &= \delta_\alpha \quad \text{if } T = c_\alpha \\ &= 0 \quad \text{if } T > c_\alpha \end{aligned}$$

Here c_α, δ_α are given by

$$\beta_{\varphi_2}(\theta_1) = P_\theta[T \le c_\alpha \mid \theta_1] - P[T = c_\alpha \mid \theta_1] + \delta_\alpha P[T = c_\alpha \mid \theta_1]$$

which is the power of the test φ_2 at θ_1. This implies that $c_\alpha = k_\alpha'$ and $\gamma_\alpha' = \delta_\alpha$ and $\varphi_2' \equiv \varphi_2$. However $\beta_{\varphi_2'}(\theta_2) \ge \beta_{\varphi_2'}(\theta_1)$ and thus $\beta_{\varphi_2}(\theta_1) \le \beta_{\varphi_2}(\theta_2)$. In case T is continuous r.v. $P[T = k_\alpha \mid \theta] = 0$ and the proofs are simpler. The UMP test given by φ_1 is called as a right tailed test and φ_2 is called as left tailed test. The proofs of the remaining two cases are similar and are left to reader as an exercise.

EXAMPLE 8.4.3. Let $(X_1, \ldots, X_n)$ be i.i.d. Pareto wih pdf $f(x, \lambda) = \frac{\lambda}{x^{\lambda+1}}$, $x > 1$, $\lambda > 0$. As seen before $\{f(x, \lambda), \lambda > 0\}$ is one parameter exponential family with $u(\lambda) = -(\lambda + 1)$ and $T = \sum_{i=1}^{n} \log x_i$ as minimal sufficient statistic with $g_n(t, \lambda) = \frac{\lambda^n}{\Gamma(n)} e^{-\lambda t} \cdot t^{n-1}$, $t > 0$, $\lambda > 0$. Now $u'(\lambda) = -1$ and $u(\lambda)$ is decreasing and as T is continuous r.v. the UMP level α test for testing $H_0 : \lambda = 1$ vs $H_A : \lambda < 1$ is given by $\varphi(x) = 1$ if $T > t_\alpha$ and zero otherwise where t_α is given by $\int_{t_\alpha}^{\infty} g_n(t, 1) = \alpha$ or $1 - G_n(t_\alpha) = \alpha$ where $G_n(u)$ is the d.f. of $G(n, 1)$ r.v. Thus $t_\alpha = \eta_{n,1-\alpha}$ the 100 $(1 - \alpha)$% point of $G(n, 1)$. The power function of this test is given by

$$\int_{\eta_{n,1-\alpha}}^{\infty} g_n(t, \lambda)\, dt = 1 - G_n(\lambda\, \eta_{n,1-\alpha}) = \beta_\varphi(\lambda).$$

Now $\frac{\partial \beta_\varphi(\lambda)}{\partial \lambda} = -\, g_n(\lambda \eta_{n,1-\alpha})\, \eta_{n,1-\alpha} < 0$ as $\eta_{n,1-\alpha} > 0$. Therefore β_φ is decreasing. On the other hand if $H_0 : \lambda = 1$ and $H_A : \lambda > 1$ then MP level α test is given by $\varphi'(x) = 1$ if $T < k'_\alpha$ and zero otherwise where $G_n(k'_\alpha) = \alpha$ or $k'_\alpha = \eta_{n,\alpha}$ i.e. M.P. level α test is left tailed. The power function of this test is given by $\beta_{\varphi'}(\lambda) = G_n(\lambda \eta_{n,\alpha})$ and is increasing in λ.

EXAMPLE 8.4.4. Consider $(X_1, \ldots, X_n)$ as i.i.d. $b(1, \theta)$. Then $\{L(x, \theta), \theta \in (0, 1)\}$ is one parameter exponential family with $u(\theta) = [\log \theta - \log (1 - \theta)]$ and $T = \sum X_i$ as minimal sufficient statistic. Here $u'(\theta) = \frac{1}{\theta(1-\theta)} > 0$ and therefore $u(\theta)$ is increasing. Suppose we want to test $H_0 : \theta = \theta_0$ against $H_A : \theta > \theta_0$ then UMP level α test is given by

$$\begin{aligned} \varphi(x) &= 1 \quad \text{if } T > r_0 \\ &= \gamma_\alpha \quad \text{if } T = r_0 \\ &= 0 \quad \text{if } T < r_0 \end{aligned}$$

where r_0, γ_α are given by

$$\sum_{r_0+1}^{n} \binom{n}{r} \theta_0^r (1 - \theta_0)^{n-r} + \gamma_\alpha \binom{n}{r_0} \theta_0^{r_0} (1 - \theta_0)^{n-r_0} = \alpha.$$

The power function of this test is given by

$$\beta_\varphi(\theta) = \sum_{r_0+1}^{n} \binom{n}{r} \theta^r (1 - \theta)^{n-r} + \gamma_\alpha \binom{n}{r_0} \theta^{r_0} (1 - \theta_0)^{n-r_0} \text{ for } \theta > \theta_0$$

The reader can verify that $\beta_\varphi(\theta)$ is increasing in θ.

Exercise 8.4. (i) (Inverse binomial sampling). In a production process rather than taking n items and finding out number of items which are defective, we use the inverse sampling procedure. Here we continue testing items until m_0 defective items are found where m_0 is a pre-assigned number. The total number of items thus examined would be $n + m_0$ where n is the number of non-defective items in the sample. In usual binomial sampling the number of items examined is fixed and number defective items is random. For inverse binomial sampling the number of defective items is fixed and number of items examined is random. The pmf is $P[X = x] = \binom{x + m_0 - 1}{x} \theta^{m_0}(1-\theta)^x$, $x = 0, 1, 2, \ldots, 0 < \theta < 1$. Show that the distribution belongs to one parameter exponential family with $u(\theta)$ decreasing and where x is minimal sufficient. For testing $H_0 : \theta = .1$ against $H_A : \theta > .1$ find UMP level $\alpha = .05$ test and its power function when $m_0 = 2$. Also do the similar exercise for $m_0 = 1$ and $m_0 = 3$ and compare the power functions of all the three UMP tests.

(ii) Using the solution in Exercise 8.3 (iii) obtain UMP level α tests for (a) $H_0 : \theta = \theta_0$, $H_A : \theta < \theta_0$ and (b) $H_0 : \theta = \theta_0$, $H_A : \theta > \theta_0$. Verify that their power functions are monotone.

(iii) Let $(X_1, \ldots, X_n)$ be i.i.d. with pdf $f(x, \theta) = \theta/x^2$ for $0 < \theta < x < \infty$. Using results on Pitman family (Ch. 2, Sec. 6), we use $t = x_{(1)}$, the minimal sufficient statistic with pdf

$$g(t, \theta) = n\theta^n/t^{n+1}, \ \theta < t < \infty.$$

Obtain UMP level α test for testing $H_0 : \theta = \theta_0$ against $H_A : \theta = \theta_1 < \theta_0$. After having obtained the results from the first principles verify that using the transformation $y = \frac{1}{x}$ we have $y \sim U\left(0, \frac{1}{\theta}\right)$ and results from Examples 8.3.5 and 8.4.2 can be used directly.

(iv) Let $(X_1, \ldots, X_n)$ be i.i.d. $G(\lambda, 1)$ with pdf $f(x, \lambda) = \frac{1}{\Gamma(\lambda)} e^{-x} x^{\lambda - 1}$, $x > 0, \lambda > 0$. Obtain UMP level α test for testing $H_0 : \lambda = 1$ against $H_A : \lambda > 1$.

(v) Let $(X_1, \ldots, X_n)$ be i.i.d. with pdf $f(x, \theta) = \theta x^{\theta - 1}$, $0 < x < 1$, $\theta > 0$. Obtain UMP level α test for testing $H_0 : \theta = 1$ against $H_A : \theta < 1$.

We conclude this section with the observation that for one parameter exponential family the UMP test derived for simple null hypothesis $\theta = \theta_0$ against one sided alternatives $\theta > \theta_0$ (or $\theta < \theta_0$) continues to be UMP test for testing one sided composite null hypotheses $\theta \le \theta_0$ (or $\theta \ge \theta_0$) against one sided alternatives $\theta > \theta_0$ (or $\theta < \theta_0$). This is a consequence of the monotone nature of power function $\beta_\varphi(\theta)$. We illustrate the case for testing $H_0 : \theta \le \theta_0$ against $H_A : \theta > \theta_0$ and where $u(\theta)$ is increasing and leave the other three cases as an exercise to the reader.

Let $\mathbb{D}_\alpha$ denote the class of all level α tests for $H_0 : \theta \le \theta_0$ i.e.

$$\mathbb{D}_\alpha = \{\varphi \mid \beta_\varphi(\theta) \le \alpha, \ \forall \ \theta \le \theta_0\} \qquad (8.4.7)$$

Consider the boundary point $\theta = \theta_0$ of H_0 and any $\theta \in H_A$. Then by NP lemma, MP level α test for the subproblem $H_0' : \theta = \theta_0$ vs $H_A' : \theta = \theta_1$ is given by φ_1 as defined in (8.4.2) with constants k_α, γ_α given by (8.4.4). The MP level α test φ_1 does not depend on the specific choice of $\theta_1 \in H_A$, and therefore is UMP test for $H_0' : \theta = \theta_0$ against $H_A : \theta > \theta_0$. We now show that $\varphi_1 \in \mathbb{D}_\alpha$ as defined in (8.4.7). We have already seen that the power function $\beta_{\varphi_1}(\theta)$ is increasing for $\theta > \theta_0$. We now show that $\beta_{\varphi_1}(\theta)$ is increasing for

$\theta \le \theta_0$ also. Let $\theta'' < \theta' < \theta_0$ and consider testing the simple null hypothesis $H_0'' : \theta = \theta''$ against the alternative $H_A'' : \theta = \theta'$ at level $\beta_{\varphi_1}(\theta'') = 1 - P[T \le k_\alpha \mid \theta''] + \gamma_\alpha P[T = k_\alpha \mid \theta'']$. Then by N-P lemma and the fact that $u(\theta)$ is increasing, the MP level $\beta_{\varphi_1}(\theta'')$ test is given by

$$\begin{aligned} \varphi_1'' &= 1 \quad \text{if } T > k_\alpha'' \\ &= \gamma_\alpha'' \quad \text{if } T = k_\alpha'' \\ &= 0 \quad \text{if } T < k_\alpha'' \end{aligned}$$

Here constants k_α'', γ_α'' are determined by

$$\beta_{\varphi_1}(\theta'') = 1 - P[T \le k_\alpha'' \mid \theta''] + \gamma_\alpha'' P[T = k_\alpha'' \mid \theta'']$$

and therefore $k_\alpha'' = k_\alpha$ and $\gamma_\alpha'' = \gamma_\alpha$ and $\varphi_1'' \equiv \varphi_1$. The power of MP test φ_1'' at $\theta' = \beta_{\varphi_1''}(\theta')$ is not less than the level $\beta_{\varphi_1''}(\theta'')$. But $\varphi_1'' \equiv \varphi_1$ and therefore for $\theta'' < \theta' < \theta_0$ we have $\beta_{\varphi_1}(\theta'') \le \beta_{\varphi_1}(\theta')$ and $\beta_{\varphi_1}(\theta)$ is increasing for $\theta \le \theta_0$. Therefore $\varphi_1 \in \mathbb{D}_\alpha$ as defined in (8.4.7) and would be UMP level α test for $H_0 : \theta \le \theta_0$ against $H_A : \theta > \theta_0$.

It now follows that UMP level α tests derived in Examples 8.4.1, 8.4.3, 8.4.4, 8.4.5 and Exercises 8.4 (iv) and (v) are UMP level α tests for one sided composite null hypotheses.

EXAMPLE 8.4.5. Let $(X_1, \ldots, X_n)$ be i.i.d. $U(0, \theta)$. Suppose we want to test $H_0 : \theta \le \theta_0$ against $H_A : \theta > \theta_0$. Then consider the sub-problem $H_0' : \theta = \theta_0$ against $H_A' : \theta = \theta_1 > \theta_0$. From Example 8.3.5, MP level α test is given by

$$\begin{aligned} \varphi_1(t) &= 1 \quad \text{if } t = x_{(n)} > \theta_0 \\ &= \gamma(t) \quad \text{if } 0 < t < \theta_0 \end{aligned}$$

where $0 \le \gamma(t) \le 1$ is such that $\int_0^{\theta_0} \gamma(t) \dfrac{nt^{n-1}}{\theta_0^n} dt = \alpha$. As the MP test does not depend upon specific alternative $\theta_1 \in H_A$ we have φ_1 is UMP level α test. The power function for any choice of $\gamma(t)$ given by

$$\beta_{\varphi_1}(\theta) = 1 - \left(\frac{\theta_0}{\theta}\right)^n (1 - \alpha) \quad \text{for } \theta > \theta_0.$$

Since $\gamma(t)$ can be chosen suitably let $\gamma(t) \equiv \alpha$ for $0 < t < \theta_0$. Then an UMP test of level α is given by

$$\begin{aligned} \varphi_1^*(t) &= 1 \quad \text{if } t > \theta_0 \\ &= \alpha \quad \text{if } 0 < t < \theta_0. \end{aligned}$$

Now for $\theta' < \theta_0$, $\beta_{\varphi_1^*}(\theta') = \int_0^{\theta'} \dfrac{\alpha \cdot nt^{n-1}}{\theta'^n} dt = \alpha$. Therefore $\beta_{\varphi_1^*}(\theta') \le \alpha$,

$\forall\ \theta' \le \theta_0$ and φ_1^* is level α test for composite null hypotheses $H_0 : \theta \le \theta_0$ and continues to be UMP level α test for $H_0 : \theta \le \theta_0$ against $H_A : \theta > \theta_0$.

The connecting link between one parameter exponential family and Pitman family is the Monotone Likelihood Ratio (MLR) property. Let $\{L(x, \theta), \theta \in \Omega \subset R_1\}$ be the family of joint pdf of $(X_1, \ldots X_n)$. Consider θ_0, θ_1 with $\theta_1 > \theta_0$ in Ω then $\{L(x, \theta), \theta \in \Omega \subset R_1\}$ is said to have MLR property in a real valued statistic $T(x)$ if $L(x, \theta_1)/L(x, \theta_0)$ is a monotone function of $T(x)$ i.e. it is either non-increasing or non-decreasing function of $T(x)$. Note that if $T(x)$ is minimal sufficient for θ then we can consider $L(x, \theta_1)/L(x, \theta_0) = g(t, \theta_1)/g(t, \theta_0)$ and check whether this ratio or logarithm of it is a monotone function of T. It is easy to check that the one parameter exponential family has MLR property in $T = \Sigma K(x_i)$ and $U(0, \theta), \theta > 0$ has MLR property in $T = \text{Max}(X_1, \ldots, X_n)$. In the first case

$$\log L(x, \theta_1) - \log L(x, \theta_0) = (u(\theta_1) - u(\theta_0)) \Sigma K(x_i) + n(v(\theta_1) - v(\theta_0))$$

and is strictly increasing or decreasing in T according as $\frac{du}{d\theta} > 0$ or $\frac{du}{d\theta} < 0$. In $\{U(0, \theta), \theta > 0\}$ case

$$\frac{L(x, \theta_1)}{L(x, \theta_0)} = \left(\frac{\theta_0}{\theta_1}\right)^n \quad \text{if } 0 < t \le \theta_0$$

$$= \infty \quad \text{if } \theta_0 < t < \theta_1$$

which is a non-decreasing function of $t = x_{(n)}$.

Let $\{L(x, \theta), \theta \in \Omega\}$ be a family with MLR property in some statistic T. Suppose we want to test $H_0 : \theta = \theta_0$ and $H_1' : \theta = \theta_1 > \theta_0$, and where $\frac{L(x, \theta_1)}{L(x, \theta_0)}$, is say non-decreasing function of T. Then NP-lemma gives the test φ_1 given by (8.4.2). The proof that φ_1 is UMP for $H_0 : \theta = \theta_0$ vs $H_1 : \theta > \theta_0$ and that $\beta_{\varphi_1}(\theta)$ is increasing in θ is word to word same for this case as it is in case of one parameter exponential family and therefore will not be repeated here. Similarly φ_1 continues to be UMP level α test for the wider problem $H_0 : \theta \le \theta_0$ vs $H_1 : \theta > \theta_0$ and the proof is left to the reader as an exercise.

Exercise 8.4 (contd.). (vi) Show that $f(x, \theta) = \theta/x^2, 0 < \theta < x < \infty$, has MLR property in $t = x_{(1)}$ and obtain UMP test for $H_0 : \theta \ge \theta_0$ against $H_1 : \theta < \theta_0$.

(vii) For a sample of size one from the Double exponential distribution with pdf $f(x, \theta) = \frac{1}{2} \exp\{-|x - \theta|\}, x \in R_1, \theta \in R_1$, show that $\{f(x, \theta), \theta \in R_1\}$ has MLR property and obtain UMP level $\alpha = .05$ for testing $H_0 : \theta \le 0$ against $\theta > 0$.

(viii) Let X be a sample of size one from Cauchy distribution with $f(x, \mu) = \frac{1}{\pi} \frac{1}{1 + (x - \mu)^2}, x \in R_1, \mu \in R_1$. Check whether $\{f(x, \mu), \mu \in R_1\}$ has MLR property.

Section 8.5 shows that in general UMP level α test does not exist for testing a simple null hypothesis $H_0 : \theta = \theta_0$ vs $H_A : \theta \ne \theta_0$, a two sided alternative. We also give an example in which the UMP test exists for testing a simple null hypothesis against two sided composite alternatives.

8.5 Non-existence of UMP Tests

Now consider a situation in which we want to test $H_0 : \theta = \theta_0$ against $H_A : \theta \neq \theta_0$. We first show by way of an example that for a random sample of size n from say $N(\theta, 1)$, there does not exist UMP level α test for two sided alternatives.

EXAMPLE 8.5.1 Suppose if possible there exists an UMP level α test φ^* for $H_0 : \theta = \theta_0$ vs $H_A : \theta \neq \theta_0$. Consider a subproblem of testing $H_0 : \theta = \theta_0$ against $H'_A : \theta > \theta_0$. Then by Example 8.4.1, UMP test for this subproblem is given by

$$\varphi_1(\bar{x}) = 1 \text{ if } \bar{x} > \theta_0 + \frac{\xi_{1-\alpha}}{\sqrt{n}}$$

$$= 0 \quad \text{otherwise.}$$

Now as seen earlier, for any test $\varphi(x)$ there exists a test based on minimal sufficient statistic $\bar{x}$ given by $E[\varphi(x) \mid \bar{x}] = \psi(\bar{x})$ such that $\beta_\varphi(\theta) = \beta_\psi(\theta)$, $\forall\ \theta \in \Omega$. Therefore wlg we can take $\varphi^*(x)$ to be a test based on $\varphi^*(\bar{x})$. Let $\beta_{\varphi^*}(\theta)$ be its power function. Then $\beta_{\varphi^*}(\theta) \equiv \beta_{\varphi_1}(\theta)$, $\forall\ \theta > \theta_0$ because if this were not true then there exists a $\theta_1 > \theta_0$ such that either $\beta_{\varphi^*}(\theta_1) > \beta_{\varphi_1}(\theta_1)$ or $\beta_{\varphi^*}(\theta_1) < \beta_{\varphi_1}(\theta_1)$. In case the first possibility holds this contradicts the fact that $\varphi_1(\bar{x})$ is UMP for $H_0 : \theta = \theta_0$ against $H'_A : \theta > \theta_0$. In case the second possibility holds then this contradicts the fact that φ^* is UMP for H_A which includes H'_A. Therefore $\beta_{\varphi^*}(\theta) - \beta_{\varphi_1}(\theta) = 0$, $\forall\ \theta \geq \theta_0$. However in view of completeness of the family $\left\{N\left(\theta, \frac{1}{n}\right), \theta \geq \theta_0\right\}$ we must have $\varphi_1 \equiv \varphi^*$.

Next consider the problem $H_0 : \theta = \theta_0$ against $H''_A : \theta < \theta_0$. Then UMP level α test is given by $\varphi_2(\bar{x}) = 1$ if $\bar{x} < \theta_0 + \frac{\xi_\alpha}{\sqrt{n}}$ and zero otherwise. Arguing as above we first claim that $\beta_{\varphi_2}(\theta) \equiv \beta_{\varphi^*}(\theta)$ for $\theta \leq \theta_0$ and then using completeness of the family $\left\{N\left(\theta, \frac{1}{n}\right), \theta \leq \theta_0\right\}$ conclude that $\varphi_2 \equiv \varphi^*$.

This leads to a contradiction that $\varphi_1 \equiv \varphi_2$ when in fact $\varphi_1 \neq \varphi_2$ for any $\bar{x} \in R_1$. Thus there does not exist any UMP test for $H_0 : \theta = \theta_0$ against H_A: $\theta \neq \theta_0$. In course on statistical methods we come across the two-sided equal tail criterion test given by

$$\varphi_3(\bar{x}) = 1 \text{ if } \bar{x} \notin \left(\theta_0 + \frac{\xi_{\alpha/2}}{\sqrt{n}}, \theta_0 + \frac{\xi_{1-\alpha/2}}{\sqrt{n}}\right)$$

$$= 0 \quad \text{otherwise.}$$

This is a level α test for two sided problem with power function given by

$$\beta_{\varphi_3}(\theta) = 1 - \Phi((\theta_0 - \theta)\sqrt{n} + \xi_{1-\alpha/2}) + \Phi(\sqrt{n}(\theta_0 - \theta) + \xi_{\alpha/2}) \qquad (8.5.1)$$

One can verify that for $\delta > 0$

$$\beta_{\varphi_3}(\theta_0 + \delta) < \beta_{\varphi_1}(\theta_0 + \delta) \text{ and } \beta_{\varphi_3}(\theta_0 - \delta) < \beta_{\varphi_2}(\theta_0 - \delta).$$

However $\beta_{\varphi_1}(\theta) < \beta_{\varphi_3}(\theta)$ for $\theta < \theta_0$ and $\beta_{\varphi_2}(\theta) < \beta_{\varphi_3}(\theta)$ for $\theta > \theta_0$. Thus we can view φ_3 as a compromise test although it is not UMP for two sided alternatives. In fact φ_1 and φ_2 are biased tests for H_0: $\theta = \theta_0$ vs $H_A : \theta \neq \theta_0$.

Analysis given here for $N(\theta, 1)$ case can be extended to the one parameter exponential family case in a straightforward manner where $T = \sum K(x_i)$ is minimal sufficient statistic and we can restrict ourselves to tests based on T with pdf $\{g(t, \theta), \theta \in \Omega)$ which itself is a one parameter exponential family. Again sub-families $\{g(t, \theta), \theta \geq \theta_0\}$ and $\{g(t, \theta), \theta \leq \theta_0\}$ are complete. If we assume that there exists an UMP level α test for $H_0 : \theta = \theta_0$ against $H_A : \theta \neq \theta_0$ given by $\varphi^*(t)$ then we can show that $\varphi_1(t) \equiv \varphi^*(t) \equiv \varphi_2(t)$ where $\varphi_1(t)$ and $\varphi_2(t)$ are UMP level α tests for one sided alternatives $H'_A : \theta > \theta_0$ and $H''_A : \theta < \theta_0$ respectively. However $\varphi_1(t) \neq \varphi_2(t)$ for any t and thus we have a contradiction.

We note that in the above proof of non-existence of UMP level α test for two sided alternatives, completeness of subfamilies $\{g(t, \theta), \theta \geq \theta_0\}$ and $\{g(t, \theta), \theta \leq \theta_0\}$ plays a very crucial role. Now if any one of the above two subfamilies is not complete the proof will break down and there may exist UMP level α test for two sided alternatives. We show that this is indeed the case for $U(0, \theta)$ model where $t = x_{(n)}$ is minimal sufficient and as seen earlier in Chapter 3, $\{g(t, \theta), \theta \geq \theta_0\}$ is not a complete family.

EXAMPLE 8.5.2. From Examples 8.3.6 and 8.4.2 for testing $H_0 : \theta = \theta_0$ vs $H'_A : \theta > \theta_0$ the UMP level α test is given by

$$\varphi_1(t) = 1 \quad \text{if } t > \theta_0$$

$$= \gamma_1(t) \quad \text{if } 0 < t < \theta_0$$

where $\gamma_1(t)$ is chosen so that $\int_0^{\theta_0} \gamma_1(t) \frac{nt^{n-1}}{\theta_0^n} dt = \alpha$. Thus UMP test is not unique and we will exploit this nonuniqueness to determine UMP test for two sided alternatives. Consider $H_0 : \theta = \theta_0$ against H''_A: $\theta = \theta_1 < \theta_0$ then

$$\lambda(t) = (\theta_0/\theta_1)^n \quad \text{if } 0 < t < \theta_1$$

$$= 0 \quad \text{if } \theta_1 \le t < \theta_0.$$

Here an UMP level α test is given by

$$\varphi_2(t) = \gamma_2(t) \quad \text{if } 0 < t < \theta_1$$

$$= 0 \quad \text{if } \theta_1 \le t < \theta_0$$

where $\gamma_2(t)$ is such that $\int_0^{\theta_1} \gamma_2(t) \frac{nt^{n-1}}{\theta_0^n} dt = \alpha$. The power

$$\beta_{\varphi_2}(\theta_1) = \int_0^{\theta_1} \gamma_2(t) \frac{nt^{n-1}}{\theta_1^n} dt = (\theta_0/\theta_1)^n \alpha$$

if $(\theta_0/\theta_1)^n \alpha < 1$ and otherwise it would be unity as $\beta_{\varphi_2}(\theta) \le 1$ for any θ. Here the MP level α test appears to depend on particular θ_1. However if $\gamma_2(t)$ is appropriately chosen then we can get around this problem. Towards this end, let $\gamma_2(t) = 1, 0 < t < a < \theta_1$ such that

$$\int_0^a \frac{nt^{n-1}}{\theta_0^n} dt = \alpha \quad \text{or } a = \theta_0 \alpha^{1/n}$$

The power of the test φ_2' is

$$\beta_{\varphi_2'}(\theta_1) = \int_0^a \frac{nt^{n-1}}{\theta_1^n} dt \quad \text{if } a = \theta_0 \alpha^{1/n} < \theta_1$$

$$= 1 \quad \text{if } \theta_1 < \theta_0 \alpha^{1/n} = a$$

Thus

$$\varphi_2'(t) = 1 \quad \text{if } 0 < t < \theta_0 \alpha^{1/n}$$

$$= 0 \quad t \ge \theta_0 \alpha^{1/n}$$

is UMP level α test for $H_0 : \theta = \theta_0$ against $H_A'' : \theta < \theta_0$ since $\varphi_2'(t)$ is independent of particular $\theta_1 \in H_A''$.

The power function of φ_2' is given by

$$\beta_{\varphi_2'}(\theta_1) = \left(\frac{\theta_0}{\theta_1}\right)^n \alpha \quad \text{if } \theta_0 \alpha^{1/n} < \theta_1$$

$$= 1 \quad \text{if } \theta_1 < \theta_0 \alpha^{1/n}.$$

Now define φ^* to be a combination of φ_2' and φ_1 so that

$$\varphi^* = 1 \quad \text{if } 0 < t < \theta_0\, \alpha^{1/n}$$

$$= 0 \quad \text{if } \theta_0 \alpha^{1/n} < t < \theta_0$$

$$= 1 \quad \text{if } t > \theta_0$$

Then φ^* is UMP level α test and its power function is given by

$$\begin{aligned}\beta_{\varphi^*}(\theta) &= 1 \text{ if } \theta < \theta_0\alpha^{1/n} \\ &= (\theta_0/\theta)^n\alpha \quad \text{if } \theta_0\alpha^{1/n} < \theta \leq \theta_0 \\ &= 1 - (\theta_0/\theta)^n(1-\alpha) \quad \text{if } \theta > \theta_0\end{aligned}$$

When H_0 and H_A are both composite hypotheses one can show that UMP tests generally do not exist particularly when nuisance parameter is present i.e. when θ is vector valued and H_0 and H_A are statements about say θ_1 only. For example such a situation occurs in testing $H_0 : \theta = 0$ against $H_A : \theta > 0$ in $N(\theta, \sigma^2)$ family where σ^2 is unknown. Here both H_0 and H_A are composite and σ^2 is called as a nuisance parameter and θ is called as the parameter of interest. The reader familiar with standard statistical methods will observe that although UMP test does not exist in this $N(\theta, \sigma^2)$ case, the well known Student's t test which rejects H_0 if $\sqrt{n}(\bar{x})/\sqrt{S^2/(n-1)} > c$ is used in many applications with $c = t_{n-1,1-\alpha}$ the $100(1-\alpha)\%$ point of Student's t with $n-1$ degrees of freedom. Chapter 9 gives the above test and several other well known tests used in Analysis of variance, regression analysis, testing equality of proportions of two samples from binomial distributions etc. using the method of Likelihood Ratio Test (LRT).

9

Tests of Hypotheses-II

9.1 The Likelihood Ratio Test (LRT)

Consider the problem of testing $H_0 : \theta \in \Omega_0$ against $H_1 : \theta \in \Omega_1$ where both H_0 and H_1 could be composite and where x and θ could be both vector valued. The basic idea is to compare the $L_0(x, \hat{\theta}_0)$ the maximum of the likelihood of θ for $\theta \in \Omega_0$ and $L_1(x, \hat{\theta}_1)$ the maximum of the likelihood of θ for $\theta \in \Omega_1$. We make this comparison in a slightly different way by defining the likelihood ratio test statistic (LRTS)

$$\lambda(x) = \sup_{\theta \in \Omega_0} L(x, \theta) / \sup_{\theta \in \Omega} L(x, \theta) \tag{9.1.1}$$

where $\Omega = \Omega_0 \cup \Omega_1$.

Note that $0 < \lambda(x) \leq 1$ as $\max_{\theta \in \Omega} L(x, \theta) \geq \max_{\theta \in \Omega_0} L(x, \theta)$ since $\Omega_0 \subset \Omega$. The LRT based on $\lambda(x)$ follows the logic of likelihood ordering (Sec. 1.3) and assumes that small values of $\lambda(x)$ near zero indicate support to H_1. Thus LRT rejects H_0 if $\lambda(x) \leq c$ where c is chosen so that $\sup_{\theta \in \Omega_0} P[\lambda(x) \leq c \mid \theta] = \alpha$ where α is the specified upper limit to type I error or level of the test. We first give a few examples which show that the LRT based on $\lambda(x)$ leads to MP or UMP level α tests derived in the previous chapter. We also observe that if T is minimal sufficient for $\{L(x, \theta), \theta \in \Omega\}$ then we need to consider the LRT based on $\{g(t, \theta), \theta \in \Omega\}$ only, with LRTS

$$\lambda(t) = \sup_{\theta \in \Omega_0} g(t, \theta) / \sup_{\theta \in \Omega} g(t, \theta). \tag{9.1.2}$$

EXAMPLE 9.1.1. We first consider the case where $\Omega_0 = \{\theta_0\}$ and $\Omega_1 = \{\theta_1\}$ and show that in this case the LRT leads to the MP level α test given by *N-P* lemma. Note that $\sup_{\theta \in \Omega_0} L(x, \theta_0) = L(x, \theta_0)$ since $\Omega_0 = \{\theta_0\}$. Further

$$\begin{aligned} \sup_{\theta \in \Omega} L(x, \theta) &= L(x, \theta_0) \quad \text{if } L(x, \theta_0) \geq L(x, \theta_1) \\ &= L(x, \theta_1) \quad \text{if } L(x, \theta_0) < L(x, \theta_1). \end{aligned}$$

Therefore the LRTS

$$\lambda(x) = 1 \quad \text{if } L(x, \theta_0) \geq L(x, \theta_1)$$

$$= \frac{L(x, \theta_0)}{L(x, \theta_1)} \quad \text{if } L(x, \theta_0) < L(x, \theta_1).$$

We reject H_0 if $\lambda(x) \le c$ such that $P_{\theta_0}[\lambda(x) \le c] = \alpha$. If $c = 0$ then we have $P_{\theta_0}[\lambda(x) \le 0] = P_{\theta_0}[\lambda(x) = 0] = 0$ and if $c = 1$ then $P_{\theta_0}[\lambda(x) \le 1] = 1$. As in most situations of interest, we have $0 < \alpha < 1$ and we therefore take $0 < c < 1$. The LRT then rejects H_0 if $[\lambda(x) \le c]$ and $c < 1$ and is therefore given by

$$\begin{aligned} \varphi(x) &= 1 \quad \text{if } L(x, \theta_0) \le cL(x, \theta_1) \\ &= 0 \quad \text{otherwise} \end{aligned}$$

where c is such that $P_{\theta_0}[L(x, \theta_0) \le cL(x, \theta_1)] = \alpha$. This is same as the test given by N-P lemma namely

$$\begin{aligned} \varphi^*(x) &= 1 \quad \text{if } L(x, \theta_0) < cL(x, \theta_1) \\ &= \gamma \quad \text{if } L(x, \theta_0) = cL(x, \theta_1) \\ &= 0 \quad \text{otherwise} \end{aligned}$$

where (c, γ) are such that φ^* is size α test or $E_{\theta_0}[\varphi^*(x)] = \alpha$.

EXAMPLE 9.1.2. Let $(X_1, \ldots, X_n)$ be a random sample of size n from $N(\theta, 1)$. Suppose we want to test $H_0 : \theta \le \theta_0$ against $H_1 : \theta > \theta_0$ so that $\Omega = R_1$. Now $L(x, \theta) = g(\bar{x}, \theta) \cdot h(x)$ where $\bar{X} = \frac{1}{n} \Sigma X_i$ is the minimal sufficient statistic for the family $\{N(\theta, 1), \theta \in \Omega\}$ and as seen before $\hat{\theta} = \bar{x}$ and $\sup_{\theta \in \Omega} L(x, \theta) = g(\bar{x}, \bar{x}) h(x)$. For $\theta \in \Omega_0$, using techniques of Example 7.2.2 we have $\hat{\theta}_0 = \bar{x}$ if $\bar{x} \le \theta_0$ and $\hat{\theta}_0 = \theta_0$ if $\bar{x} > \theta_0$. Therefore

$$\begin{aligned} \sup_{\theta \in \Omega_0} L(x, \theta) &= g(\bar{x}, \bar{x}) h(x) \quad \text{if } \bar{x} \le \theta_0 \\ &= g(\bar{x}, \theta_0) h(x) \quad \text{if } \bar{x} > \theta_0 \end{aligned}$$

or

$$\begin{aligned} \lambda(\bar{x}) &= 1 \quad \text{if } \bar{x} \le \theta_0 \\ &= \exp\left\{-\frac{n(\bar{x} - \theta_0)^2}{2}\right\} \quad \text{if } \bar{x} > \theta_0. \end{aligned}$$

Thus we accept H_0 if $\bar{x} \le \theta_0$ and reject H_0 if $\bar{x} > \theta_0$ and $\exp\left\{-\frac{n(\bar{x} - \theta_0)^2}{2}\right\} \le c$ which is equivalent to rejecting H_0 if $\bar{x} > \theta_0 + k$ where k is chosen so that $\sup_{\theta \in \Omega_0} P[\bar{X} > \theta_0 + k] = \alpha$. However $P_\theta[\bar{X} > \theta_0 + k]$

$= 1 - \Phi[\sqrt{n}(\theta_0 - \theta + k)]$ which is an increasing function of $\theta \in (-\infty, \theta_0]$ and $\sup_{\theta \le \theta_0} P[\lambda(x) \le c]$ therefore occurs at θ_0 and is given by $1 - \phi[\sqrt{n}k]$. Thus k is determined by $1 - \Phi[\sqrt{n}k] = \alpha$ or $k = \dfrac{\xi_{1-\alpha}}{\sqrt{n}}$. The LRT is then given by

$$\varphi_1(\bar{x}) = 1 \quad \text{if } \bar{x} > \theta_0 + \frac{\xi_{1-\alpha}}{\sqrt{n}}$$

$$= 0 \quad \text{otherwise}$$

We have already seen in the previous chapter that $\varphi_1(x)$ defined above is in fact UMP level α test for testing $H_0 : \theta \le \theta_0$ vs $H_1 : \theta > \theta_0$.

EXAMPLE 9.1.3. Suppose that in Example 9.1.2 we have $H_0 : \theta = \theta_0$ and H_1: $\theta \ne \theta_0$ then

$$\lambda(x) = g(\bar{x}, \theta_0)/g(\bar{x}, \bar{x}) = \exp\left\{-\frac{n(\bar{x} - \theta_0)^2}{2}\right\}.$$

Now $\lambda(x) \le c$ is equivalent to $n(\bar{x} - \theta_0)^2 \ge k$ and we must select k such that $P_{\theta_0}[n(\bar{X} - \theta_0)^2 \ge k] = \alpha$. But under θ_0, $\bar{X} \sim N\left(\theta_0, \dfrac{1}{n}\right)$ and therefore $n(\bar{X} - \theta_0)^2 \sim \chi_1^2$ and this gives $k = \chi^2_{1,1-\alpha} = \xi^2_{1-\alpha/2}$, where ξ_α is $100\alpha\%$ point of $N(0, 1)$. Therefore LRT rejects H_0 when the $\bar{x} > \theta_0 + \dfrac{\xi_{1-\alpha/2}}{\sqrt{n}}$ or $\bar{x} < \theta_0 - \dfrac{\xi_{1-\alpha/2}}{\sqrt{n}} = \theta_0 + \dfrac{\xi_{\alpha/2}}{\sqrt{n}}$. This is the usual equal tail criterion test that we considered in Example 8.5.1, as a compromise test. Note that here UMP level α test does not exist.

We have already observed that the power function of this test is symmetric around θ_0 and that as $\theta \to \pm\infty$, power goes to 1. Further $\beta_\varphi(\theta) \ge \beta_\varphi(\theta_0) = \alpha$ and $\beta_\varphi(\theta)$ has minimum at $\theta = \theta_0$. A test φ such that

$$\sup_{\theta \in \Omega_0} \beta_\varphi(\theta) = \alpha \le \inf_{\theta \in \Omega_1} \beta_\varphi(\theta) \tag{9.1.3}$$

has the property that its power is never below the level α which specifies the maximum possible type I error. A test having this property is called as an unbiased test. The unbiasedness is a desirable property for any test since if a test is biased there exist an alternative $\theta_1 \in \Omega_1$ and $\theta_0 \in \Omega_0$ such that $\beta_\varphi(\theta_0) > \beta_\varphi(\theta_1)$ or P [Rejecting $H_0 \mid H_0$ is true] is greater than P [Rejecting $H_0 \mid H_0$ is false]. One can show that the LRT derived above is UMP test for $H_0 : \theta = \theta_0$ vs $H_1 : \theta \ne \theta_0$ within the class of all unbiased tests of level α i.e. consider a sub-class of $\mathbb{D}_\alpha$ namely $\mathbb{D}_\alpha^{(u)} = \{\varphi \mid \varphi \in \mathbb{D}_\alpha, \beta_\varphi(\theta_0) \le \operatorname{Inf}_{\theta \in \Omega_1} \beta_\varphi(\theta)\}$. We will however not derive this result which depends on using the generalized Neyman-Pearson Lemma. The interested reader should refer to Lehmann (1959) for further details.

EXAMPLE 9.1.4. Next consider the problem of testing composite null hypotheses against composite alternatives. Historically Student (1908) proposed the problem of testing $H_0 : \theta = \theta_0$ against $H_1 : \theta \neq \theta_0$ in the $N(\theta, \sigma^2)$ model where σ^2 is not specified under either H_0 or H_1. As mentioned at the end of Chapter 8, the methods books recommend the test based on well known student's statistic, $\dfrac{\sqrt{n}(\bar{x} - \theta_0)}{\sqrt{S^2/(n-1)}} = t_{n-1}$ where $\bar{x} = \dfrac{1}{n}\Sigma x_i$, and $S^2 = \Sigma (x_i - \bar{x})^2$, which rejects H_0 for large values of t_{n-1}^2.

Since $t_{n-1} \in (-\infty, \infty)$ and its distribution under H_0 is symmetric around zero Student's t-test rejects H_0 if observed value of $t_{n-1}^2 > t_{n-1,1-\alpha/2}^2$ where $t_{n-1,1-\alpha/2}$ is the $100(1 - \alpha/2)\%$ point of the Student's t distribution with $(n - 1)$ d.f. We now show that the Student's t test is indeed a LRT.

Now here $L(x, \theta, \sigma^2) = \left(\dfrac{1}{\sqrt{2\pi\sigma^2}}\right)^n \exp\left\{-\dfrac{\Sigma (x_i - \theta)^2}{2\sigma^2}\right\}$ and $\hat{\theta} = \bar{x}$, $\hat{\sigma}^2 = S^2/n$ so that

$$\sup_{\theta \in R_1, \sigma^2 > 0} L(x, \theta, \sigma^2) = \left(\frac{1}{\sqrt{2\pi\hat{\sigma}^2}}\right)^n \exp\left\{-\frac{n}{2}\right\}$$

Now under H_0 we have

$$L_0(x, \theta_0, \sigma^2) = \left(\frac{1}{\sqrt{2\pi\sigma^2}}\right)^n \exp\left\{-\frac{\Sigma (x_i - \theta_0)^2}{2\sigma^2}\right\}$$

$$\hat{\sigma}_0^2 = \frac{1}{n}\Sigma (x_i - \theta_0)^2$$

and

$$\sup_{\theta = \theta_0, \sigma^2 > 0} L(x, \theta, \sigma^2) = \left(\frac{1}{\sqrt{2\pi\hat{\sigma}_0^2}}\right)^n \exp\left\{-\frac{n}{2}\right\}$$

Thus $\lambda(x) = \left(\dfrac{\hat{\sigma}^2}{\hat{\sigma}_0^2}\right)^{n/2}$.

Now LRT rejects H_0 if $\lambda(x) \leq c$ which is equivalent to $\dfrac{\hat{\sigma}^2}{\hat{\sigma}_0^2} \leq c_1$. But $n\hat{\sigma}_0^2 = \Sigma (x_i - \theta_0)^2 = n\hat{\sigma}^2 + n(\bar{x} - \theta_0)^2$. Thus $\lambda(x) \leq c_1$ is equivalent to $\dfrac{n\hat{\sigma}^2 + n(\bar{x} - \theta_0)^2}{\hat{\sigma}^2} \geq k_1$, that is, $\dfrac{n(\bar{x} - \theta_0)^2}{\hat{\sigma}^2} \geq k_2$ or $t_{n-1}^2 \geq k_3$ as $\hat{\sigma}^2 = \dfrac{S^2}{n} = \dfrac{S^2}{n-1}\left(\dfrac{n-1}{n}\right)$. Now $P[t_{n-1}^2 \geq k_3 \mid \theta_0, \sigma^2]$ does not depend on σ^2 and therefore we have $k_3 = t_{n-1,1-\alpha/2}^2$, and Student's t test and LRT test are equivalent.

Historically whether the observed sample men $\bar{x}$ is significantly different from the hypothetical population mean θ_0 was tested by the absolute deviation $|\bar{x} - \theta_0|$ or $(\bar{x} - \theta_0)^2$ measured in units of standard deviation of $\bar{x}$, namely σ^2/n if σ^2 was known.

Thus for known σ^2 the test statistic was $|Z_n| = \left|\frac{\sqrt{n}(\bar{x} - \theta_0)}{\sigma}\right|$ and using normality of Z_n, the null hypotheses was rejected if $|Z_n| \geq \xi_{1-\alpha/2}$ or $Z_n^2 \geq \xi_{1-\alpha/2}^2 = \chi_{1,1-\alpha}^2$. When σ^2 is not known, it was recommended that we estimate σ^2 by S^2/n and use $Z_n' = \frac{\sqrt{n}(\bar{x} - \theta_0)}{\sqrt{S^2/n}}$ and reject H_0 if $|Z_n'| \geq \xi_{1-\alpha/2}$ which is valid for large samples in view of the fact that $Z_n' \xrightarrow{d} N(0, 1)$. If n is small, Student (1908) recommended using $t_{n-1} = \frac{\sqrt{n}(\bar{x} - \theta_0)}{\sqrt{S^2/(n-1)}}$ and obtained the exact sampling distribution of t_{n-1} and its critical points. The most important property of student's statistic t_{n-1} is that its distribution under H_0 does not depend on the unknown nuisance parameter σ^2. This further implies that $P_{\theta_0}[t_{n-1}^2 \geq t_{n-1,1-\alpha/2}^2] = \alpha$ for any $\sigma^2 > 0$ or if $\beta(\theta, \sigma^2)$ denotes the power function of the test, then $\beta(\theta_0, \sigma^2) = \alpha, \forall\, \sigma^2 > 0$. This property of the power function of a test is called as similarity. The corresponding critical region is called as a similar critical region and the corresponding test as a similar test.

Remark 9.1.2. In general, suppose $H_0 : \theta \in \Omega_0$ is composite hypotheses where $\theta = (\theta_1, \ldots, \theta_m)'$. Then a test φ is called a similar test of size α, if $\beta_\varphi(\theta) = \alpha, \forall\, \theta \in \Omega_0$. By using advanced techniques one can show that the Student's test based on t_{n-1} is UMP test within the class of all similar size α tests defined by $S_\alpha = \{\varphi \mid \beta_\varphi(\theta_0, \sigma^2) = \alpha\}$. We will not pursue the problem of obtaining UMP tests within the class of similar size α tests but refer to Lehmann (1959) for details.

We close this section with some additional examples to demonstrate how the approach based on LRT leads to some of the well known tests that the reader may have come across in the methods course.

EXAMPLE 9.1.5. Consider a situation where we have k-samples from k normal population with possible different means $(\mu_1, \ldots, \mu_k)$ and common variance σ^2. Our object is to test the homogeneity of the k samples i.e. $H_0 : \mu_1 = \mu_2 = \ldots = \mu_k$ and H_1 is $\mu_i \neq \mu_j$ for at least one pair (i, j). The common variance σ^2 is unspecified under H_0 as well as H_1.

We then have the following setup.

$$X_{ij} = \mu_i + \varepsilon_{ij},\ j = 1, 2, \ldots, n_i,\ i = 1, 2, \ldots, k$$

where $\{\varepsilon_{ij}\}$ are i.i.d. $N(0, \sigma^2)$. Here $\theta = (\mu_1, \mu_2, \ldots, \mu_k, \sigma^2)'$ and under H_0, $\theta = (\mu, \ldots, \mu, \sigma^2)'$ and $\Omega_0 \cup \Omega_1 = \Omega = R_k \times R_+$. The likelihood of the data (x_{ij}), $j = 1, 2, \ldots, n_i$, $i = 1, 2, \ldots, k$ is given by

$$L(x, \mu_1, \ldots, \mu_k, \sigma^2) = \prod_{i=1}^{k} \left(\frac{1}{\sqrt{2\pi\sigma^2}} \right)^{n_i} \exp \left\{ - \sum_{j=1}^{n_i} \frac{(x_{ij} - \mu_i)^2}{2\sigma^2} \right\}$$

The MLEs are $\hat{\mu}_i = \bar{x}_i$ and $\hat{\sigma}^2 = \frac{1}{N} \sum_{i=1}^{k} \sum_{j=1}^{n_i} (x_{ij} - \bar{x}_i)^2$ where $N = \sum_{i=1}^{k} n_i$ so that

$$\sup_{\theta \in \Omega} L(x, \theta) = \left(\frac{1}{\sqrt{2\pi\hat{\sigma}^2}} \right)^N \exp\{-N/2\}$$

The MLEs under H_0 are given by

$$\hat{\mu} = \frac{1}{N} \sum_{i=1}^{k} \sum_{j=1}^{n_i} x_{ij} = \bar{x}, \text{ the grand mean}$$

and $$\hat{\sigma}_0^2 = \frac{1}{N} \sum_{i=1}^{k} \sum_{j=1}^{n_i} (x_{ij} - \bar{x})^2$$

so that $$\sup_{\theta \in \Omega_0} L(x, \theta) = \left(\frac{1}{\sqrt{2\pi\hat{\sigma}_0^2}} \right)^N \exp\left\{ -\frac{N}{2} \right\}.$$

The LRTS $\lambda(x) = \left(\frac{\hat{\sigma}^2}{\hat{\sigma}_0^2} \right)^{N/2}$ and $\lambda(x) \le c$ is equivalent to $\left(\frac{\hat{\sigma}_0^2}{\hat{\sigma}^2} \right)^N \ge c_1$.

We now use the well known decomposition of sums of squares (SS) namely

$$\sum_{i=1}^{k} \sum_{j=1}^{n_i} (x_{ij} - \bar{x})^2 = \sum_{i=1}^{k} \sum_{j=1}^{n_i} (x_{ij} - \bar{x}_i)^2 + \sum_{i=1}^{k} n_i (\bar{x}_i - \bar{x})^2$$

that is, Total SS = Within samples SS + Between Samples SS, which gives $N\hat{\sigma}_0^2 = N\hat{\sigma}^2 + \sum n_i(\bar{x}_i - \bar{x})^2$. Therefore $\frac{\hat{\sigma}_0^2}{\hat{\sigma}^2} \ge c$ is equivalent to $\frac{\sum_{i=1}^{k} n_i(\bar{x}_i - \bar{x})^2}{\sum_{i=1}^{k} \sum_{j=1}^{n_i} (x_{ij} - \bar{x}_i)^2} \ge c_3$. Now under H_0, using Cochran's (1934) theorem one can show that the numerator is distributed as χ^2_{k-1} and the denominator is χ^2_{N-k} and they are independent. For relevant distribution theory reference may be made to Rao (1973) and Kendall and Stuart Vol. II (1967).

Thus the LRT in this case is equivalent to the well known F test in the one way analysis of variance, given by $\frac{\chi^2_{k-1}}{k-1} \Big/ \frac{\chi^2_{N-k}}{N-k}$. Note that in this case $\Omega_{H_0} = R_1 \times R_+$, $\Omega = \Omega_{H_0} \cup \Omega_{H_1} = R_k \times R_+$ and H_0 and H_1 are both composite. The F test is also a similar test in that its power function under H_0 satisfies $\beta_F(\mu, \sigma^2) = \alpha$, $\forall\ (\mu, \sigma^2)' \in \Omega_{H_0}$. Again one can show that the F test derived above is UMP level α test within the class of similar level α tests. For $k = 2$ the LRT is equivalent to the well known two sample $t_{n_1+n_2-2}$ test which rejects H_0 if

$$\frac{(\bar{x}_1 - \bar{x}_2)^2}{s_p^2\left(\frac{1}{n_1} + \frac{1}{n_2}\right)} > t^2_{n_1+n_2-2,.1-\alpha/2}$$

where $s_p^2 = \dfrac{(n_1 - 1)s_1^2 + (n_2 - 1)s_2^2}{n_1 + n_2 - 2}$ is the pooled unbiased estimator of σ^2 and s_1^2 and s_2^2 are unbiased estimators of σ^2 from the two samples respectively.

So far we have been lucky in that the LRT led us to a test statistic whose distribution under H_0 is well known and even well tabulated for small samples as well. In the next section we will consider the famous Pearson χ^2 test of goodness of fit and show that it is a large sample approximation to the LRT for this problem. We will then show how similar techniques could be used to obtain large sample tests in many other situations.

Exercise 9.1. (i) Let $(X_1, \ldots, X_n)$ be i.i.d. exponential with mean θ. Obtain LRT for testing $H_0 : \theta \le 1$ against $H_1 : \theta > 1$ and show that it coincides with the UMP test for the problem. Obtain LRT to test H_0: $\theta = 1$ against $H_1 : \theta \ne 1$. Show that it is equivalent to the test which rejects H_0 if $T = \sum X_i \notin (a, b)$ where (a, b) are such that $G_n(b) - G_n(a) = 1 - \alpha$. Show that if $ag_n(a) = bg_n(b)$ where G_n and g_n denote the d.f. and p.d.f. of $G(n, 1)$ r.v., then the LRT is an unbiased test.

(ii) Show how results of the above exercise can be used to test $H_0 : \lambda \ge 1$ vs $H_1 : \lambda < 1$ in the Pareto distribution with pdf $f(x, \lambda) = \dfrac{\lambda}{x^{\lambda+1}}$, $x \ge 1$, $\lambda > 0$.

(iii) Consider the simple regression problem

$$y_i = \alpha + \beta x_i + \varepsilon_i,\ i = 1, 2, \ldots, n$$

where $(x_1, \ldots, x_n)$ are fixed constants with $\sum(x_i - \bar{x})^2 > 0$. Obtain the LRT for testing H_0: $\beta = 0$ vs $H_1 : \beta \ne 0$ when $\{\varepsilon_i\}_1^n$ are i.i.d. $N(0, \sigma^2)$. How will you modify the above LRT if H_1 is $\beta > 0$.

(iv) Let $(X_1, \ldots, X_{n_1})$ and $(Y_1, Y_2, \ldots, Y_{n_2})$ be two independent random samples from Poisson distributions with means λ_1 and λ_2 respectively. Obtain LRT for testing the $H_0 : \lambda_1 = \lambda_2$ against $H_1 : \lambda_1 \ne \lambda_2$ and also for $H_0 : \lambda_1 = \lambda_2$ against $H_1 : \lambda_1 > \lambda_2$. The two above LRTs are examples of similar tests.

(v) Refer to Example given in Sec. 7.5 where $P(GG) = \theta^2$, $P(Gg) = \theta(1 - \theta) = P(gG)$ and $P(gg) = (1 - \theta)^2$. Obtain LRT (i) for testing $\theta = \frac{1}{2}$ against $H_1 : \theta \ne 1/2$ and (ii) for testing $\theta = 3/4$ against $H_1 : \theta \ne 3/4$.

The first null hypothesis corresponds to equiprobable cells where as the second null hypothesis corresponds to Mendelian hypothesis, with frequencies proportional to (9:3:3:1).

9.2 Likelihood Ratio Tests for Multinomials

Consider a situation when we have a random sample of size n on a multinomial distribution in k-cells with observed cell frequencies $(n_1, n_2, \ldots, n_k)'$ such that $\sum n_i = n$. The well known χ^2 test of goodness of fit due to K. Pearson (1900) uses the test statistic $\chi^2 = \sum_{i=1}^{k} (O_i - E_i)^2/E_i$ where O_i are observed cell frequencies and E_i are expected cell frequencies under $H_0 : p_i = p_{i0}$, $i = 1, 2, \ldots, k$. We reject H_0 if the observed value of χ^2 exceeds the critical value. $\chi^2_{k-1, 1-\alpha}$, the 100(1 – α)% point of the chi-square distribution with $(k - 1)$ degrees of freedom (d.f.).

We show that χ^2 test is an approximation to LRT for testing $H_0 : p_i = p_{i0}$, $i = 1, 2, \ldots, k$ against $H_1 : p_i \neq p_{i0}$ for at least one $i = 1, 2, \ldots, k$.

Since H_0 is a simple null hypothesis

$$\operatorname*{Sup}_{p \in \Omega_0} L(x, p) = L(x, p_0) = \frac{n!}{\prod_{i=1}^{k} n_i!} \prod (p_{i0})^{n_i}$$

Now

$$L(x, p) = \frac{n!}{\prod_{i=1}^{k} n_i!} \prod_{i=1}^{k} p_i^{n_i}$$

and

$$\sup_{p \in \Omega} L(x, p) = L(x, \hat{p}) = \frac{n!}{\prod_{i=1}^{k} n_i!} \prod_{i=1}^{k} (\hat{p}_i)^{n_i}$$

where $\hat{p}_i = \frac{n_i}{n}$. We leave it as an exercise to the reader to show that $\hat{p} = (\hat{p}_1, \ldots, \hat{p}_k)'$ is infact the MLE of $p = (p_1, \ldots, p_k)'$ where $p_i > 0$, $i = 1, 2, \ldots, k$, $\sum p_i = 1$. We have already seen in Example 6.3.3 that $(\hat{p}_1, \ldots, \hat{p}_{k-1})'$ is CAN for $(p_1, \ldots, p_{k-1})'$ with asymptotic variance covariance matrix $\frac{1}{n} \Lambda_{k-1}$ where

$$\lambda_{ii} = p_i(1 - p_i) \text{ and } \lambda_{ij} = - p_i p_j.$$

The LRTS then is given by

$$\lambda(x) = \prod_{i=1}^{k} (p_{i0})^{n_i} / \prod_{i=1}^{k} (\hat{p}_i)^{n_i} \tag{9.2.1}$$

or $\log \lambda(x) = \sum_{i=1}^{k} n_i \log \frac{p_{i0}}{\hat{p}_i}$. Using $n_i = n\hat{p}_i$ we have

$$-2 \log \lambda(x) = 2n \sum \hat{p}_i \log \frac{\hat{p}_i}{p_{i0}} \tag{9.2.2}$$

We now show that RHS of (9.2.2) is equal to Pearson χ^2 plus a remainder term which converges in probability to zero. Put

$$y_i = \sqrt{n}(\hat{p}_i - p_{i0}) \text{ or } \hat{p}_i = \frac{y_i}{\sqrt{n}} + p_{i0},\ i = 1, 2, \ldots, k \text{ then}$$

$$-2\log\lambda(x) = 2n\sum_{i=1}^{k}\left\{\left(\frac{y_i}{\sqrt{n}} + p_{i0}\right)\log\left(1 + \frac{y_i}{p_{i0}\sqrt{n}}\right)\right\} \tag{9.2.3}$$

Since $\hat{p}_i \xrightarrow{P} p_{i0}$ under H_0, $\dfrac{y_i}{\sqrt{n}\,p_{i0}} \xrightarrow{P} 0$ under H_0, and we can use expansion of $\log(1 + x)$ for $|x| < 1$. Therefore

$$\log\left(1 + \frac{y_i}{p_{i0}\sqrt{n}}\right) = \frac{y_i}{p_{i0}\sqrt{n}} + \frac{y_i^2}{p_{i0}\,n} + 0\left(\frac{1}{n}\right).$$

But $\sum y_i = 0$ and after summing up over i we obtain

$$-2\log\lambda(x) = 2n\sum\left\{\frac{1}{2n}\frac{y_i^2}{p_{i0}^2} + 0\left(\frac{1}{n}\right)\right\} = \sum_{i=1}^{k}(n\hat{p}_i - np_{i0})^2/np_{i0} + \varepsilon_n$$

$$= \chi^2 + \varepsilon_n$$

where $\varepsilon_n \xrightarrow{P} 0$ under H_0. As seen in Example 6.3.3, Pearson χ^2 has asymptotic chi-square distribution with $(k - 1)$ d.f. We emphasize here that $\hat{p} = (\hat{p}_1, \ldots, \hat{p}_{k-1})'$ is CAN for $p = (\hat{p}_1, \ldots, \hat{p}_{k-1})'$ with asymptotic variance covariance matrix

$$\frac{1}{n}\Lambda_{k-1} = \frac{1}{n}I_{k-1}^{-1}(p) \text{ where } \lambda_{ii} = p_i(1 - p_i) \text{ and } \lambda_{ij} = -p_i p_j,$$

and $$\chi^2 = (\hat{p} - p_0)'[n\,\Lambda_{k-1}^{-1}(p_0)](\hat{p}_0 - p_0)$$

$$= n(\hat{p} - p_0)'\,I_{k-1}(p_0)(\hat{p} - p_0)$$

Now consider a situation in which H_0 can be composite. For example such a situation arises in a 2×2 contingency table or in general $r \times s$ contingency table where H_0 is the hypotheses of row and column independence.

Let $A_1, \ldots, A_r$ be a partition of the sample space according to a factor A at r levels and let $B_1, \ldots, B_s$ be a partition of the same sample space according to a factor B at s levels. Then $C_{ij} = A_i \cap B_j$, $i = 1, 2, \ldots, r$, $j = 1, 2, \ldots, s$ is a partition of the sample space and let p_{ij} denote the cell probabilities and n_{ij} observed frequencies in a random sample of size n on the corresponding multinomials. Let $p_{i.} = \sum_{j=1}^{s} p_{ij} = P(A_i)$ and $p_{.j} = \sum_{i=1}^{r} p_{ij} = P(B_j)$. Then the hypotheses of independence of the classification factors A and B at all levels corresponds to the relations $p_{ij} = P(A_i \cap B_j) = P(A_i)\, P(B_j) = p_{i.} \times p_{.j}$, $i = 1, 2, \ldots, r$, $j = 1, 2, \ldots, s$.

Now, likelihood of the sample when p_{ij} are such that $\Sigma\Sigma p_{ij} = 1$, $p_{ij} > 0$ is given by

$$L(n, p) = \frac{n!}{\prod_{i,j} n_{ij}!} \prod_{i,j} (p_{ij})^{n_{ij}} \tag{9.2.4}$$

and that under the null hypotheses of independence is given by

$$L_0(n, p) = \frac{n!}{\prod_{i,j} n_{ij}!} \prod_{i,j} (p_{.i}\, p_{.j})^{n_{ij}} \tag{9.2.5}$$

The MLE's $\hat{p}_{ij} = \dfrac{n_{ij}}{n}$ and $\sup_{p \in \Omega} \log L(n, p) = \text{const.} + \sum_i \sum_j n_{ij} \log \hat{p}_{ij}$.

Similarly MLEs of $p_{i.}$ and $p_{.j}$ are $\dfrac{n_{i.}}{n} = \sum_j n_{ij}/n$ and $\dfrac{n_{.j}}{n} = \sum_i n_{ij}/n$ which gives

$$\sup_{p \in \Omega_0} \log L_0(n, p) = \text{const.} + \sum_i n_{i.} \log \hat{p}_{i.} + \sum_j n_{.j} \log \hat{p}_{.j}.$$

The logarithm of LRTS is

$$\log \lambda(n) = \Sigma n_{i.} \log \hat{p}_{i.} + \Sigma n_{.j} \log \hat{p}_{.j} - \Sigma\Sigma n_{ij} \log \hat{p}_{ij}$$

which can be simplified to

$$-2 \log \lambda(n) = 2n \sum_i \sum_j \frac{n_{ij}}{n} (\log \hat{p}_{ij} - \log \hat{p}_{i.}\, \hat{p}_{.j}) \tag{9.2.6}$$

Now following the techniques used in derivation of Pearson χ^2 as an approximation to LRTS, we can show that

$$-2 \log \lambda(n) \simeq \sum_{i=1}^{r} \sum_{j=1}^{k} \left(n_{ij} - \frac{n_{i.} n_{.j}}{n} \right)^2 \Bigg/ \frac{n_{i.} n_{.j}}{n} + 0(1/n).$$

However we will consider a more basic technique. Consider Taylor series expansion of log b about log a which is given by

$$\log b = \log a + (b - a)\frac{1}{a} - \frac{1}{2}(b - a)^2 \frac{1}{a^2} + 0(|\, b - a\,|^{2+\delta})$$

or $\quad a(\log a - \log b) = -(b - a) + \dfrac{1}{2}(b - a)^2 \dfrac{1}{a} + 0(|\, b - a\,|^{2+\delta})\, a.$

Take $a = \hat{p}_{ij}$ and $b = \hat{p}_{i.}\, \hat{p}_{.j}$ and expand each term on RHS of (9.2.6). Then as $\Sigma\Sigma\ \hat{p}_{ij} = 1 = \Sigma\ \hat{p}_{i.} = \Sigma\ \hat{p}_{.j}$ we have RHS of (9.2.6)

$$\sum_i \sum_j \left(\frac{n_{ij}}{n} - \frac{n_{i.}n_{.j}}{n}\right)^2 \Big/ \frac{n_{i.}}{n}\frac{n_{.j}}{n} = \frac{\Sigma\left(n_{ij} - \dfrac{n_{i.}n_{.j}}{n}\right)^2}{n_{i.}n_{.j}} + \varepsilon_n$$

where $\varepsilon_n \xrightarrow{p} 0$ under H_0. The reader will recall that the methods course states that $\sum_{i,j} \frac{(O_{ij} - E_{ij})^2}{E_{ij}}$ has a χ^2 distribution with $(r-1)(s-1)$ d.f. and not $(rs-1)$ d.f. In estimating $p_{i.}$ and $p_{.j}$ under H_0 we are estimating $(r-1) + (s-1)$ parameters, whereas under $H_0 \cup H_1$ we are estimating $(rs-1)$ parameters. As a consequence of Pearson-Fisher Theorem quoted below, the Pearson χ^2 has d.f. given by $(rs-1) - (r-1) - (s-1) = (r-1)(s-1)$. Originally Pearson (1900) indicated that for testing independence in a 2×2 table the number of d.f. for χ^2 is $4 - 1 = 3$. However as pointed by Yule and Greenwood (1915) and Bowley (1920) this led to certain inconsistencies. Fisher (1922) indicated that the correct number of d.f. in this case is not three but $1 = (2-1)(2-1)$. No body at that time thought that K. Pearson could be wrong and it appears that Fisher had difficulty in publishing this result. However Fisher (1928) showed that if the multinomial distribution in k cells has probabilities $p_i(\theta)$ depending on a vector parameter $\theta = (\theta_1, \ldots, \theta_m)'$ then the $\chi^2 = \sum_{i=1}^{k} (n_i - np_i(\hat{\theta}))^2 / np_i(\hat{\theta})$ has a χ^2 distribution with $(k-1) - m$ d.f. when $\hat{\theta}$ is MLE of θ. Thus the d.f. $(k-1)$ for the simple $H_0 : p_i = p_{io}$, $i = 1, 2, \ldots, k$ are reduced by m, the number of parameters estimated under H_0. This is the well known Pearson-Fisher theorem. There were a few gaps in the proof given by Fisher but these were filled up by Cramer (1946) and Birch (1964). We will not go into details here, but strongly recommend the readers to refer to the original papers of Fisher, particularly Author's notes written by Fisher himself in Shewhart (1950) where references to Yule and Greenwood (1915) and Bowley (1920) are listed fully.

We now consider an example to illustrate the Pearson-Fisher Theorem.

EXAMPLE 9.2.1. Consider example 7.5.2 due to Fisher (1954) which considers a multinomial distribution is 4 cells with cell probabilities

$$p_1(\theta) = \frac{2+\theta}{4},\ p_2(\theta) = p_3(\theta) = \frac{1-\theta}{4} \text{ and } p_4(\theta) = \theta/4,\ 0 < \theta < 1$$

where θ denotes the linkage factor between two varieties of maize classified as Sugary and Starchy with colors Green and White. In Example 7.5.2 for Carver's data we hae seen that MLE $\hat{\theta} = .035712$ and therefore the expected cell frequencies under the null hypotheses that the model is true are given by

$$n\left(\frac{2+\hat{\theta}}{4}\right) = 1943.60,\ n\left(\frac{1-\hat{\theta}}{4}\right) = n\left(\frac{1-\hat{\theta}}{4}\right) = 920.65,\ \frac{n\hat{\theta}}{4} = 34.10$$

where $n = 3819$.

On the other hand $n\hat{p}_1 = 1977 = n\hat{p}_2 = 906, n\hat{p}_3 \ 904$ and $n\hat{p}_4 = 32$. This gives

$$\sup_{p\in\Omega} L(n, p) = \frac{n!}{\pi(n_i!)}\pi(\hat{p}_i)^{n_i} \text{ and}$$

$$\sup_{\theta\in\Theta_0} L(n, p(\theta)) = \frac{n!}{\pi(n_i!)}\pi(\hat{p}_i(\hat{\theta}))^{n_i}$$

so that

$$-2\log\lambda(n) = 2n\sum_{i=1}^{4}\hat{p}_i \log\cdot \hat{p}_i/p_i(\hat{\theta})$$

$$= 2\sum_{i=1}^{4} n_i \log\frac{n_i}{np_i(\hat{\theta})} = .6660$$

On the other hand Pearson $\chi^2 = \sum_{i=1}^{4}(O_i - E_i)^2/E_i = 1.2375$. We observe that even for $n = 3819$ a fairly large sample size, the approximation to $-2\log\lambda(n)$ by Pearson χ^2 is not very close.

Since $k = 4$ and $m = 1$, Pearson χ^2 test statistic has 2 d.f. Suppose $\alpha = .05$ then we reject the model specified by H_0 if observed χ^2 value is greater than the critical value $\chi^2_{2,1-\alpha} = 5.99147$. In this case therefore we do not reject H_0 either on the basis of Pearson χ^2 or LRTS.

If Pearson χ^2 test statistic is regarded as a measure of discrepancy between the observed multinomial distribution and the hypothetical multinomial, then probability of obtaining a value of Pearson χ^2 as large or larger than the observed value can be regarded as a measure of goodness of fit of the model with the observed data. Nearer this value is to zero more strongly we should suspect H_0 and nearer this value is to one, stronger is the support in favour of model specified by H_0. For the Carver's data the observed Pearson χ^2 is 1.2375 and $P[\chi^2_2 \geq 1.2375] = \int_{1.2375}^{\infty} f_2(u)\,du$ where $f_2(u)$ is the pdf of χ^2_2 which is same as that of an exponential distribution with mean 2. Thus the above probability is given by $\frac{1}{2}\int_{1.2375}^{\infty} e^{-u/2}\,du = .5386$.

From Abramowitz and Stegun (1964) (Table 26.7) by interpolation one obtains the value .5388 which is quite close to the exact value. The above probability is called as the observed level of significance of the Pearson χ^2 statistic and in the general case it would be

$$P[\chi^2_{k-1-m} > \text{observed value of Pearson } \chi^2]$$

If this probability is less than α, the specified level of significance or maximum tolerable type I error, then we reject H_0 otherwise we accept H_0.

The Pearson Fisher theorem can also be used to test the model prescribed by a continuous distribution. This is achieved by grouping the data into number of cells with cell probabilities $p_i(\theta) = \int_{c_i} f(x, \theta)\, dx$. If the Pearson χ^2 test statistic rejects H_0; i.e. $p_i = p_i(\theta)$, $i = 1, 2, \ldots, k$ then this implies rejection of the model specified by the class of pdf given by $\{f(x, \theta), \theta \in \Omega\}$. Note that here both x and θ can be vector valued. On the other hand if we have a discrete random variable with $P[X = x_i] = p_i(\theta)$, $i = 1, 2, \ldots, k$ then each c_i can be taken as x_i or we can still have a grouped data. We illustrate this technique by examples.

EXAMPLE 9.2.2. Consider the data obtained by Rutherford Chadwick and Ellis (1920) reported earlier in Chapter 1. Suppose we want to test the hypotheses that the model is described by the Poisson pmf with parameter $\lambda > 0$ unspecified. Following Cramer (1946) (Table 30.4.1) we obtain:

x	n_i	$np_i(\hat{\lambda})$
0	57	54.399
1	203	210.523
2	383	407.361
3	525	525.496
4	532	508.418
5	408	393.575
6	273	253.817
7	139	140.325
8	45	67.882
9	27	29.189
≥ 10	16	17.075

The MLE $\hat{\lambda} = 3.87$ and Pearson $\chi^2 = 12.885$. The d.f. of Pearson χ^2 are $11 - 1 - 1 = 9$ since under H_0 only one parameter is estimated. The critical value of χ^2_9 at $\alpha = .05$ is 16.919 and we accept H_0 or say that Poisson distribution model holds. The observed level of significance is obtained as

$$p \text{ value} = P[\chi^2_9 \geq 12.885] \simeq .17$$

Thus although the data supports the hypotheses of Poisson distribution, this support is not as strong as that for H_0 in the following example.

EXAMPLE 9.2.3. In his famous novel "Jurassic Park" Michael Crichton (1990) presents a statistical analysis of the data on the heights in cms of 68 procompsognathids bred in the Park and released in three batches at six months intervals. The statistical analysis done by Malcolm, the information scientist of the team shows that the table presented below has frequency polygon which very closely resembles a perfectly symmetric normal ditribution. Malcolm argues that this resemblence constitutes a proof that the dinosaurs were breeding at random because only then the distribution of their heights will produce a normal curve. On the other hand since three batches were released at six months interval the data on heights should more look like a mixture of three different distributions if there was no random breeding. We will examine this analysis using Pearson-Fisher theorem and application of Pearson χ^2 to a grouped data for testing H_0 : Data is a random sample from $N(\mu, \sigma^2)$, both (μ, σ^2) unspecified.

Table 9.2.1. Height of 68 "Compsos" (cm)

Height	O_i	$np_i(\hat{\mu}, \hat{\sigma}^2) = E_i$
≤ 29	4	5.91
≤ 30	4	4.39
≤ 31	6	5.95
≤ 32	9	7.64
≤ 33	10	8.71
≤ 34	10	8.85
≤ 35	9	9.08
≤ 36	6	6.44
≤ 37	4	4.92
> 37	6	6.06
	68	68.00

The MLE are $\hat{\mu} = 33.15$ and $\hat{\sigma}^2 = 3.05$. Pearson $\chi^2 = 1.4184$ and is to be tested at $10 - 1 - 2 = 7$ d.f. as two parameters μ and σ^2 are estimated. Now $\chi^2_{7,.95} = 14.067$ and we do not reject H_0. The p-value of the observed χ^2 is given by $P[\chi^2_7 \geq 1.4184] \approx .9824$ which very strongly supports the hypotheses of normality. Hence the incontrovertible conclusion is that dinosaurs are breeding at random. This later turns out to be the fact and there is a plausible explanation of how this could happen due to certain possible properties of DNA sequences used in breeding "Compsos" in the laboratory. The data is fictious but the ingenuity of the novelist and his use of statistical thinking is very noteworthy.

We refer to Cramer (1946, Sec. 30.2) and Fisz (1963, Example 12.4.2) and Rao (1973, Sec. 6b) for some real life data (grouped) to test normality by using Pearson χ^2 statistic and Pearson-Fisher Theorem.

Exercise 9.2. (i) As per the Mendelian hypotheses $H_0 : p_1 = 9/16$ $p_2 = 3/16$, $p_3 = 3/16$ and $p_4 = 1/16$ where the four cells correspond to 2×2 classification of peas according to (Round and Angular) × (Yellow and Green). For the data $n_1 = 315$, $n_2 = 108$, $n_3 = 101$

and $n_4 = 32$, calculate Pearson χ^2 and show that the p-level is between 90% to 95%. The Mendelian Hypothesis is strongly supported by the data. Should one then suspect data as the fit is too good to be true? Thus for χ^2 test, should one suspect very small value of χ^2 also? The common view is to believe in the honesty of the concerned experimenters and reject the model specified by H_0 only for large values of Pearson χ^2.

(ii) Consider a multinomial distribution in 4 cells with $p_1(\theta) = \theta^2$, $p_2(\theta) = \theta(1 - \theta) = p_3(\theta)$ and $p_4(\theta) = (1 - \theta)^2$ (Refer Section 7.5). Obtain Pearson χ^2 test for this model. Suppose for a data of 1000 observations the observed cell frequency are 558, 192, 178, 72. What is the p-level in support of H_0 that the model is true?

(iii) Consider a 2 × 2 table discussed by Yule and Greenwood (1915) to study the effect of antityphoid and anticholera inoculations on incidence of these diseases. The data is taken from Kendall and Stuart, Vol. 2 [(1967) Example 33.1].

	Not attacked	Attacked	Total
Innoculated	276	3	279
Not innoculated	473	66	539
	749	89	818

Calculate Pearson χ^2 and test (at $\alpha = .5$) the hypotheses of independence between the two attributes namely being innoculated and not attacked. What is the p-level of the observed Pearson χ^2?

9.3 Large Sample Tests

Consider testing $H_0 : \theta = \theta_0$ against $H_1 : \theta \neq \theta_0$ on the basis of a random sample of size n with pdf belonging to one parameter Cramer family given by $\{f(x, \theta), \theta \in \Omega \subset R_1\}$ then

$$-2 \log \lambda(x) = 2[\log L(x, \hat{\theta}) - \log L(x, \theta_0)].$$

Now expand $\log L(x, \theta_0)$ around $\hat{\theta}$ by Taylor series to obtain

$$\log L(x, \theta_0) = \log L(x, \hat{\theta}) + (\theta_0 - \hat{\theta})\left(\frac{\partial \log L}{\partial \theta}\right)_{\theta=\hat{\theta}}$$

$$+ \frac{(\theta_0 - \hat{\theta})^2}{2}\left(\frac{\partial^2 \log L}{\partial \theta^2}\right)_{\theta=\hat{\theta}} + \varepsilon_n$$

where $\varepsilon_n = 0(|\theta_0 - \hat{\theta}|^{2+\delta})$. Noting that $\left(\frac{\partial \log L}{\partial \theta}\right)_{\hat{\theta}} = 0$ we have

$$-2 \log \lambda(x) = -(\theta_0 - \hat{\theta})^2\left(\frac{\partial^2 \log L}{\partial \theta^2}\right)_{\theta=\hat{\theta}} + \varepsilon_n$$

$$= nI(\theta_0)(\theta_0 - \hat{\theta})^2 \cdot \frac{-1}{nI(\theta_0)}\left(\frac{\partial^2 \log L}{\partial \theta^2}\right)_{\theta=\hat{\theta}} + \varepsilon_n$$

$$= UV + \varepsilon_n$$

where $U = nI(\theta_0)(\theta_0 - \hat{\theta})^2 \xrightarrow{d} \chi_1^2$ and $V = \frac{1}{nI(\theta_0)}\left(-\frac{\partial^2 \log L}{\partial \theta^2}\right)_{\hat{\theta}} \xrightarrow{p} 1$ in view of Cramer-Huzurbazar Theorem and $\varepsilon_n \xrightarrow{p} 0$.

Thus $-2 \log \lambda \xrightarrow{d} \chi_1^2$ under H_0. Therefore, LRT rejects H_0 if $-2 \log \lambda(x) > \chi^2_{1,1-\alpha} = \xi^2_{1-\alpha/2}$ where ξ_α is $100\alpha\%$ point of the standard normal distribution.

The result can be generalized to m-parameter Cramer family to show that

$$-2 \log \lambda(x) = U_m V_m + \varepsilon_n$$

where

$$U_m = n(\hat{\theta} - \theta_0)' I(\theta_0)(\hat{\theta} - \theta_0) \tag{9.3.1}$$

and

$$V_m = I^{-1}(\theta_0)\left(\left(\frac{1}{n}\frac{\partial^2 \log L}{\partial \theta_r \partial \theta_s}\right)\right)_{\hat{\theta}}.$$

As $\sqrt{n}(\hat{\theta} - \theta_0) \xrightarrow{d} N^{(m)}(0, I^{-1}(\theta_0))$ we have $U_m \xrightarrow{d} \chi^2_m$. Further $V_m \xrightarrow{p} I_{m\times m}$ the identity matrix of dimension m. Thus $-2 \log \lambda(x) \xrightarrow{d} \chi^2_m$ and we reject $H_0 : \theta = \theta_0$ if $-2 \log \lambda(x) > \chi^2_{m,1-\alpha}$. The approximate p-level is given by $P[\chi^2_m > -2 \log \lambda(x)]$.

A large sample test equivalent to the LRT derived above can be obtained by using the fact that under H_0, $\hat{\theta} \sim \text{AN}^{(m)}\left(\theta_0, \frac{1}{n} I^{-1}(\theta_0)\right)$. Thus if observed deviation between $\hat{\theta}$ and hypothesized value θ_0 is large we have an occurrence of rare event and we reject H_0. The deviation between $\hat{\theta}$ and θ_0 is however not measured by the usual norms $\Sigma(\hat{\theta}_i - \theta_0)^2$ or $\max_i |\hat{\theta}_i - \theta_{i0}|$ but by the norm defined by

$$U_m = n(\hat{\theta} - \theta_0)' I(\theta_0)(\hat{\theta} - \theta_0).$$

Wald (1943) proposed that instead of using $I(\theta_0)$ one could use its estimator given by $I(\hat{\theta})$ i.e. approximate $-2 \log \lambda(x)$ by $n(\hat{\theta} - \theta_0)' I(\hat{\theta})(\hat{\theta} - \theta_0)$ and we reject H_0 if $n(\hat{\theta} - \theta_0)' I(\hat{\theta})(\hat{\theta} - \theta_0) > \chi^2_{m,1-\alpha}$. Here $\hat{\theta}$ is MLE of θ under $H_0 \cup H_1$.

The LRT as well as Wald's test requires explicit evaluation of MLE $\hat{\theta}$ and the information matrix $I(\theta)$ at θ_0 or $\hat{\theta}$. One can obtain a large sample test based on the score functions $\frac{\partial \log L}{\partial \theta} = \left(\frac{\partial \log L}{\partial \theta_1}, \ldots, \frac{\partial \log L}{\partial \theta_m} \right)'$, using the fact that under the null hypotheses $H_0: \theta = \theta_0$, $\frac{1}{\sqrt{n}} \left(\frac{\partial \log L}{\partial \theta} \right)_{\theta=\theta_0}$ is $AN^{(m)}(0, I(\theta_0))$ and

$$Q(\theta_0) = \frac{1}{n} \left(\frac{\partial \log L}{\partial \theta} \right)'_{\theta=\theta_0} I^{-1}(\theta_0) \left(\frac{\partial \log L}{\partial \theta} \right)_{\theta=\theta_0} \tag{9.3.2}$$

has asymptotic χ^2_m distribution, and we reject H_0 if $Q(\theta_0) > \chi^2_{m,1-\alpha}$. This is known as a score test and was proposed by Rao (1947). In the score test one can use $\left(\left(\frac{\partial^2 \log L}{\partial \theta_r \partial \theta_s} \right) \right)^{-1}_{\theta=\theta_0}$ instead of $\frac{1}{n} I^{-1}(\theta_0)$ to obtain

$$Q_1(\theta_0) = \left(\frac{\partial \log L}{\partial \theta} \right)'_{\theta=\theta_0} \left(\left(\frac{\partial^2 \log L}{\partial \theta_r \partial \theta_s} \right) \right)^{-1}_{\theta=\theta_0} \left(\frac{\partial \log L}{\partial \theta} \right)_{\theta=\theta_0} \tag{9.3.3}$$

The score test given by $Q_1(\theta_0)$ does not require explicit evaluation of θ and $I^{-1}(\theta_0)$. The asymptotic distribution of the test statistic $Q(\theta_0)$ or $Q_1(\theta_0)$ is still χ^2_m under H_0 and we reject H_0 if the observed value of the test statistic exceeds $\chi^2_{m,1-\alpha}$.

EXAMPLE 9.3.1. Let $(x_1, \ldots, x_n)$ be a random sample of size n from Cauchy distribution with location μ. Then

$$\log L = -n \log \pi + \sum_{i=1}^{n} \frac{1}{[1 + (x - \mu)^2]}$$

$$\frac{\partial \log L}{\partial \mu} = \sum_{i=1}^{n} \frac{2(x_i - \mu)}{1 + (x_i - \mu)^2}$$

We also know that $I(\mu) = \frac{1}{2}$. Suppose we want to test $H_0 : \mu = 0$ against $H_1 : \mu \neq 0$, then $-2 \log \lambda$, the LRT statistic is approximated by $\frac{n}{2} (\hat{\mu})^2$ where $\hat{\mu}$ is MLE of μ and we reject H_0 if $\frac{n}{2} (\hat{\mu})^2 > \chi^2_{1,1-\alpha}$. This requires explicit evaluation of $\hat{\mu}$ which depends on the method of scoring for parameter given in (7.6.3). If instead we use (9.3.2) we reject H_0 if $\frac{2}{n} \left(\frac{\partial \log L}{\partial \mu} \right)^2_{\mu=0} > \chi^2_{1,1-\alpha}$ i.e. $\frac{2}{n} \left\{ \sum \frac{2(x_i)}{(1 + x_i^2)} \right\}^2 > \chi^2_{1,1-\alpha}$ which avoids explicit evaluation of $\hat{\mu}$. Since $I(\mu) = \frac{1}{2}$ we do not recommend using $Q_1(\mu_0)$ given by (9.3.3) since it requires evaluating

$$\left[-\frac{\partial^2 \log L}{\partial \mu^2}\right]^{-1}_{\mu=0} = \left\{\sum_{i=1}^{n} \frac{2(1-x_i^2)}{(1+x_i^2)^2}\right\}^{-1}.$$

EXAMPLE 9.3.2. Consider a two parameter Gamma distribution with pdf given by $f(x, \alpha, \lambda) = \frac{\alpha^{\lambda}}{\Gamma(\lambda+1)} e^{-\alpha x} x^{\lambda-1}$, $x > 0$, $\alpha > 0$, $\lambda > 0$. Suppose $H_0 : \alpha = 1, \lambda = 1$ i.e. under H_0 we have a standard exponential distribution then we have

$$\frac{\partial \log L}{\partial \alpha} = -\Sigma x_i + n\frac{\lambda}{\alpha} \cdot \left(\frac{\partial \log L}{\partial \alpha}\right)_{\alpha=1, \lambda=1} = -\Sigma x_i + n$$

$$\frac{\partial \log L}{\partial \lambda} = -n\psi(\lambda) + n \log \alpha + \Sigma \log x_i,$$

$$\left(\frac{\partial \log L}{\partial \lambda}\right)_{\alpha=1, \lambda=1} = -n\psi(1) + \Sigma \log x_i$$

where $\psi(\lambda) = \frac{d}{d\lambda} \log \Gamma(\lambda)$. The information matrix is

$$I_{\alpha\alpha} = n\frac{\lambda}{\alpha^2}, \; I_{\alpha\lambda} = I_{\lambda\alpha} = -\frac{n}{\alpha^2} \text{ and } I_{\lambda\lambda} = n\psi'(\lambda)$$

Hence the test statistic $Q(\theta_0)$ given by (9.3.2) now becomes

$$[n - \Sigma x_i, \Sigma \log x_i - n\psi(1)] \begin{bmatrix} n & -n \\ -n & n\psi'(1) \end{bmatrix} \begin{bmatrix} n - \Sigma x_i \\ \Sigma \log x_i - n\psi(1) \end{bmatrix}$$

From Abromowitz and Stegun (1968) $\psi(1) = -\gamma$ (Euler's constant) and $\psi'(1) = \pi^2/6$ and for observed data we can calculate $Q(\theta_0)$.

Here also MLE $(\hat{\alpha}, \hat{\lambda})$ has to be obtained by iterative procedure given in (7.6.5).

We recommend to reader to carry out a small simmulation experiment and compare values of the test statistics $Q(\theta_0)$ and U_m and compare the labour involved in computing them.

In the above discussion we considered the situation where $H_0 : \theta = \theta_0$ was a simple hypothesis. More often than not we are interested in the null hypotheses which specifies only a few components of θ say $H_0 : \theta_1 = \theta_{10}$, $\theta_k = \theta_{k0}$ and where the remaining components of θ are unspecified and act as nuisance parameters. We have seen examples of this type of problems in Sections 9.1 and 9.2. To illustrate the situation consider testing $H_0 : \lambda = 1$

with α unspecified in Example 9.3.2 considered above. This is similar to testing $H_0 : \theta = \theta_0$, σ^2 unspecified in $N(\theta, \sigma^2)$ model for which we have shown that LRT is equivalent to the well known one-sample Student's t test. We thus first consider the problem where $(x_1, \ldots, x_n)$ is random sample of size n from a population with pdf belonging to the m-parameter Cramer family $\{f(x, \theta^{(1)}, \theta^{(2)}), (\theta^{(1)}, \theta^{(2)}) \in \Omega_k \times \Omega_{m-k} \subset R_m\}$ and $H_0 : \theta^{(1)} = \theta_0^{(1)}$ with $\theta^{(2)}$ acting as a nuisance parameter. Under H_0, $L_0(x, \theta_0^{(1)}, \theta^{(2)})$ is the likelihood of the sample for fixed x and $\theta^{(2)} \in \Omega_{m-k}$. Let $\hat{\theta}_0^{(2)}$ be the MLE of $\theta^{(2)}$ when $\theta^{(1)} = \theta_0^{(1)}$ and let $(\hat{\theta}^{(1)}, \hat{\theta}^{(2)})'$ be the MLE of $\theta = (\theta^{(1)}, \theta^{(2)})'$ under $H_0 \cup H_1$. Then LRTS is given by

$$-2 \log \lambda = 2\{\log L(x, \hat{\theta}^{(1)}, \hat{\theta}^{(2)}) - \log L(x, \theta_0^{(1)}, \hat{\theta}_0^{(2)})\} \quad (9.3.4)$$

It can be shown that the asymptotic distribution of $-2 \log \lambda(x)$ as given in (9.3.4) is χ_k^2. We will not go into a detailed proof of the result but present a heuristic argument below. The interested reader may refer to Rao (1973), and Wald (1943).

Consider $\theta_0 = (\theta_0^{(1)}, \theta_0^{(2)}) \in \Omega_{H_0}$. Then RHS of (9.3.4) can be written as

$$-2 \log \lambda(x) = 2\,[\log L(x, \hat{\theta}^{(1)}, \hat{\theta}^{(2)}) - \log L(x, \theta_0^{(1)}, \theta_0^{(2)})]$$
$$-2\,[\log L(x, \theta_0^{(1)}, \theta_0^{(2)}) - \log L(x, \theta_0^{(1)}, \theta_0^{(2)})]$$

On expansion in Taylor series around $(\hat{\theta}^{(1)}, \hat{\theta}^{(2)})$ in the first bracket and $(\theta_0^{(1)}, \hat{\theta}_0^{(2)})$ in the second bracket we have

$$-2 \log \lambda(x) = n(\hat{\theta} - \theta_0)' I(\theta_0)(\hat{\theta} - \theta_0) - n(\hat{\theta}_0^{(2)} - \theta_0^{(2)})' I_{22}(\theta_0)(\hat{\theta}_0^{(2)} - \theta_0^{(2)}) \quad (9.3.5)$$

where information matrix is partitioned as

$$I(\theta_0) = \left|\begin{array}{c|c} I_{11}(\theta_0) & I_{12}(\theta_0) \\ (k \times k) & \\ \hline I_{21}(\theta_0) & I_{22}(\theta_0) \\ & (m-k)(m-k) \end{array}\right|$$

Under $(\theta_0^{(1)}, \theta_0^{(2)}) \in \Omega_{H_0}$, the asymptotic distribution of the first term on RHS of (9.3.5) is χ_m^2 and of the second term is χ_{m-k}^2. Thus $-2 \log \lambda(x)$ for large n has asymptotic distribution as that of $\chi_m^2 - \chi_{m-k}^2$. It is thus tempting to guess that such a difference should behave like χ_k^2 in view of the additive property of independent chi-square variables, namely $\chi_m^2 = \chi_k^2 + \chi_{m-k}^2$. If Cochran's Theorem (1934) on quadratic forms in normal variables hold, [See also Cramer (1946)] then we can conclude that $\chi^2_m - \chi^2_{m-k}$ is indeed χ^2_k. A delicate argument given by Wilks (1938) shows that the Cochran's theorem holds for the asymptotic distribution and indeed one can show that $-2 \log \lambda(x) \xrightarrow{d} \chi^2_k$.

We can generalize this result further where the null hypotheses states that $\theta = (\theta_1, \ldots, \theta_m)'$ satisfies k functional relations $\psi_1(\theta) = c_1, \ldots, \psi_k(\theta) = c_k$ where $\psi_1, \ldots, \psi_k$ are such that there exists function $\psi_{k+1}(\theta), \ldots, \psi_m(\theta)$ so that

$\left| \dfrac{\partial(\psi_1, \ldots, \psi_m)}{\partial(\theta_1, \ldots, \theta_m)} \right| \neq 0$ for any $\theta \in \Omega$ i.e. $\psi(\theta) = (\psi_1, \ldots, \psi_m)'$ is a one to one transformation.

The question of obtaining a large sample test for $H_0 : \psi_i(\theta) = c_i$ where c_i are known constants $i = 1, 2, \ldots, k$, can now be solved by using a reparametrization $\theta \to \psi$ which is one-to-one and differentiable. Let $\psi = (\psi_1, \ldots, \psi_m)' \in \Psi$ and let $\{L_1(x, \psi), \psi \in \Psi \subset R_m\}$ be the transformed family of pdfs. Then H_0 specifies first k co-ordinates of ψ to be $(c_1, \ldots, c_k)'$ and $\psi_{k+1}, \ldots, \psi_m$ are now nuisance parameters. We then have the LRT in terms of MLEs of ψ under H_0 and $H_0 \cup H_1$. Using invariance of MLE under non-singular one-to-one transformation, by transforming back to θ from ψ, the LRTS and $-2 \log \lambda(x)$ would be asymptotically χ^2_k under H_0. We illustrate this with a problem for test of homogeneity for several exponential distributions.

EXAMPLE 9.3.3. Consider testing homogeneity of m exponential populations with samples of size n_i each so that

$$L(x, \sigma_1, \ldots, \sigma_m) = \prod_{i=1}^{m} \left\{ \frac{1}{\sigma_i^{n_i}} \exp \left\{ \sum_{j=1}^{n_i} x_{ij} / \sigma_i \right\} \right\}$$

and $H_0 : \sigma_1 = \sigma_2 = \ldots = \sigma_m$ (= σ unspecified).

Routine calculations show that under H_0, $\hat{\sigma} = \dfrac{1}{N} \sum_{i=1}^{m} \sum_{j=1}^{n_i} x_{ij} = \dfrac{T}{N}$, where $N = \sum n_i$ and $\hat{\sigma}_i = \dfrac{1}{n_i} \sum_{j=1}^{n_i} x_{ij} = \dfrac{T_i}{n_i}$. Thus, LRT is given by $-2 \log \lambda(x) = 2\{N \log \hat{\sigma} - \sum_{i=1}^{m} n_i \log \hat{\sigma}_i\}$ which is asymptotically χ^2_{m-1}. Since under H_0 we have $m - 1$ relations specified, the LRTS is χ^2_{m-1}. Another equivalent way to specify the degrees of freedom for $-2 \log \lambda(x)$ is to subtract the number of parameters estimated under H_0 (1 in this case) from the number of parameters estimated under $H_0 \cup H_1$ (m in this case).

EXAMPLE 9.3.4. Consider a system with two components (C_1, C_2) such that to start with failure time distributions of C_1 and C_2 are both exponential with failure rates α and β respectively. If C_1 fails first the system continues to work until C_2 fails but the failure time distribution of C_2 now becomes exponential with failure rate β'. Similarly if C_2 fails first then the system continues to work until C_1 fails but the failure time distribution of C_1 now becomes exponential with failure rate α'. Let (X, Y) denote the failure times of components C_1 and C_2 then the joint pdf of (X, Y) was obtained by Freund (1961) and is given by

$$\log f = I_A(x, y)\,[\log \alpha\beta' - \log \alpha'\beta] + \log \alpha'\beta - \alpha' x - (\alpha + \beta - \alpha')y$$
$$+ (\alpha' + \beta' - \alpha - \beta)\,(x - y)\, I_A(x, y) \qquad (9.3.6)$$

where $A = \{(x, y) \mid 0 \le x < y\}$ and $I_A(x, y)$ is the indicator function of the set A. One can show that the above pdf depending on $\theta = (\alpha, \beta, \alpha', \beta')'$ belongs to a four parameter exponential family with Fisher information matrix

$$I(\theta) = \text{diag}\left[\frac{1}{\alpha(\alpha+\beta)}, \frac{1}{\beta(\alpha+\beta)}, \frac{\beta}{(\alpha')^2(\alpha+\beta)}, \frac{\alpha}{(\beta')^2(\alpha+\beta)}\right]$$

The minimal sufficient statistic is four dimensional and is given by $\Sigma I_A(x, y) = n_1$ = number of observations with $x < y$ i.e. where C_1 failed first, together with Σx, Σy, $\Sigma (x - y) I_A(x, y)$. The MLE's for a random sample of size n on $f(x, y, \alpha, \beta, \alpha', \beta')$ are given by

$$\hat{\alpha} = \frac{n_1}{\Sigma \min(x, y)}, \quad \hat{\beta} = \frac{n_2}{\Sigma \min(x, y)}.$$

$$\hat{\alpha}' = \frac{n_2}{\Sigma x - \Sigma \min(x, y)}, \quad \hat{\beta}' = \frac{n_1}{\Sigma y - \Sigma \min(x, y)}$$

We urge the reader to work out details which are simple but interesting. The MLE's are CAN for $(\alpha, \beta, \alpha', \beta')'$ with variance covariance matrix $\frac{I^{-1}(\theta)}{n}$ given by

$$\frac{1}{n}\,\text{diag}\left[\alpha(\alpha+\beta), \beta(\alpha+\beta), \frac{\alpha'^2(\alpha+\beta)}{\beta}, \frac{\beta'^2(\alpha+\beta)}{\alpha}\right]$$

Freund (1961) had obtained the MLE but did not obtain $I(\theta)$ or $\frac{I^{-1}(\theta)}{n}$ or did not note that the pdf in (9.3.6) is a four parameter exponential family. These results were derived by Hanagal and Kale (1992), in which they consider testing hypotheses of symmetry $H_0 : \alpha = \beta$ and $\alpha' = \beta'$. Then the pdf under H_0 is given by

$$f_0(x, y, \theta_1, \theta_2) = \theta_1\theta_2 \exp\{-\theta_2 \mid x - y \mid - 2\theta_1 \min(x, y)\} \qquad (9.3.7)$$

with $\hat{\theta}_1 = \frac{n}{2\,\Sigma \min(x, y)}$ and $\hat{\theta}_2 = \frac{n}{\Sigma \mid x - y \mid}$ and the asymptotic variance covariance matrix of $(\hat{\theta}_1, \hat{\theta}_2)'$ as diag$\left(\frac{\theta_1^2}{n}, \frac{\theta_2^2}{n}\right)$, where $\theta_1 = \alpha = \beta$ and $\theta_2 = \alpha' = \beta'$.

Since explicit MLE's are available

$\lambda = L(x, y, \hat{\alpha}, \hat{\beta}, \hat{\alpha}, \hat{\beta}')/L_0(x, y, \hat{\theta}_1, \hat{\theta}_2)$ can be computed.

LRT will reject H_0 if $-2\log\lambda > \chi^2_{2,1-\alpha}$. The degrees of freedom for $-2\log\lambda$ is two, as under $H_0 \cup H_1$, we estimate four parameters whereas under H_0 we estimate two parameters.

Hanagal and Kale (1992) suggest using the fact that

$$\begin{bmatrix} \hat{\alpha} - \hat{\beta} \\ \hat{\alpha}' - \hat{\beta}' \end{bmatrix} \sim AN^{(2)}\left[\begin{pmatrix} \alpha - \beta \\ \alpha' - \beta' \end{pmatrix}, \frac{\alpha + \beta}{n} \text{ diag}\left(\alpha + \beta, \frac{\alpha'^2}{\beta} + \frac{\beta'^2}{\alpha}\right)\right]$$

and construct the test statistic

$$T_1 = \frac{n(\hat{\alpha} - \hat{\beta})^2}{(\hat{\alpha} + \hat{\beta})^2} + \frac{n(\hat{\alpha}' - \hat{\beta}')^2}{(\hat{\alpha} + \hat{\beta})\left[\frac{\hat{\alpha}'^2}{\hat{\beta}} + \frac{\hat{\beta}'^2}{\hat{\alpha}}\right]}$$

and reject H_0 if $T_1 > \chi^2_{2,1-\alpha}$.

The test statistic T_1 is a good large sample approximation to $-2\log\lambda$ and follows Wald's suggestion to use $I^{-1}(\hat{\theta})$ where $\hat{\theta}$ is MLE of θ under $H_0 \cup H_1$.

Hanagal and Kale (1992) also give a test for $H_0 : \alpha = \alpha'$ and $\beta = \beta'$ which is equivalent to independence of x and y in the Freund model.

Suppose we are testing $H_0 : \theta_i = \theta_{i0}$, $i = 1, 2, \ldots, k$ on the basis of a sample of size n when the pdf belongs to the m parameter Cramer family. Then one can obtain a test for H_0 against $H_1 : \theta_i \neq \theta_{i0}$ for some i based on a CAN estimator $\hat{\theta} = (\hat{\theta}_1, \ldots, \hat{\theta}_m)$ with asymptotic variance covariance matrix Λ. Partition the matrix Λ as $\begin{bmatrix} \Lambda_{11} & \Lambda_{12} \\ \Lambda_{21} & \Lambda_{22} \end{bmatrix}$ then

$$\hat{\theta}^{(1)} = (\hat{\theta}_1, \ldots, \hat{\theta}_k) \sim AN^{(k)}\left(\theta^{(1)}, \frac{\Lambda_{11}}{n}\right).$$

Following Wald (1943) let

$$W = n(\hat{\theta}^{(1)} - \theta_0^{(1)})'\Lambda^{11}(\hat{\theta})(\hat{\theta}^{(1)} - \theta_0^{(1)}) \tag{9.3.8}$$

A large sample test based on W rejects H_0 if observed $W > \chi^2_{k,1-\alpha}$. In general if $\hat{\theta}$ is MLE, the test would have better power performance than any test based on other CAN estimators. It is important to note that in constructing the quadratic form (9.3.8) we use $\hat{\theta}$ which is CAN for the parameter θ under $H_0 \cup H_1$. In this we are following Student (1908). To illustrate this point consider testing $H_0 : \mu = \mu_0$ against $H_1 : \mu \neq \mu_0$ in $N(\mu, \sigma^2)$ model where σ^2 is a nuisance parameter. Then consider $\hat{\theta} = (\bar{x}, S^2/n)'$ with $\frac{1}{n}\Lambda = \text{diag}\left(\frac{\sigma^2}{n} \frac{2\sigma^4}{n}\right)$. Wald statistic given by (9.3.8) is $W = n(\bar{x} - \mu_0)^2/\frac{S^2}{n} \sim \chi_1^2$ for large n. Nuisance parameter σ^2 is estimated by $\hat{\sigma}^2 = S^2/n$ which is CAN for σ^2 under H_0 as well as H_1. Suppose we use a CAN estimator of σ^2 under H_0 only say $\hat{\sigma}_0^2 = \Sigma\,(x_i - \mu_0)^2/n$. Then modified test statistic is given by $W' = \dfrac{n(\bar{x} - \mu_0)^2}{\Sigma\,(x_i - \mu_0)^2/n}$ which is also χ_1^2 for large n under H_0. However as $\Sigma\,(x_i - \mu_0)^2 = \Sigma\,(x_i - \bar{x})^2 + n(\bar{x} - \mu_0)^2$, $W' < W$ and there will be several samples for which $W' < \chi^2_{1,1-\alpha} < W$. This shows that the power of W would be larger than the power of W'. However one cannot guarantee that this kind of situation will occur in every case.

In general we recommend that in constructing test based on a CAN estimator with variance covariance matrix Λ which involves nuisance parameter one should evaluate Λ at $\hat{\theta}$ the MLE or any other CAN estimator of the parameter θ under $H_0 \cup H_1$. We illustrate this technique by way of a few examples.

EXAMPLE 9.3.5. Consider testing equality of probability of occurrence of an event E in two independent Bernoulli series of trials. Thus the likelihood of the data is

$$L(x, y, \theta_1, \theta_2) = \theta_1^{\Sigma x_i}(1 - \theta_1)^{n_1 - \Sigma x_i}\ \theta_2^{\Sigma y_i}(1 - \theta_2)^{n_2 - \Sigma y_i}$$

and we have a two parameter exponential family with the MLE's $\hat{\theta}_1 = \frac{n_{11}}{n_1}$, $\hat{\theta}_2 = \frac{n_{21}}{n_2}$ which are CAN with a variance covariance matrix, $\Lambda = \text{diag}\left(\frac{\theta_1(1 - \theta_1)}{n_1}, \frac{\theta_2(1 - \theta_2)}{n_2}\right)$. Here n_{11}, n_{21} denotes the number of successes in the two samples respectively. Note that for CANness to hold we must have $n_1 \to \infty$ as well as $n_2 \to \infty$. It immediately follows that

$$\hat{\theta}_1 - \hat{\theta}_2 \sim AN\left(\theta_1 - \theta_2, \frac{\theta_1(1 - \theta_1)}{n_1} + \frac{\theta_2(1 - \theta_2)}{n_2}\right).$$

Now for $H_0 : \theta_1 = \theta_2$ we reject H_0 if

$$T_0 = (\hat{\theta}_1 - \hat{\theta}_2)^2 \Big/ \left(\frac{\hat{\theta}_1(1 - \hat{\theta}_1)}{n_1} + \frac{\hat{\theta}_2(1 - \hat{\theta}_2)}{n_2}\right) > \chi^2_{1,1-\alpha} = \xi^2_{1-\alpha/2}$$

if the alternatives are two sided. If H_1 is $\theta_1 > \theta_2$ then we reject H_0 when

$$T_1 = (\hat{\theta}_1 - \hat{\theta}_2) \Big/ \left\{ \frac{\hat{\theta}_1(1-\hat{\theta}_1)}{n_1}, \frac{\hat{\theta}_2(1-\hat{\theta}_2)}{n_2} \right\}^{1/2} > \xi_{1-\alpha}$$

Note that we use the estimator of Var $(\hat{\theta}_1 - \hat{\theta}_2)$ which is consistent under $H_0 \cup H_1$. A consistent estimator of Var $(\hat{\theta}_1 - \hat{\theta}_2)$ under H_0 is $\hat{\theta}(1-\hat{\theta})\left(\frac{1}{n_1} + \frac{1}{n_2}\right)$ where $\hat{\theta} = \frac{n_{11} + n_{21}}{n_1 + n_2}$ the MLE of θ under H_0 and some books on statistical methods recommend its use.

We have considered variance stabilization transformation for CAN estimators due to Bartlett (1947) in Exercise (3) of Section 6.2. One can very effectively use such a transformation for testing hypotheses. For example in the Bernoulli series of trials the MLE of θ, $\hat{\theta} = \bar{x}$ has variance $\frac{\theta(1-\theta)}{n}$. It can be easily verified that the transformation $\psi(\theta) = \sin^{-1}\sqrt{\theta}$ is such that $\psi(\bar{x}) \sim AN\left(\sin^{-1}\sqrt{\theta}, \frac{1}{4n}\right)$. Thus in the above example we can use the test statistic.

$$T_2 = \left[\sin^{-1}\left(\frac{n_{11}}{n_1}\right)^{1/2} - \sin^{-1}\left(\frac{n_{21}}{n_2}\right)^{1/2}\right]^2 \Big/ \frac{1}{4n_1} + \frac{1}{4n_2} > \chi^2_{1,1-\alpha}$$

in case alternatives are two sided. For one sided alternatives $\theta_1 > \theta_2$ we reject H_0 if

$$\sin^{-1}\left(\frac{n_{11}}{n_1}\right)^{1/2} - \sin^{-1}\left(\frac{n_{21}}{n_2}\right)^{1/2} \Big/ \left(\frac{1}{4n_1} + \frac{1}{4n_2}\right)^{1/2} > \xi_{1-\alpha}.$$

Fisher-Yates (1963) have tabulated the inverse sine function but now a days most hand held calculators have this function built in.

Exercise 9.3. (1) Analogous to the comparison of proportions from two independent samples on binomial distribution consider two independent samples of size n_1 and n_2 on two exponential distributions with means σ_1 and σ_2, respectively. Obtain LRT to test $H_0 : \sigma_1 \geq 2\sigma_2$ against $H_1 : \sigma_1 < 2\sigma_2$. Here MLEs $\hat{\sigma}_1 = \bar{x}_1$ and $\hat{\sigma}_2 = \bar{x}_2$ are CAN with asymptotic variance covariance matrix diag $\left(\frac{\sigma_1^2}{n_1}, \frac{\sigma_2^2}{n_2}\right)$. Construct a test for the above problem based on $\bar{x}_1 - 2\bar{x}_2$ which is $AN\left(\sigma_1 - 2\sigma_2, \frac{\sigma_1^2}{n_1}, \frac{4\sigma_2^2}{n_2}\right)$. Observing that the variance stabilization transformation is logarithmic construct a large sample test based on $[\log \hat{\sigma}_1 - \log (2\hat{\sigma}_2)]$ for the above hypotheses.

(2) Let (x_i, y_i) $i = 1, 2, \ldots, n$ be a random sample of size n from a bivariate normal distribution with parameter $\theta = (\mu_1, \mu_2, \sigma_1^2, \sigma_2^2, \rho)'$. Obtain MLE of θ. Construct a

large sample test for testing $H_0 : \rho \geq \rho_0$ against $\rho < \rho_0$ based on MLE $\hat{\rho} = r$, the Pearson correlation coefficient of the sample with asymptotic variance $\frac{(1-\rho^2)^2}{n}$. Show that Fisher's transformation $\psi(r) = \frac{1}{2} \log \frac{1+r}{1-r}$ is a variance stabilizing transformation and give a large sample test based on this transformation.

9.4 Consistency of a Large Sample Test

Consider the problem of testing $H_0 : \theta \leq \theta_0$ vs $H_A : \theta > \theta_0$ when we have a random sample of size n from $N(\theta, \sigma_0^2)$ where σ_0^2 is known. Then the LRT, which is also UMP test for the problem, has test function $\varphi_1(\bar{x}) = 1$ if $\bar{x} > \theta_0 + \frac{\sigma_0}{\sqrt{n}} \xi_{1-\alpha}$ and zero elsewhere. The power function of this test is given by

$$\beta_{\varphi_1}(\theta) = 1 - \Phi\left[\sqrt{n}\frac{(\theta_0 - \theta)}{\sigma_0} + \xi_{1-\alpha}\right] \tag{9.4.1}$$

For any $\theta \in \Omega_{H_1}$ i.e. $\theta > \theta_0$ $\lim_{n\to\infty} \beta_{\varphi_1}(\theta) = 1$, i.e. the power of test φ_1 approaches one as $n \to \infty$ for any alternative hypothesis. This property is known as the consistency of the test and formally we define a test $\varphi \in \mathbb{D}_\alpha$ to be a consistent test if $\beta_\varphi(\theta) \to 1$ as $n \to \infty$ for $\forall\ \theta \in \Omega_{H_1}$. Note that here Ω_{H_1} could be two sided and further θ could be vector valued. For example in the above problem suppose we are testing $H_0 : \theta = \theta_0$ vs $H_1 : \theta \neq \theta_0$ then LRT test of size α, which is also known to be UMP Unbiased, rejects H_0 if $\frac{n(\bar{x} - \theta_0)^2}{\sigma_0^2} > \chi^2_{1,1-\alpha} = \xi^2_{1-\alpha/2}$. As the reader can verify the power function of this test is given by

$$\beta_2(\theta) = 1 - \Phi\left[\sqrt{n}\,\frac{(\theta_0 - \theta)}{\sigma_0} + \xi_{1-\alpha/2}\right] + \Phi\left[\sqrt{n}\,\frac{(\theta_0 - \theta)}{\sigma_0} - \xi_{1-\alpha/2}\right]$$

For any $\theta \neq \theta_0$, $\lim_{n\to\infty} \beta_2(\theta) = 1$ and the LRT is consistent.

If σ^2 is unknown then for the one sided problem $H_0 : \theta \leq \theta_0$ against $H_1 : \theta > \theta_0$ the LRT rejects H_0 if $\sqrt{n}(\bar{x} - \theta_0)/(S^2/(n-1))^{1/2} > t_{n-1,1-\alpha}$. The power function of this test is given by

$$\beta_3(\theta, \sigma^2) = P[\sqrt{n}(\bar{x} - \theta_0)/(S^2/(n-1))^{1/2} > t_{n-1,1-\alpha} \mid (\theta, \sigma^2)]. \tag{9.4.2}$$

Under H_1 the test statistic has a non-central t-distribution with $(n-1)$ d.f. and non-centrality parameter $\delta^2 = (\theta - \theta_0)^2/\sigma^2$ and the power function can be evaluated at any point $(\theta, \sigma^2) \in \Omega_{H_1}$. For details we refer to Kendall and Stuart Vol. 2 (1967) and other references contained therein. To obtain $\lim_{n\to\infty} \beta_3(\theta, \sigma^2)$, the RHS of (9.4.2) can be written as

$$P\left[(\bar{x}-\theta_0)/(S^2/(n-1))^{1/2} > \frac{t_{n-1,1-\alpha}}{\sqrt{n}} \,\middle|\, \theta, \sigma^2\right].$$

Observe that as $n \to \infty$, $\bar{x} \xrightarrow{p} \theta$ and $(S^2/(n-1))^{1/2} \xrightarrow{p} \sigma$. Further $t_{n-1,1-\alpha} \to \xi_{1-\alpha}$ and $\frac{1}{\sqrt{n}} \to 0$ therefore the r.v. in bracket converges in probability to $(\theta - \theta_0)/\sigma$ and we have

$$\begin{aligned}\lim_{n\to\infty} \beta_3(\theta, \sigma^2) &= 0 \quad \text{if } \theta < \theta_0\\ &= \alpha \quad \text{if } \theta = \theta_0\\ &= 1 \quad \text{if } \theta > \theta_0.\end{aligned}$$

If we have two sided alternatives $\theta \neq \theta_0$ then LRT rejects H_0 if $\frac{n(\bar{x}-\theta_0)^2}{S^2/(n-1)} > t^2_{n-1,1-\alpha/2}$ and using similar argument as above we can show that for any $\theta \neq \theta_0$ $\lim_{n\to\infty} \beta_4(\theta, \sigma^2) = 1$.

In fact the above argument can be generalized to show that the LRT in general is consistent. Consider testing $H_0 : \theta_i = \theta_{i0}$, $i = 1, 2, \ldots, k$ with $\theta_{k+1}, \ldots, \theta_m$ as nuisance parameters. As seen in the previous section, the LRT rejects H_0 if $-2 \log \lambda > \chi^2_{k,1-\alpha}$ and Wald's test as an approximation to LRT rejects H_0 if

$$W = n(\hat{\theta}^{(1)} - \theta_0^{(1)})' I_{11}(\hat{\theta})(\hat{\theta}^{(1)} - \theta_0^{(1)}) > \chi^2_{k,1-\alpha}.$$

The power function of the Wald test is given by

$$\begin{aligned}\beta_w(\theta) &= P_\theta[W > \chi^2_{k,1-\alpha}]\\ &= P_\theta[(\hat{\theta}^{(1)} - \theta_0^{(1)})' I_{11}(\hat{\theta})(\hat{\theta}^{(1)} - \theta_0^{(1)}) > \chi^2_{k,1-\alpha}/n].\end{aligned}$$

Now

$$(\hat{\theta}^{(1)} - \theta_0^{(1)})' I_{11}(\hat{\theta})(\hat{\theta}^{(1)} - \theta_0^{(1)}) \xrightarrow{p} (\theta^{(1)} - \theta_0^{(1)})' I_{11}(\dot{\theta}_0)(\theta^{(1)} - \theta_0^{(1)}) = \delta^2.$$

As $\delta^2 > 0$ and as $\frac{1}{n}\chi^2_{k,1-\alpha} \to 0$ for any $\theta^{(1)} \neq \theta_0^{(1)}$ we have $\lim_{n\to\infty} \beta_w(\theta) = 1$ for any $\theta \in \Omega_{H_1}$ and Wald's test is consistent which implies that LRT is also consistent. We note that the asymptotic distribution of Wald's test statistic is a non-central χ^2 with k degrees of freedom and non-centrality parameter δ^2 as defined above. For specific values of n, sufficiently large, the value of $\beta(\theta)$ at a given point $\theta \in \Omega_{H_1}$ can be obtained. For more details we again refer to Kendall and Stuart, Vol. 2 (1967) and references contained therein. We can also argue in a similar way to claim that the tests based on CAN estimators discussed at the end of Section 9.3 are consistent. The detailed proofs are not attempted here as this is a first course in parametric inference. On the other hand we discuss the implications of the consistency of a test in practice as well as in theory assuming that arbitrarily large sample are available.

EXAMPLE 9.4.1. Consider the problem discussed at the beginning of this section. Suppose at θ we want that the test has power β or equivalently type II error is $(1 - \beta)$ then using (9.4.1), we have the equation defining sample size n given by

$$1 - \Phi\left(\sqrt{n}\,\frac{(\theta - \theta_0)}{\sigma_0} + \xi_{1-\alpha}\right) = \beta \text{ which on simplification gives}$$

$$n_0 = \left[\frac{\sigma_0^2(\xi_{1-\beta} - \xi_{1-\alpha})^2}{(\theta_0 - \theta)^2}\right] + 1.$$

This determination of n_0 is very similar to determination of minimum sample size required for a consistent estimator T to achieve a degree of accuracy specified by $(\varepsilon; \delta)$ which has been illustrated in Examples 5.1.1 and 5.4.1.

Let $\theta = \theta_0 + k\sigma_0$ where k measures departure of θ from θ_0 in the units of standard deviation σ_0. Then $n_0 = \left[\frac{(\xi_{1-\beta} - \xi_{1-\alpha})^2}{k^2}\right] + 1$. Observe that for fixed (α, β), n_0 decrease as k increases indicating that larger departures from θ_0 are easier to detect with smaller sample sizes, But we need large samples to detect small departures from θ_0. Following table gives values of n_0 for different combinations β = .90, .95, .99 and k = .1, .3, .5, 1 when α = .05.

Values of n_0

$k \backslash \beta$	.90	.95	.99
.1	857	1083	1578
.3	96	121	176
.5	35	44	64
1.0	9	11	16

In a more realistic situation where σ^2 is unknown, the problem of determining minimum sample size n_0 for level α = .05 test to have power β = .90, .95, .99 at specific values of $\frac{\theta - \theta_0}{\sigma}$ is much more difficult. Again we refer to Section 24.35 in Kendall and Stuart, Vol. 2 (1967) and references contained therein to obtain the power function of the Student's t test at specific values of $(\theta - \theta_0)/\sigma = k$ for given n. One can obtain values of n_0 by assuming n_0 to be large enough so that the distribution of t_{n-1} is $N(0, 1)$

under H_0 and under the alternatives it is $N\left(\dfrac{\theta - \theta_0}{\sigma}, 1\right)$. This approximation will lead to the table of values of n_0 given above.

Next consider an example in which we have mean and the variance are the functions of the same parameter such as in the case of exponential distribution with mean θ.

EXAMPLE 9.4.2. Let $(X_1, \ldots, X_n)$ be i.i.d. exponential with mean θ and let $H_0 : \theta \geq \theta_0$ and $H_A : \theta < \theta_0$. The LRT of level α which is also UMP, rejects H_0 for small values of ΣX_i and has power function given by $\beta(\theta) = F_n\left(\dfrac{\theta_0}{\theta}\eta_{n,\alpha}\right)$ where $\eta_{n,\alpha}$ is the $100\alpha\%$ point of a $G(n, 1)$ r.v. and $F_n(u)$ is its d.f. Suppose we want to determine n_0 such that $F_n\left(\dfrac{\theta_0}{\theta}\eta_{n,\alpha}\right) = \beta$ at $\dfrac{\theta_0}{\theta} = k$. Here as $\theta < \theta_0$, $k > 1$ and n is determined by $k\eta_{n,\alpha} = F_n^{-1}(\beta) = \eta_{n,\beta}$ and tables of incomplete Gamma functions can be used to determine n_0 by using trial and error method. If we assume that n_0 is sufficiently large so that $\hat{\theta} = \bar{x} \sim AN(\theta, \theta^2/n)$ then we reject H_0 if $\bar{x} < \theta_0 + \dfrac{\theta_0}{\sqrt{n}}\xi_\alpha$, $\beta(\theta) \simeq \Phi[\sqrt{n}(k-1) + k\xi_\alpha]$. Hence n_0 is determined by the equation $\sqrt{n}(k-1) + k\xi_\alpha = \xi_\beta$ or

$$n_0 = \left[\frac{(\xi_\beta - k\xi_\alpha)}{(k-1)}\right]^2 + 1.$$

Following table give the values of n_0 for $\alpha = .05$, $\beta = .90, .95, .99$ and $k = 1.1, 1.2, 1.3, 1.5, 2.0$.

Values of n_0

$k\backslash\beta$	.90	.95	.99
1.1	955	1194	1711
1.2	266	328	428
1.5	57	67	92
2.0	21	24	32

The problem of determination of minimum samples size with guaranteed power at the specific alternative is a difficult one as it involves equations or inequalities (mostly nonlinear) which are difficult to solve. However this is an important problem and we refer interested reader to Desu and Raghavrao (1990) and references contained there in.

We point out that for $\alpha = .05$ if we require power $= .99$ then it implies that the type I error $= .05 \geq$ type II error $= .01$. This is contradictory to our initial assumption that type I error is more serious than type II error.

In fact for any consistent test this situation will occur for any $\alpha \in (0, 1)$ since as $n \to \infty$, power $\to 1$ and therefore type II error $\to 0$. The above paradoxical situation arises because of the fact that in bounding type I error by α we have not taken into account n, the sample size. An alternative then is to select the level α_n depending on n, the sample size available. Consider the following example to illustrate the point.

EXAMPLE 9.4.3. We consider testing $H_0 : \theta = \theta_0$ against $H_1 : \theta = \theta_1 > \theta_0$ in $N(\theta, \sigma_0^2)$ model with σ_0^2 known and where n can be chosen sufficiently large. By (9.4.1)

$$\beta_n(\theta_1) = 1 - \Phi\left[\frac{\sqrt{n}(\theta_0 - \theta_1)}{\sigma_0} + \xi_{1-\alpha_n}\right] \tag{9.4.3}$$

We now require that $1 - \beta_n(\theta_1)$ = Type II error $\geq \alpha_n$ = Type I error. Thus for given large n select α_n such that $1 - \beta_n(\theta_1) = \alpha_n = \beta_n(\theta_0)$ i.e.

$$\Phi\left[\frac{\sqrt{n}(\theta_0 - \theta_1)}{\sigma_0} + \xi_{1-\alpha_n}\right] = \alpha_n = \beta_n(\theta_0)$$

or

$$\sqrt{n}\left(\frac{(\theta_0 - \theta_1)}{\sigma_0}\right) + \xi_{1-\alpha_n} = \xi_{\alpha_n} = -\xi_{1-\alpha_n}.$$

Now, as $\theta_1 > \theta_0$, $\xi_{1-\alpha_n}$ is therefore given by

$$\xi_{1-\alpha_n} = \frac{\sqrt{n}}{2}\left(\frac{(\theta_0 - \theta_1)}{\sigma_0}\right).$$

For example if $n = 400$ and $\dfrac{\theta_0 - \theta_1}{\sigma_0} = 1$ then $\xi_{1-\alpha_n} = 10$ or the level $\alpha_n = 1 - \Phi(10)$. Using the approximation $1 - \Phi(x) \simeq \dfrac{1}{x}\dfrac{1}{\sqrt{2\pi}} e^{-x^2/2}$ for large x or using tables of Abromowitz and Stegun (1964), we obtain $1 - \Phi(10) = 10^{-23.119}$. For $n = 100$, similar calculations lead to $\alpha_n = 1 - \Phi(5) = 10^{-6.543}$. We refer to Kunte and Gore (1992) for more discussion on this point. They also point out that similar problems occur in tests of goodness of fit. In fact it has been observed that very large samples would reject any precise null hypothesis H_0 at the conventional levels $\alpha = .01, .05, .10$.

When H_1 is composite, $\theta > \theta_0$, α_n would depend both on n and $\theta \in \Omega_{H_1}$. By fixing a power at a particular alternative θ_1 such that $1 - \beta_\varphi(\theta_1) \geq \alpha_n$ we can determine α_n as a function of n and θ_1. However for any $\theta \in (\theta_0, \theta_1)$ as $n \to \infty$, $1 - \beta_\varphi(\theta) \to 0$ and we have the paradoxical situation that the type II error at $\theta \in (\theta_0, \theta_1)$ falls below the level of significance α_n. The paradox also appears even when n is fixed but $\theta \in \Omega_{H_1}$ is sufficiently large as $1 - \beta_\varphi(\theta) \to 0$ as $\theta \to \infty$. Thus the choice of α, the upper limit to the type I error has to depend on the sample size n as well as the alternatives.

The conventional levels $\alpha = .01$, $\alpha = .05$ or $\alpha = .10$ were introduced by Fisher in developing tests of significance which did not refer to specific alternatives but depended on a test statistic T whose distribution under H_0 is completely known. If the observed value of T belongs to the tails of its distribution then we reject H_0. The tails were defined by levels $\alpha = .01$, $\alpha = .05$ or $\alpha = .10$. Whether we should use both the tails or only right tail or only left tail of the distribution depends on the context and indirectly involves H_1. Fisher insisted that in scientific investigations alternatives to H_0 are difficult to specify and power considerations or type II error is not relevant. In cases where alternatives are specified as in SQC problems, Fisher agreed that the type II error can be considered although he would prefer to view these as problems in estimation.

The Neyman-Pearson theory does not provide any guidelines as to how to choose α. Further once α is fixed it recommends randomization on the boundary $E_2 = \{x \mid L(x, \theta_1) = k_\alpha L(x, \theta_0)\}$. Most scientists and engineers do not like the idea of making decision based on randomization when the data has been collected after a careful experimentation. The real meaning of necessity of randomization on E_2 is the fact that some additional data is required before the decision could be made.

Such modifications to fixed sample size procedures have lead to two stage or in general multistage and sequential estimation and testing procedures. The path breaking work of Wald (1947, 1950) deserves a special mention. Wald generalized the Neyman-Pearson approach to Decision Theoretic Statistics and Sequential Sampling. We will not pursue this line of thought, but instead refer to some well known texts such as Ferguson (1967), Kendall and Stuart Vol. II (1967), Rohatgi (1976) and Zacks (1971) among others.

9.5 Convex Combination of Two Types of Errors

Consider the simplest problem of testing $H_0 : \theta = \theta_0$ against $H_1 : \theta = \theta_1$ on the basis of the sample $x = (x_1, \ldots, x_n)$. As seen before let $E[\varphi(x)] = \beta_\varphi(\theta)$ be the power function of a test φ. Then the two types of errors are given $\beta_\varphi(\theta_0)$, $1 - \beta_\varphi(\theta_1)$ respectively. Let $0 < w < 1$ denote the relative weight of the type I error so that the average error is given by

$$\begin{aligned} R(\varphi, w) &= w\beta_\varphi(\theta_0) + (1 - w)\,(1 - \beta_\varphi(\theta_1)) \\ &= (1 - w) + \int [wL(x, \theta_0) - (1 - w)L(x, \theta_1)]\varphi(x)\,dx \end{aligned}$$

where the integration is over $S_0 \cup S_1$, the union of support of $L(x, \theta_0)$ and $L(x, \theta_1)$. For given weights w and $(1 - w)$ the test φ^* would be optimum if $R(\varphi^*, w) \le R(\varphi, w)$ for any test function $\varphi(x)$. To minimize $R(\varphi, w)$ for variations in φ observe that as $0 \le \varphi(x) \le 1$, if $[wL(x, \theta_0) - (1 - w)L(x, \theta_1)]$

< 0 then $\varphi^*(x) = 1$ and $\varphi^*(x) = 0$ if $[w[(x, \theta_0) - (1 - w)L(x, \theta_1)] > 0$. On the set $E_2 = \{x \mid wL(x, \theta_0) = (1 - w)L(x, \theta_1)\}$, $\varphi^*(x)$ can be defined arbitrarily as the integral over E_2 is zero and does not contribute to $R(\varphi, w)$. Thus the best test which minimizes average error for fixed weights w is given by

$$\left.\begin{aligned} \varphi_w^*(x) &= 1 \quad \text{if } L(x, \theta_1) > \frac{w}{1-w} L(x, \theta_0) \\ &= \gamma(x) \quad \text{if } L(x, \theta_1) = \frac{w}{1-w} L(x, \theta_0) \\ &= 0 \quad \text{if } L(x, \theta_1) < \frac{w}{1-w} L(x, \theta_0) \end{aligned}\right\} \tag{9.5.1}$$

The structure of $\varphi^*(x)$ is similar to that of MP test given by N-P lemma with $k = \dfrac{w}{1-w}$ or $w = \dfrac{k}{1+k}$.

Now in N-P lemma, to obtain MP test we take $\gamma(x) = \gamma$ on the set E_2, such that $E[\varphi_0(x) \mid \theta_0] = \alpha$ and without loss of generality we can take $\gamma(x) = \gamma$ on the set E_2 for the test $\varphi^*(x)$. In case $P(E_2) = 0$ under both H_0 and H_1, we have a one to one correspondence between the MP test $\varphi_0(x)$ of level α and $\varphi^*(x)$ which minimizes $R(\varphi, w)$ for given w. Thus in choosing α one implicitly fixes w and in choosing w one implicitly fixes α, at least in the case where $P_{\theta_0}(E_2) = 0 = P_{\theta_1}(E_2)$ which is the case when we have a continuous X with pdf $\{L(x, \theta), \theta \in \Omega\}$ with the support of L not depending on θ. In the further discussion we will assume this to be the case and therefore take

$$\left.\begin{aligned} \varphi_w^*(x) &= 1 \quad \text{if } L(x, \theta_1) > \frac{w}{1-w} L(x, \theta_0) \\ &= 0 \quad \text{otherwise.} \end{aligned}\right\} \tag{9.5.2}$$

Let the α corresponding to $k = \dfrac{w}{1-w}$ be denoted by α_w and $\varphi_0(x)$ the MP level α_w test is in one to one correspondence with $\varphi_w^*(x)$.

One can interprete the weights w and $(1 - w)$ also as probabilities measuring uncertainty about the values of θ. Thus we can regard $w = P(\theta = \theta_0)$ and $(1 - w) = P(\theta = \theta_1)$ and then $R(\varphi, w)$ is the average error of test procedure φ when the distribution over $\{\theta_0, \theta_1\}$ is given by weights $\{w, 1 - w\}$.

The basic idea of expressing uncertainty about the parameter value θ by a probability distribution over possible values of θ is due to Bayes (1763). Suppose that we have k possible mutually exclusive and exhaustive causes or hypotheses $H_1, H_2, \ldots, H_k$ indexed by $\theta_1, \ldots, \theta_k$ respectively and the probability of obtaining data when the hypotheses H_j i.e. say $\theta = \theta_j$ holds, is given by $P(D \mid H_j) = L(D \mid \theta_j)$. Further if prior probability distribution over $\{\theta_1, \ldots, \theta_k\}$ is given by $\{w_1, \ldots, w_k\}$ then a straight forward calculation shows that

$$P(H_j \mid D) = L(D \mid \theta_j)w_j / \Sigma\, L(D \mid \theta_j)w_j \tag{9.5.3}$$

which is the well known Bayes Theorem first proved by Laplace (1774). When $k = 2$ with possible values θ_0, θ_1 with prior probabilities w and $1 - w$ respectively, we have

$$\left.\begin{aligned} P(\theta_0 \mid x) &= \frac{L(x, \theta_0)w}{L(x, \theta_0)w + L(x, \theta_1)(1 - w)} \\ P(\theta_1 \mid x) &= \frac{L(x, \theta_1)(1 - w)}{L(x, \theta_0)w + L(x, \theta_1)(1 - w)} \end{aligned}\right\} \tag{9.5.4}$$

$\{P(\theta_0 \mid x), P(\theta_1 \mid x)\}$ is called as the posterior distribution over $\{\theta_0, \theta_1\}$ when the data x is observed. According to Bayes the statistical inference consists of change from prior distribution $\{w, 1 - w\}$ over $\{\theta_0, \theta_1\}$ to posterior distribution $\{P(\theta_0 \mid x), P(\theta_1 \mid x)\}$ when the data x is observed.

We define the Bayes test for H_0 against H_1 as

$$\begin{aligned} \varphi_B(x) &= 1 \quad \text{if } P(\theta_1 \mid x) > P(\theta_0 \mid x) \\ &= 0 \quad \text{otherwise.} \end{aligned}$$

However $P(\theta_1, x) > P(\theta_0 \mid x)$ is equivalent to $L(x, \theta_1) > \frac{w}{1 - w} L(x, \theta_0)$ and $\varphi_B(x)$ is equivalent to $\varphi^*(x)$ which minimizes $R(\varphi, w)$ and corresponds to MP level α_w test. Thus Neyman-Pearson approach of controlling type I error at a level α and then minimizing type II error is equivalent to minimizing average error $R(\varphi, w)$ with weights w and $(1 - w)$ which in turn is equivalent to Bayes test for H_0 against H_1 when prior distribution over $\{\theta_0, \theta_1\}$ is given by $\{w, 1 - w\}$. Of course if $P(E_2) \neq 0$ under either θ_0 or θ_1 this will create a few problems of details in establishing such equivalence and we will not pursue this aspect further.

The above approach of assuming a prior probability distribution on the parameter space of interest is originally due to Bayes (1763). Bayes solution to the problem of estimation of θ in the model $\{L(x, \theta), \theta \in \Omega\}$ was to quantify uncertainity about θ by a probability distribution called prior distribution of θ. Then treating θ as an unobservable r.v., we define the joint distribution of (x, θ) given by $L(x, \theta)\, p(\theta)$. Using standard arguments of conditional probability, Bayes suggested obtaining posterior distribution of θ given x, definded by

$$p(\theta \mid x) = L(x, \theta)\ p(\theta) \Big/ \int_\Omega L(x, \theta)\, p(\theta)\, d\theta \tag{9.5.5}$$

in case θ is continuous. If θ is discrete then the posterior distribution of θ for the data x is given by

$$p(\theta_i \mid x) = L(x, \theta_i)\ p(\theta_i) / \Sigma\ L(x, \theta_r)\, p(\theta_r) \tag{9.5.6}$$

The above method was proposed by Bayes to solve the following problem which in his own words was stated as follows.

"Given the number of times in which an unknown event has happened and failed; Required the chance that the probability of its happening in a single trial lies somewhere between any two degrees of probability that can be named".

Note that an 'unknown event' means known event E whose occurrence in a single trial has unknown probability $P(E) \in (0, 1)$ Observe that in Required part, Bayes uses a different word 'chance' than the probability i.e. Chance $[a < P(E) < b]$ rather than $P(a < P(E) < b)$. Although later in definitions given in Sec. 1 of the paper, Bayes states that by chance he means same as probability. Assuming Bernoulli series of trials as model and uniform prior distribution for θ over (0, 1) Bayes shows that

$$\text{Chance } [a < P(E) < b] = \int_a^b p^m (1-p)^n \, dp \Big/ \int_0^1 p^m (1-p)^n \, dp \qquad (9.5.7)$$

when E has occurred m times and failed to occur n times in a Bernoulli series of $(m + n)$ trials.

Laplace [(1774), (1812), (1820)] took lead from Bayes and proved Bayes theorem and formulae (9.5.5) and (9.5.6). Given the status of Laplace as "the second Newton" in European mathematical world, the Bayes-Laplace method soon established itself and won many great followers and came to be known as the method of inverse probability. For example Gauss showed that under local uniform prior for θ, the mean of the normal distribution, the posterior distribution of θ is $N(\bar{x}, \sigma^2/n)$. Gauss then called $\bar{x}$ as the maximum (posterior) probability estimator of θ since the posterior distribution of θ given $\bar{x}$ has the unique mode at $\bar{x}$.

Fisher (1912) justified his method of MLE using the inverse probability argument although Fisher (1922) later adds a disclaimer "I must indeed plead guilty in my original statement of the method of MLE to having based my argument on inverse probability". The change in the view point of Fisher was due to criticisms of Boole (1958), Venn (1866, 1876) and the fact that Bayes-Laplace method of inverse probability required specification of prior distribution which introduces a subjective element in scientific inference.

However Bayes-Laplace theory is undoubtedly attractive and fits well into Descartes-Bacon philosophy of rational empiricism which assumes that one can learn from empirical data obtained by careful experimentation. Even though Fisher later became a critic of inverse probability he gives due credit, (Fisher, 1956), to Bayes for systematically tackling the problem of inductive inference for the first time when the logicians and philosophers of those times regarded that the inductive inference is not possible. Fisher (1930) tried to provide solution to the problem of inductive inference based on fiducial probability distributions and confidence intervals without assuming

θ to be a random variable with known prior distribution but assuming some well defined structure on the parameter space Ω. We will not discuss the issues further here but emphasize that the last word has not been said and perhaps will never be said in regard to the solution of the problem of inductive inference of which the parametric inference is an important special case.

Chapter 10, dealing with the confidence interval estimation, will show that similar problems do crop up there also.

10
Interval Estimation

10.1 Background

Let $(x_1, \ldots, x_n)$ be a random sample on X with pdf from $\{f(x, \theta), \theta \in \Omega \subset R_1\}$. In the problem of estimation our search was for an estimator $T(x)$ of θ or $\psi(\theta)$ having some desirable propeties such as MVUness or CANness. In the problem of testing of hypotheses $H_0 : \theta \in \Omega_{H_0}$ against $H_1 \in \Omega_{H_1}$, we searched for a test procedure defined by a test function $\varphi(x)$ with type I error bounded above by a specified level α and having some desirable properties such as UMPness, Unbiasedness, Similarity, Consistency etc. In the problem of interval estimation of θ we are searching for a random interval $(T_1(x), T_2(x))$ such that for any $\theta \in \Omega$

$$P[T_1(x) \leq \theta \leq T_2(x) \mid \theta] \geq 1 - \gamma, \quad \forall\ \theta \in \Omega \qquad (10.1.1)$$

where γ is a specified constant. The constant on the RHS is called as the level of the confidence interval (CI). If $P[T_1(x) \leq \theta \leq T_2(x) \mid \theta]$ does not depend on θ then this probability is called as the size of CI given by $(T_1(x), T_2(x))$. We illustrate this by an example.

EXAMPLE 10.1.1 Let $(x_1, \ldots, x_n)$ be a random sample from $N(\theta, 1)$ with $\theta \in R_1$ and consider $T_1(x) = \bar{x} - a$ and $T_2(x) = \bar{x} + a$ where $a > 0$. Then

$$P[T_1(x) \leq \theta \leq T_2(x) \mid \theta] = P[-a \leq \bar{X} - \theta \leq a \mid \theta]$$

Under each θ, $\bar{X} - \theta \sim N(0, 1/n)$ and therefore

$$P[\bar{X} - a \leq \theta \leq \bar{X} + a \mid \theta] = \Phi[\sqrt{n}a] - \Phi[-\sqrt{n}a] = 2\Phi[\sqrt{n}a] - 1.$$

Hence $(\bar{x} - a, \bar{x} + a)$, for a given $a > 0$, provides a CI of size $2\Phi(\sqrt{n}a) - 1$. Observe that the size of the CI, $(\bar{x} - a, \bar{x} + a)$, as a takes values over $(0, \infty)$ ranges over $(0, 1)$ for any fixed n. Therefore given γ, we can determine a_γ such that size of the CI is $1 - \gamma$ and the corresponding a_γ is given by

$$a_\gamma = \frac{1}{\sqrt{n}}\, \Phi^{-1}(1 - \gamma/2) = \xi_{1-\gamma/2}/\sqrt{n} \qquad (10.1.2)$$

Any value of $a \geq a_\gamma$ will provide CI of level $1 - \gamma$ and size larger than $1 - \gamma$ but larger the value of a, larger will be the length of CI which is $2a$. In the limit we have size $\to 1$ as $a \to \infty$ and we have a CI of size one but with length infinite. On the other hand if $a < a_\gamma$ then $(\bar{x} - a, \bar{x} + a)$ will not be a CI of level $1 - \gamma$. This shows that our object is to find a CI of given level

$1 - \gamma$ which is such that its expected length $= E(T_2(x) - T_1(x))$ is as small as possible.

Suppose we consider CI the type $T_1(x) = \bar{x} + c$ and $T_2(x) = \bar{x} + d$ with $-\infty < c < d < \infty$. Then the expected length is $(d - c)$ and $P[\bar{X} + c \leq \theta \leq \bar{X} + d \mid \theta] = \Phi(\sqrt{n}d) - \Phi(\sqrt{n}c)$.

We want to determine c and d such that $(d - c)$ is minimized subject to condition that $\Phi(\sqrt{n}d) - \Phi(\sqrt{n}c) = 1 - \gamma$.

Using Lagrange's method of undetermined multipliers the equations determining c and d are

$$1 + \lambda\, \varphi(\sqrt{n}\, d)\, \sqrt{n} = 0 = 1 + \lambda\, \varphi(\sqrt{n}c)\sqrt{n} \qquad (10.1.3)$$

This immediately gives $\varphi(\sqrt{n}d) = \varphi(\sqrt{n}c)$. As $c < d$, in view of symmetry of $\varphi(x)$ around zero we must have $c = -d$. Hence we select d such that $\Phi(\sqrt{n}d) - \Phi(-\sqrt{n}d) = 1 - \gamma$ or

$$d = \frac{1}{\sqrt{n}}\, \Phi^{-1}\left(1 - \frac{\gamma}{2}\right) = \xi_{1-\gamma/2}/\sqrt{n}$$

EXAMPLE 10.1.2 Let (X_1, X_2) be a random sample of size two from an exponential distribution with mean θ. Then as $x_1 \geq 0$ and $x_2 \geq 0$ we have $x_1 \leq x_1 + x_2$ and we can take $T_1(x) = x_1$ and $T_2(x) = x_1 + x_2$. If $P(X_1 < \theta < X_1 + X_2)$ does not depend on θ and is constant say $1 - \gamma$ then $(x_1, x_1 + x_2)$ would be a CI for θ of size $1 - \gamma$. Now let

$$1 - \gamma(\theta) = \iint\limits_{x_1<\theta<x_1+x_2} \frac{1}{\theta^2}\, e^{-(x_1+x_2)/\theta}\, dx_1\, dx_2 \qquad (10.1.3)$$

making a transformation $u_1 = \dfrac{x_1}{\theta}$, $u_2 = \dfrac{x_2}{\theta}$ we have

$$1 - \gamma(\theta) = \iint\limits_{u_1<1<u_1+u_2} e^{-(u_1+u_2)}\, du_1\, du_2$$

which is independent of θ. One can easily check that $1 - \gamma(\theta) = e^{-1}$. Hence $(x_1, x_1 + x_2)$ is a CI of size e^{-1} and level less than or equal to e^{-1}.

Note that $u_1 = \dfrac{x_1}{\theta}$ or $u_2 = \dfrac{x_2}{\theta}$ are not statistics as these are functions of observations and the parameter. Fisher (1935) called a function $T(x, \theta)$ of observations and the parameter as a 'pivot' or pivotal quantity if $P[T(x, \theta) \leq a \mid \theta] = G(a, \theta)$ does not depend on θ for any $\theta \in \Omega$ and $a \in R_1$. Note that $(\bar{X} - \theta)$ in Example 10.1.1 being $N(0, 1/n)$ is a pivot. Similarly, in Example 10.1.2, $u_1 = \dfrac{x_1}{\theta}$, $u_2 = \dfrac{x_2}{\theta}$ are pivots as these are independent standard exponentials and $u_3 = \dfrac{x_1 + x_2}{\theta}$ under each θ has $G(2, 1)$ distribution with pdf $g_2(u) = u\, e^{-u}, u \geq 0$ and therefore is a pivotal quantity which can be used to construct CI.

EXAMPLE 10.1.2 (*contd.*). To illustrate the use of pivotal quantity u_3 to construct a CI for θ, observe that for any a, b, $0 < a < b$, we have $a < u_3 < b$ is equivalent to $\frac{x_1 + x_2}{b} < \theta < \frac{x_1 + x_2}{a}$. We therefore take $T_1(x) = \frac{x_1 + x_2}{b}$ and $T_2(x) = \frac{x_1 + x_2}{a}$. Then $(T_1(x), T_2(x))$ is a CI for θ of size $1 - \gamma = \int_a^b g_2(u_3)\, du_3 = G_2(b) - G_2(a)$. Length of the CI of the above type is $(x_1 + x_2)\left(\frac{1}{a} - \frac{1}{b}\right)$ which is a r.v. Hence we try to determine (a, b) such that $G_2(b) - G_2(a) = 1 - \gamma$ and $(x_1 + x_2)\left(\frac{1}{a} - \frac{1}{b}\right)$ is as small as possible. However as $(x_1 + x_2)$ is always positive, this amounts to minimizing $\left(\frac{1}{a} - \frac{1}{b}\right)$ subject to constraint that $0 < a < b$ and $G_2(b) - G_2(a) = 1 - \gamma$. Using Lagrange's method, equations defining a, b are given by $b^3 e^{-b} = a^3 e^{-a}$ and $G_2(b) - G_2(a) = 1 - \gamma$. If instead of a sample of size two we have a sample of size n then we search for CI for θ of size $1 - \gamma$ of the type $\left(\frac{\Sigma x_i}{b}, \frac{\Sigma x_i}{a}\right)$ such that $G_n(b) - G_n(a) = 1 - \gamma$. One can show that shortest length CI in the above class of CI has a and b defined by $a^{n+1}\, e^{-a} = b^{n+1}\, e^{-b}$ which can be determined using tables given by Tate and Klett (1959) for standard values of $1 - \gamma = .90, .95$, etc. or by numerical methods.

10.2 Shortest Expected Length CI

The general problem of determining shortest expected length CI of size $1 - \gamma$ can be posed as follows. Let C_γ denote the class of CIs of level $1 - \gamma$. Determine $(T_1^*(x), T_2^*(x)) \in C_\gamma$ such that

$$E_\theta(T_2^*(x) - T_1^*(x)) \leq E_\theta(T_2(x) - T_1(x)),\ \forall\ \theta \in \Omega$$

for any $(T_1(x), T_2(x)) \in C_\gamma$. This is analogous to determining $T^* \in U_\psi$ such that $M_T - M_{T^*}$ is nnd, $\forall\ \theta \in \Omega$, $\forall\ T \in U_\psi$ or determining $\varphi^* \in D_\alpha$ such that $\beta_{\varphi^*}(\theta) \geq \beta_\varphi(\theta)$, $\forall\ \theta \in \Omega$, $\forall \varphi \in D_\alpha$. We have already observed that the optimal estimators and tests exist in some specific models only. The problem of obtaining optimum CI of given level $1 - \gamma$ is more complex and we will restrict ourselves to the CIs obtained by using a pivotal quantity $u(x, \theta)$ with p.d.f. $h(u)$ and d.f. $H(u)$ under each $\theta \in \Omega$. We therefore consider an interval (a, b) such that

$$P\,(a \leq u(x, \theta) \leq b \mid \theta) = H(b) - H(a) = 1 - \gamma \tag{10.2.1}$$

Select $u(x, \theta)$ such that the inequality $a \leq u(x, \theta) \leq b$ can be inverted to obtain $T_1(x) \le \theta \le T_2(x)$ for each x and every $\theta \in \Omega$. Then $(T_1(x), T_2(x))$ is a CI of size $1 - \gamma$. The upper limit $T_2(x)$ and the lower limit $T_1(x)$ of the CI would depend on (a, b). We then select (a, b) satisfying (10.2.1) such that the expected length of the corresponding CI is shortest. Such an interval is called as shortest expected length CI (SELCI). We will usually consider the pivotal quantity $u(x, \theta)$ based on minimal sufficient statistic for the family $\{L(x, \theta), \theta \in \Omega\}$.

Note that in Example 10.1.1, $\bar{X} - \theta$ is a pivotal quantity based on minimal sufficient statistic $\bar{X}$ and in Example 10.1.2, $\Sigma X_i/\theta$ is a pivotal quantity based on minimal sufficient statistic ΣX_i. To illustrate the technique further we consider some more examples.

Example 10.2.1 Let $(X_1, \ldots, X_n)$ be a random sample from $U(0, \theta)$ so that $X_{(n)}$ is minimal sufficient statistic with pdf

$$g(x_{(n)}, \theta) = \frac{n\,x_{(n)}^{n-1}}{\theta^n},\ 0 < x_{(n)} < \theta$$

An obvious choice for a pivotal quantity $u(x_{(n)}, \theta)$ is $\dfrac{x_{(n)}}{\theta}$ with its pdf $h(u) = nu^{n-1}$, $0 < u < 1$.

Now consider $\{a < u(x_{(n)}, \theta) < b\} = \left\{a < \dfrac{x_{(n)}}{\theta} < b\right\} = \left\{\dfrac{x_{(n)}}{b} < \theta < \dfrac{x_{(n)}}{a}\right\}$.

We have to determine (a, b) such that $P(a < u(x, \theta) < b \mid \theta) = b^n - a^n = 1 - \gamma$ and $E\left[X_{(n)}\left(\dfrac{1}{a} - \dfrac{1}{b}\right)\right] = \dfrac{n\theta}{n+1}\left(\dfrac{1}{a} - \dfrac{1}{b}\right)$ is minimum or $Q = \left(\dfrac{1}{a} - \dfrac{1}{b}\right)$ is minimum.

As $b^n - a^n = 1 - \gamma$, treating a as a function of b, we have

$$nb^{n-1} - na^{n-1}\frac{da}{db} = 0 \text{ i.e. } \frac{da}{db} = \frac{b^{n-1}}{a^{n-1}}$$

Therefore

$$\frac{\partial Q}{\partial b} = -\frac{1}{a^2}\frac{da}{db} + \frac{1}{b^2} = -\frac{1}{a^2}\frac{b^{n-1}}{a^{n-1}} + \frac{1}{b^2} = \frac{a^{n+1} - b^{n+1}}{b^2\, a^{n-1}}.$$

As $0 \le a < b \le 1$ we have $\dfrac{dQ}{db} < 0$ and the minimum of Q occurs at the maximum possible value of b, namely $b = 1$. This gives $a = \gamma^{1/n}$ and SELCI is given by $\left(x_{(n)}, \dfrac{x_{(n)}}{\gamma^{1/n}}\right)$.

Example 10.2.2 Consider a Pareto distribution with pdf $f(x, \theta) = \dfrac{\theta}{x^{\theta+1}}$, $x > 1, \theta > 0$. Then $Y = \log X$ is exponential with mean $\dfrac{1}{\theta}$ and ΣY_i is minimal sufficient statistic with distribution $G\left(n, \frac{1}{\theta}\right)$. An obvious choice of pivotal quantity is $u(y, \theta) = \theta \Sigma y_i$ with $h(u) = \frac{1}{\Gamma(n)} u^{n-1} e^{-u}$, $u > 0$.

Consider $\{a < u(y, \theta) < b\} = \left\{\frac{a}{\Sigma y_i} < \theta < \frac{b}{\Sigma y_i}\right\}$ so that we have to select (a, b) such that $H_n(b) - H_n(a) = 1 - \gamma$ and $\frac{1}{\Sigma y_i}(b - a)$ is minimum. Note that $E\left(\frac{1}{\Sigma Y_i}\right)$ exists only when $n \geq 2$. However, since $\Sigma y_i > 0$ for any $n \geq 1$, we select (a, b) such that $Q = (b - a)$ is minimum subject to the condition that $H_n(b) - H_n(a) = 1 - \gamma$. Again treating a as a function of b, we have $h_n(b) - h_n(a) \frac{da}{db} = 0$. Therefore we must select (a, b) such that $e^{-b} b^{n-1} = e^{-a} a^{n-1}$ and $H_n(b) - H_n(a) = 1 - \gamma$.

We note that if $(T_1(x), T_2(x))$ is a CI of size $1 - \gamma$ for θ and if ψ is an increasing function of θ with $\frac{d\psi}{d\theta} > 0$ then $(\psi(T_1(x)), \psi(T_2(x)))$ is a CI of size $1 - \gamma$ for $\psi(\theta)$. If ψ is decreasing i.e. $\frac{d\psi}{d\theta} < 0$ then $(\psi(T_2(x)), \psi(T_1(x)))$ is a CI of size $1 - \gamma$ for $\psi(\theta)$. This follows from the fact that in case ψ is increasing

$$\{x \mid T_1(x) \leq \theta \leq T_2(x)\} = \{x \mid \psi(T_1(x)) \leq \psi(\theta) \leq \psi(T_2(x))\} \quad (10.2.2)$$

and if ψ is decreasing the RHS of (10.2.2) becomes $\{x \mid \psi(T_2(x)) \leq \psi(\theta) \leq \psi(T_1(x))\}$. However this CI may not be SELCI for $\psi(\theta)$ even though $(T_1(x), T_2(x))$ is SELCI.

To illustrate consider Example 10.1.2, where we have shown that $\left(\frac{\Sigma y_i}{b}, \frac{\Sigma y_i}{a}\right)$ is a CI for θ of size $1 - \gamma = G_n(b) - G_n(a)$ for any (a, b) and is the SELCI of size $1 - \gamma$ if a and b satisfy $e^{-a} a^{n+1} = e^{-b} b^{n+1}$. Consider $\psi(\theta) = \frac{1}{\theta}$, the failure rate of the exponential distribution. Then as $\frac{d\psi}{d\theta} = -\frac{1}{\theta^2} < 0$ we have ψ decreasing and $\left(\frac{a}{\Sigma y_i}, \frac{b}{\Sigma y_i}\right)$ is a CI of size $1 - \gamma$. However the SELCI for $\frac{1}{\theta}$ is not same as that for θ since for $\psi(\theta) = \frac{1}{\theta}$, the constants a, b must satisfy $G_n(b) - G_n(a) = 1 - \gamma$ and $e^{-b} b^{n-1} = e^{-a} a^{n-1}$.

Next suppose that θ is vector valued, $\theta = (\theta_1, \ldots, \theta_m)'$ and we want to obtain CI for say θ_1. If we can find a pivotal quantity $u(x, \theta_1)$ with pdf $h(u)$ and d.f. $H(u)$ which does not depend on θ then using the above technique we can obtain SELCI for θ_1 of size $1 - \gamma$. For example for a random sample of size n from $N(\mu, \sigma^2)$, as seen earlier $(\bar{X}, S^2)'$ is minimal sufffficient for $(\mu, \sigma^2)'$.

To obtain SELCI for μ we can consider $u(x, \mu) = \sqrt{n}(\bar{X} - \mu)/\left\{\frac{S^2}{n-1}\right\}^{1/2}$ as a pivot with pdf given by that of a Student's t with $(n - 1)$ d.f. One can show that $\bar{x} \pm t_{n-1,1-\gamma/2} \frac{S}{\sqrt{n(n-1)}}$ is SELCI for μ of size $1 - \gamma$. Similarly a pivotal for σ^2 is given by $u(x, \sigma^2) = \frac{S^2}{\sigma^2}$ which is χ^2_{n-1} and SELCI for σ^2 of size $1 - \gamma$ is given by $\left(\frac{S^2}{b}, \frac{S^2}{a}\right)$ where a and b are determined by

$$H_{n-1}(b) - H_{n-1}(a) = 1 - \gamma \text{ and } a^2 h_{n-1}(a) = b^2 h_{n-1}(b).$$

For details see Guenther (1969).

This section restricts to pivotal quantities $u(x, \theta)$ which depend on observations through the sufficient statistic T. This is a consequence of the crucial role that minimal sufficient statistic T plays in MVU estimation. Recall that Rao-Blackwell Theorem allows us to restrict our search for MVUE within the class of unbiased estimators of $\psi(\theta)$ which are functions of minimal sufficient statistic T. Similarly for any testing problem $\theta \in \Omega_{H_0}$ vs $\theta \in \Omega_1$ for any test function $\varphi(x)$ there exists a test function $E[\varphi(x) \mid t] = \varphi^*(t)$ such that the power functions of φ and φ^* are identical $\forall\ \theta \in \Omega$. Therefore we can restrict our attention to test functions depending on observations only through the minimal sufficient statistic T. As the following example shows this is not necessarily true for the problem of CI.

EXAMPLE 10.2.3 We have already seen in Example 10.1.2 that for a random sample of size two from exponential distribution with mean θ, $T_1(x) = x_1$ and $T_2(x) = x_1 + x_2$ provides us a CI of size e^{-1}. The expected length of this CI is θ. Now consider $E(x_1 \mid x_1 + x_2) = \frac{x_1 + x_2}{2}$ to obtain CI $\left(\frac{x_1 + x_2}{2}, x_1 + x_2\right)$. The expected length of this CI, obtained from $(T_1(x), T_2(x))$ by conditioning w.r.t. minimal sufficient statistic $x_1 + x_2$ is also θ. Now $P\left(\frac{X_1 + X_2}{2} < \theta < X_1 + X_2 \mid \theta\right) = P\left(\frac{u}{2} < 1 < u\right)$ where $h(u) = ue^{-u}$, $u \geq 0$.

Thus, size of the CI $\left(\frac{x_1 + x_2}{2}, x_1 + x_2\right)$ is given by $\int_1^2 ue^{-u}du = 2e^{-1} - 3e^{-2}$ which is less than e^{-1}, the size of CI given by $(x_1, x_1 + x_2)$. Thus the CI obtained by conditioning w.r.t. minimal sufficient statistic has same expected length but smaller size. However, as the next example shows we have a situation where conditioning w.r.t. minimal sufficient statistic increases the size for the same length.

EXAMPLE 10.2.4 Let $(x_1, \ldots, x_n)$ be a sample of size n from $N(\theta, 1)$ consider CI given by $(x_1 - a, x_1 + a)$, $a > 0$ having length $2a$ and size $\Phi(a) - \Phi(-a) = 2\Phi(a) - 1$.

Now $E(x_1 \mid \bar{x}) = \bar{x}$ and thus conditioning w.r.t. $\bar{x}$, leads to the CI given by $(\bar{x} - a, \bar{x} + a)$ which has same length $2a$ but its size is $2\Phi(\sqrt{n}a) - 1$ which is much larger than that of $(x_1 - a, x_1 + a)$.

The restriction of choosing a pivot for θ based on minimal sufficient statistic is thus some what arbitrary. Further, in examples considered above $\{L(x, \theta), \theta \in \Omega\}$ was such that it was easy to determine a pivotal quantity. In general it is not very clear as how to construct a pivotal quantity even in a simple case such as Bernoulli series of trials. This being a first course we sill not pursue the problem any further, but consider in the next section how this difficulty can be avoided by using a pivotal quantity for θ based on a CAN estimator of θ so that we can obtain SELCI in the large samples at least.

Exercise 10.2.1 (1) Refer Example 10.2.3. Show that the CI of size e^{-1} obtained by conditioning on $(x_1 + x_2)$ is not the SELCI within the CIs of the type $\left(\frac{x_1 + x_2}{b}, \frac{x_1 + x_2}{a}\right)$. The SELCI is given by $a \simeq 1.19$, $b \simeq 4.45$ which has much smaller expected length than θ.

(2) Obtain SELCI of size $1 - \gamma$ for θ when we have a random sample of size n from

$$f(x, \theta) = \frac{1}{2\theta} \exp\left\{-\frac{|x|}{\theta}\right\}, \; x \in R_1, \theta > 0.$$

(3) Let $(x_1, \ldots, x_n)$ be a random sample from the pdf $f(x, \mu) = \exp\{-(x - \mu)\}$, $x \geq \mu$, $\mu \in R_1$. Taking the pivot $u(x_{(1)}, \mu) = n(x_{(1)} - \mu)$ obtain SELCI for μ of size $1 - \gamma$.

(4) Consider two parameter exponential distribution with pdf

$$f(x, \mu, \sigma) = \frac{1}{\sigma} \exp\left\{\frac{-(x - \mu)}{\sigma}\right\}, \; x \leq \mu, \mu \in R_1, \sigma > 0.$$

As seen earlier $(X_{(1)}, \sum_{2}^{n} (X_{(i)} - X_{(1)})'$ is minimal sufficient for $(\mu, \sigma)'$. Obtain SELCI for μ based on $u_1(x_{(1)}, \mu) = \dfrac{n(x_{(1)} - \mu)}{\sum_{2}^{n} (x_{(i)} - x_{(1)})}$ and that for σ based on $\sum_{2}^{n} (x_{(i)} - x_{(1)})/\sigma$.

10.3 Large Sample Confidence Intervals

Let T be a CAN estimator of θ so that $T \sim AN\left(\theta, \frac{\sigma_T^2(\theta)}{n}\right)$. Then one can construct an asymptotic pivotal quantity $u(T, \theta) = \dfrac{(T - \theta)\sqrt{n}}{\sigma_T(\theta)} \sim N(0, 1)$ for which $P[-a \leq u \leq a] \simeq 2\Phi(a) - 1$, $\forall\ \theta \in \Omega$. Now, the inequality $-a \leq \dfrac{(T - \theta)\sqrt{n}}{\sigma_T(\theta)} \leq a$ is equivalent to

$$\theta - \frac{\sigma_T(\theta)}{\sqrt{n}} a \leq T \leq \theta + \frac{\sigma_T(\theta)}{\sqrt{n}} a \qquad (10.3.1)$$

If these inequalities can be inverted and are equivalent to $T_1 \le \theta \le T_2$ then (T_1, T_2) would be called as an asymptotic CI (ACI) of size $2\Phi(a) - 1$. Taking $a = \xi_{1-\gamma/2}$ we have ACI of size $1 - \gamma$ and this ACI would have shortest expected length among the class of ACI of the type $(T + c, T + d)$ based on asymptotic pivotal $u(T, \theta)$ and size $1 - \gamma$. Generally one should choose a CAN estimator T with $\sigma_T^2(\theta)$ as small as possible. Under usual regularity conditions this leads to the MLE and $u(\hat{\theta}, \theta) = \sqrt{nI(\theta)}\ (\hat{\theta} - \theta)$.

EXAMPLE 10.3.1 Let $(X_1, X_2, \ldots, X_n)$ be a random sample from Cauchy distribution with location θ. Then $I(\theta) = \frac{1}{2}$ and $u(\hat{\theta}, \theta) = \sqrt{n/2}\ (\hat{\theta} - \theta)$ and

$$\{-a \le \sqrt{n/2}\ (\hat{\theta} - \theta) \le a\} = \{\hat{\theta} - \sqrt{2/n}\ a \le \theta \le \hat{\theta} + \sqrt{2/n}\ a\}$$

Taking $a = \xi_{1-\gamma/2}$, $\hat{\theta} \pm \sqrt{2/n}\ \xi_{1-\gamma/2}$ is ACI for θ of size $1 - \gamma$.

EXAMPLE 10.3.2 Let $(X_1, \ldots, X_n)$ be a random sample from Laplace distribution with pdf $f(x, \theta) = \frac{1}{2} \exp\{-|x - \theta|\}$, $x \in R_1, \theta \in R_1$. Then the sample mean $\bar{X}$ and the sample median M_n are both CAN for θ with asymptotic variances $\frac{2}{n}$ and $\frac{1}{n}$, respectively. Therefore $(\bar{x} \pm \sqrt{2/n}\ \xi_{1-\gamma/2})$ and $(M_n \pm \xi_{1-\gamma/2}/\sqrt{n})$ are both ACI of size $1 - \gamma$. Note that the ACI based on M_n, the sample median, which is also the MLE of θ has shorter length, than that based on $\bar{X}$.

In both the above examples the asymptotic variances $\sigma_T^2(\theta)/n$ were independent of θ and inversion of the inequality $-a \le \sqrt{n}(T - \theta)/\sigma_T(\theta) \le a$ was very easy. We now consider situations in which $\sigma_T^2(\theta)$ depends on θ.

EXAMPLE 10.3.3 Let $(X_1, X_2, \ldots, X_n)$ be a random sample from an exponential distribution with mean θ then $\hat{\theta} = \bar{X} \sim AN(\theta, \theta^2/n)$ so that $AV(\hat{\theta})$ depends on θ. Let $u(\bar{x}, \theta) = n\left(\frac{\bar{x} - \theta}{\theta}\right)$ be the pivotal quantity and consider inequalities

$$-a \le \sqrt{n}\left(\frac{\bar{x} - \theta}{\theta}\right) \le a \tag{10.3.2}$$

A straightforward calculation leads to the inequalities,

$$\theta - \frac{a}{\sqrt{n}}\theta \le \bar{x} \le \theta + \frac{a}{\sqrt{n}}\theta$$

which leads to ACI

$$\frac{\bar{x}}{1 + a/\sqrt{n}} \le \theta \le \frac{\bar{x}}{1 - a/\sqrt{n}} \tag{10.3.3}$$

Taking $a = \xi_{1-\gamma/2}$ we have ACI of size $1 - \gamma$.

Next consider a situation where we have $\sigma_T^2(\theta)$ depends on θ in such a way that inversion of inequalities in (10.3.1) is not as easy as in the above case.

EXAMPLE 10.3.4 Let $(X_1, \ldots, X_n)$ be a random sample from Poisson distribution with parameter λ. Then, as seen before, $\bar{X} \sim AN(\lambda, \lambda/n)$ and we take $u(\bar{x}, \lambda) = \sqrt{n}(\bar{x} - \lambda)/\sqrt{\lambda}$ as the asymptotic pivot. Let $a = \xi_{1-\gamma/2}$, then ACI of size $1 - \gamma$ is given by inverting $-a \le u(\bar{x}, \lambda) \le a$ which is equivalent to $\frac{n(\bar{x} - \lambda)^2}{\lambda} \le a^2$ or

$$\bar{x}^2 - \lambda\left(2\bar{x} + \frac{a^2}{n}\right) + \lambda^2 \le 0 \qquad (10.3.4)$$

RHS of (10.3.4) is a quadratic in λ and can be factored as $[\lambda - \lambda_1(x)]\,[\lambda - \lambda_2(x)]$ where $\lambda_1(x) < \lambda_2(x)$ are the roots of equation $\lambda^2 - \lambda(2\bar{x} + a^2/n) + \bar{x}^2 = 0$. These are given by

$$\left.\begin{aligned} \lambda_1(x) &= \frac{(2\bar{x} + a^2/n) - \{(2\bar{x} + a^2/n)^2 - 4\bar{x}^2\}^{1/2}}{2} \\ \lambda_2(x) &= \frac{(2\bar{x} + a^2/n) - \{(2\bar{x} + a^2/n)^2 - 4\bar{x}^2\}^{1/2}}{2} \end{aligned}\right\} \qquad (10.3.5)$$

Note that $(2\bar{x} + a^2/n)^2 - 4\bar{x}^2 = 4\bar{x}\frac{a^2}{n} + \frac{a^4}{n^2} > 0$ for all $\bar{x}$ since $\bar{x} \ge 0$ w.p. 1 under each $\lambda > 0$ and therefore $\lambda_1(x)$ and $\lambda_2(x)$ are both real. Thus $(\lambda_1(x), \lambda_2(x))$ is ACI of size $1 - \gamma$ for λ.

It can be shown that for a random sample of size n on $b(1, \theta)$ the ACI of size $1 - \gamma$ for θ is given by $(\theta_1(x), \theta_2(x))$ where

$$\left.\begin{aligned} \theta_1(x) &= \frac{(2n\bar{x} + a^2) - \{4a^2 n\bar{x}(1 - \bar{x}) + a^4\}^{1/2}}{2(a^2 + n)} \\ \theta_2(x) &= \frac{(2n\bar{x} + a^2) + \{4a^2 n\bar{x}(1 - \bar{x}) + a^4\}^{1/2}}{2(a^2 + n)} \end{aligned}\right\} \qquad (10.3.6)$$

In case of Poisson or binomial models the inversion of inequalities was not that complicated since $\sigma_T^2(\theta)$ was such that it led to a quadratic equation in θ. However if $\sigma_T^2(\theta)$ is a complicated function of θ then the inversion of the basic inequalities would be much harder. Further in case where θ is vector valued say $(\theta_1, \ldots, \theta_m)'$ and T is CAN for θ_1, $\sigma_T^2(\theta)$ may depend on nuisance parameters $\theta_2, \ldots, \theta_m$ and $u(T, \theta) = \sqrt{n}(T - \theta_1)/\sigma_T(\theta)$ may not be an asymptotic pivot. We get around both of these problems by the technique of studentizing the asymptotic pivot $u(T, \theta)$, i.e. estimating $\sigma_T(\theta)$ by a consistent estimator of $\sigma_T(\theta)$. Observe that for $N(\mu, \sigma^2)$ case with σ^2 unknown, the studentized version of $u(\bar{x}, \mu) = \sqrt{n}(\bar{x} - \mu)/\sigma$ is given by Student's t_{n-1} with σ substituted by its consistent estimator $\left\{\frac{S^2}{n-1}\right\}^{1/2}$. Note that if $\hat{\theta}$ is consistent for θ and $\sigma_T(\theta)$ is continuous then $\sigma_T(\hat{\theta}) \xrightarrow{p} \sigma_T(\theta)$. Therefore $\sqrt{n}(T - \theta_1)/\sigma_T(\hat{\theta}) \xrightarrow{d} N(0, 1)$ and is an asymptotic pivotal quantity. The inequalities

$$-a \le \sqrt{n}\,(T - \theta_1)/\sigma_T(\hat{\theta}) \le a$$

can now be inverted very easily to obtain

$$T - \frac{a}{\sqrt{n}}\,\sigma_T(\hat{\theta}) \le \theta \le T + \frac{a}{\sqrt{n}}\,\sigma_T(\hat{\theta}) \tag{10.3.7}$$

For $a = \xi_{1-\gamma/2}$, $T \pm \frac{a}{\sqrt{n}}\,\sigma_T(\hat{\theta})$ gives us ACI of size $1 - \gamma$ for θ_1.

Thus studentized asymptotic pivot helps us in proposing a reasonable solution when nuisance parameters are present as well as when the basic inequalities are difficult to invert. We illustrate these situations by suitable examples.

EXAMPLE 10.3.5 Consider one way ANOVA model with $x_{ij} = \mu_i + \varepsilon_{ij}$, $i = 1, 2, \ldots, k$ and $j = 1, 2, \ldots, n$. Let ε_{ij} be i.i.d. $N(0, \sigma^2)$ so that $\theta = (\mu_1, \ldots, \mu_k, \sigma^2)'$. Then as seen before the MLEs are given by $\hat{\mu}_i = \bar{x}_i$, $i = 1, 2, \ldots, k$ and $\hat{\sigma}^2 = \frac{1}{k}\sum_{i=1}^{n} S_i^2/n$, where $S_i^2 = \sum_j (x_{ij} - \bar{x}_i)^2$. The asymptotic variance covariance matrix of MLEs is given by diag $\left(\frac{\sigma^2}{n}, \ldots, \frac{\sigma^2}{n}, \frac{2\sigma^4}{n}\right)$. Consider a linear function $\theta = \sum_{i=1}^{n} l_i\mu_i$. Then $\sum l_i\,\bar{X}_i \sim AN(\sum l_i\mu_i, \sum l_i^2\,\sigma^2/n)$, since σ^2 is unknown we studentize and consider the studentized asymptotic pivot given by $(\sum l_i\bar{x}_i - \sum l_i\mu_i)/\left\{\frac{\hat{\sigma}^2}{n}\sum l_i^2\right\}^{1/2}$ so that the ACI of size $1 - \gamma$ for $\sum li\mu_i$ is given by

$$\sum l_i\bar{x}_i \pm \frac{\xi_{1-\gamma/2}}{\sqrt{n}}(\sum l_i^2)^{1/2}\,\hat{\sigma}.$$

In the above set up we can drop the assumption of normality of errors and assume instead that $\{\varepsilon_{ij}\}$ are i.i.d. with $E(\varepsilon_{ij}) = 0$ Var $(\varepsilon_{ij}) = \sigma^2$. Then by MV CLT, $\sum l_i\bar{X}_i \sim AN\left(\sum l_i\mu_i, \frac{\sigma^2}{n}\sum l_i^2\right)$ and $\hat{\sigma}^2 = \frac{1}{k}\sum_{i=1}^{k}\frac{S_i^2}{n} \xrightarrow{p} \sigma^2$ as $n \to \infty$. We can further relax the assumption of same variance within each block. Thus now $E(\varepsilon_{ij}) = 0$, $\{\varepsilon_{ij}\}$ are mutually independent but are i.i.d. only within each block with Var $(\varepsilon_{ij}) = \sigma_i^2, i = 1, 2, \ldots, k$. Then $(\bar{x}_1, \bar{x}_2, \ldots, \bar{x}_k)'$ is CAN for $(\mu_1, \mu_2, \ldots, \mu_k)$ with asymptotic variance covariance matrix given by diag $\left(\frac{\sigma_1^2}{n}, \ldots, \frac{\sigma_k^2}{n}\right)$. Then

$$\Sigma\, l_i \bar{X}_i \sim AN\left(\Sigma\, l_i \mu_i, \sum_{i=1}^{k} l_i^2 \frac{\sigma_i^2}{n}\right)$$

Now $\hat{\sigma}_i^2 = \frac{1}{n}\Sigma\,(x_{ij} - \bar{x}_i)^2 \xrightarrow{P} \sigma_i^2$, $i = 1, 2, \ldots, k$. Therefore studentized asymptotic pivot for $\Sigma\, l_i\mu_i$ is

$$(\Sigma\, l_i \bar{x}_i - \Sigma\, l_i \mu_i) \Big/ \left\{\sum_{i=1}^{k} l_i^2 \frac{\hat{\sigma}_i^2}{n}\right\}^{1/2}$$

and $\Sigma\, l_i \bar{x}_i \pm \xi_{1-\gamma/2} \left\{\Sigma\, l_i^2 \frac{\hat{\sigma}_i^2}{n}\right\}^{1/2}$ is ACI of size $1 - \gamma$ for $\Sigma\, l_i\mu_i$. In particular for $k = 2$, we have

$$\bar{x}_1 - \bar{x}_2 \pm \xi_{1-\gamma/2} \left\{\frac{\hat{\sigma}_1^2}{n} + \frac{\hat{\sigma}_2^2}{n}\right\}^{1/2} \quad \text{is ACI for } \mu_1 - \mu_2.$$

We can also relax the assumption of the same sample size within each block i.e. take $j = 1, 2, \ldots, n_i$, $i = 1, 2, \ldots, k$. Then assuming that each $n_i \to \infty$ we have

$$\Sigma\, l_i \bar{X}_i \sim AN\left(\Sigma\, l_i \mu_i, \Sigma\, l_i^2 \frac{\hat{\sigma}_i^2}{n}\right) \text{ and } \Sigma\, l_i \bar{x}_i \pm \xi_{1-r/2} \left(\Sigma\, l_i^2 \frac{\hat{\sigma}_i^2}{n_i}\right)^{1/2}$$

is ACI for $\Sigma\, l_i\mu_i$ of size $1 - \gamma$, where $\hat{\sigma}_i^2 = \frac{1}{n_i}\sum_{j=1}^{n_1}(x_{ij} - \bar{x}_i)^2$.

When situation is such that the parameter is real valued we construct the asymptotic pivotal quantity $u(T, \theta) = \sqrt{n}(T - \theta)/\sigma_T(\tilde{\theta})$ where $\tilde{\theta}$ is consistent for θ and $\sigma_T(\tilde{\theta}) \xrightarrow{P} \sigma_T(\theta)$. We could use $\tilde{\theta} = T$ and in general we use the MLE $\hat{\theta}$ and the asymptotic pivot $u(\tilde{\theta}, \theta) = \sqrt{nI(\hat{\theta})}\,(\hat{\theta} - \theta)$. Then ACI for θ of size $1 - \gamma$ is given by $\hat{\theta} \pm \xi_{1-\gamma/2} \frac{1}{\sqrt{nI(\hat{\theta})}}$. The ACIs obtained by using $I(\hat{\theta})$ instead of $I(\theta)$ generally difer by terms $0(1/\sqrt{n})$ as can be seen from the following examples.

EXAMPLE 10.3.3 (*contd.*) We now use asymptotic pivot $\frac{\sqrt{n}(\bar{x} - \theta)}{\bar{x}}$ instead of $\frac{\sqrt{n}(\bar{x} - \theta)}{\theta}$. Then the ACI for θ is given by

$$\bar{x}(1 - a/\sqrt{n}) \leq \theta \leq \bar{x}(1 + a/\sqrt{n}) \tag{10.3.8}$$

From (10.3.3) the lower limit of ACI is

$$\frac{\bar{x}}{1+\frac{a}{\sqrt{n}}} = \bar{x}\left\{1 - \frac{a}{\sqrt{n}} + 0\left(\frac{1}{\sqrt{n}}\right)\right\}.$$

Similarly the upper limit from (10.3.3) is

$$\frac{\bar{x}}{\left(1+\frac{a}{\sqrt{n}}\right)} = \bar{x}\left\{1 + \frac{a}{\sqrt{n}} + 0\left(\frac{1}{\sqrt{n}}\right)\right\}$$

Therefore ACI given by (10.3.8) and (10.3.3) differ by $0(1/\sqrt{n})$.

EXAMPLE 10.3.4 *(contd.)* Here the studentized version of the asymptotic pivot $u(\bar{x}, \lambda) = \dfrac{\sqrt{n}(\bar{x}-\lambda)}{\sqrt{\lambda}}$ is given by $\dfrac{\sqrt{n}(\bar{x}-\lambda)}{\sqrt{\bar{x}}}$ and ACI for λ of size $1-\gamma$ is given by

$$\bar{x} - \frac{a}{\sqrt{n}}\sqrt{\bar{x}} \leq \lambda \leq \bar{x} + \frac{a}{\sqrt{n}}\sqrt{\bar{x}} \tag{10.3.9}$$

where $a = \xi_{1-\gamma/2}$.

Now lower limit of ACI based on $u(\bar{x}, \lambda)$ given in (10.3.5) is

$$\lambda_1(x) = \frac{2\bar{x} + \frac{a^2}{n} - \left\{4\bar{x}\frac{a^2}{n} + \frac{a^4}{n^2}\right\}^{1/2}}{2} = \bar{x} + \frac{a^2}{2n} - a\,\{\bar{x}/n\}^{1/2}\left\{1 + \frac{a^2}{4\bar{x}n}\right\}^{1/2}$$

$$= \bar{x} - a\,\{\bar{x}/n\}^{1/2} + 0(1/\sqrt{n})$$

Similarly one can show that the upper limit $\lambda_2(x) = \bar{x} + a(\bar{x}/n)^{1/2} + 0(1/\sqrt{n})$. We leave it to the reader to show that for the ACI for θ in $b(1, \theta)$ model, the ACI obtained by studentization is given by

$$\bar{x} - a\left\{\frac{\bar{x}(1-\bar{x})}{n}\right\}^{1/2} \leq \theta \leq \bar{x} + a\left\{\frac{\bar{x}(1-\bar{x})}{n}\right\}^{1/2} \tag{10.3.10}$$

and the upper and lower limits $\theta_1(x)$ and $\theta_2(x)$ given in (10.3.6) differ from those given above by $0(1/\sqrt{n})$.

Let $T \sim AN(\theta, \sigma_T^2(\theta)/n)$ and ψ is such that $\dfrac{d\psi}{d\theta} \neq 0$ and continuous. Then since ψ is continuous with non-vanishing derivative ψ is strictly increasing or is strictly decreasing. Under this condition $\{T_1(x) \leq \theta \leq T_2(x)\}$ is equivalent to $\{\psi(T_1(x)) \leq \psi(\theta) \leq \psi(T_2(x))\}$ if $\dfrac{d\psi}{d\theta} > 0$ and if $\dfrac{d\psi}{d\theta} < 0$ it is equivalent to $(\psi(T_2(x)) \le \psi(\theta) \le \psi(T_1(x))\}$. Hence for ψ with $\frac{d\psi}{d\theta} > 0$, $(\psi(T_1(x), \psi(T_2(x)))$ is ACI of size $1 - \gamma$ for $\psi(\theta)$. If $\frac{d\psi}{d\theta} < 0$, then $(\psi(T_2(x)), \psi(T_1(x)))$ would provide an ACI for $\psi(\theta)$ of size $1 - \gamma$. On the other hand using Theorem 6.1.1 we have if $T \sim AN\left(\theta, \frac{\sigma_T^2(\theta)}{n}\right)$ then $\psi(T) \sim AN\left(\psi(\theta), \frac{\sigma_T^2(\theta)}{n}\left(\frac{d\psi}{d\theta}\right)^2\right)$. After studentizing, the ACI for $\psi(\theta)$ of size $1 - \gamma$ is given by $\psi(T) \pm \frac{\xi_{1-\gamma/2}}{\sqrt{n}}\left[\sigma_T^2(\theta)\left(\frac{d\psi}{d\theta}\right)^2\right]^{1/2}_{\theta=T}$. Generally the two ACI's for $\psi(\theta)$ differ by $0(1/\sqrt{n})$. We illustrate this by way of an example.

EXAMPLE 10.3.4 (*Contd.*) We have already seen that $\bar{x} \pm a(\bar{x}/n)^{1/2}$ and $(\lambda_1(x), \lambda_2(x))$ given by (10.3.5) are both ACI of size $1 - \gamma$ for λ. Consider $\psi(\lambda) = e^{-\lambda}$. Then $\frac{d\psi}{d\lambda} = -e^{-\lambda} < 0$ and ψ is decreasing and $(e^{-\lambda_2(x)}, e^{-\lambda_1(x)})$ as well as $\left(\exp\left\{-\bar{x} - a\left(\frac{\bar{x}}{n}\right)^{1/2}\right\}, \exp\left\{-\bar{x} + a\left(\frac{\bar{x}}{n}\right)^{1/2}\right\}\right)$ would be ACI of size $1 - \gamma$ for $e^{-\lambda}$. Now consider the CAN estimator $\bar{e}^{\bar{x}}$ for $e^{-\lambda}$ with $AV(\bar{e}^{\bar{x}}) = \frac{\lambda}{n}e^{-2\lambda}$. A consistent estimator of $AV(\bar{e}^{\bar{x}}) = \frac{\bar{x}}{n}e^{-2\bar{x}}$. Therefore ACI for $e^{-\lambda}$ is now given by

$$\bar{e}^{\bar{x}} \pm \frac{a}{\sqrt{n}}\bar{e}^{\bar{x}}\sqrt{\bar{x}} = \bar{e}^{\bar{x}}\left(1 \pm a\left(\frac{\bar{x}}{n}\right)^{1/2}\right).$$

Now consider

$$\exp\left\{-\bar{x} \pm a\left(\frac{\bar{x}}{n}\right)^{1/2}\right\} = \bar{e}^{\bar{x}}\exp\left\{\pm a\left(\frac{\bar{x}}{n}\right)^{1/2}\right\} = \bar{e}^{\bar{x}}\left\{1 \pm a\left(\frac{\bar{x}}{n}\right)^{1/2} + 0(1/\sqrt{n})\right\}$$

Therefore the ACI of size $1 - \gamma$ given by all the three methods discussed here differ from each other by $0(1/\sqrt{n})$.

There is yet one more method by which we can obtain ACI based on variance stabilization transformations due to Bartlett (1937) discussed in Exercise 6.2 (3). We illustrate it by the Poisson example considered above.

EXAMPLE 10.3.4 (*Contd.*) For Poisson model the variance stabilizing transformation is the square root transformation given by $\varphi(\bar{x}) = \{\bar{x}\}^{1/2}$, leading to $\{\bar{x}\}^{1/2} \sim AN\left(\sqrt{\lambda}, \frac{1}{4n}\right)$. Therefore ACI for $\sqrt{\lambda} = \theta$ is given by $\left(\{\bar{x}\}^{1/2} \pm \frac{a}{2\sqrt{n}}\right)$. To obtain ACI or $\lambda = \theta^2$ we observe that $\frac{d\lambda}{d\theta} = 2\theta > 0$ which leads to ACI given by $\left(\sqrt{\bar{x}} \pm \dfrac{a}{2\sqrt{n}}\right)^2 = \bar{x} \pm a\sqrt{\bar{x}}/\sqrt{n} + 0\left(\dfrac{1}{\sqrt{n}}\right)$. Similarly for $e^{-\lambda} = e^{-\theta^2} = \psi(\theta)$ we have $\dfrac{d\psi}{d\theta} = e^{-\theta^2}(-2\theta) < 0$ so that the ACI for $e^{-\lambda} = e^{-\theta^2}$ is now given by

$$\left(\exp\left\{-\left(\{\bar{x}\}^{1/2} + \frac{a}{2\sqrt{n}}\right)^2\right\},\ \exp\left\{-\left(\{\bar{x}\}^{1/2} - \frac{a}{2\sqrt{n}}\right)^2\right\}\right)$$

which is upto $0(1/\sqrt{n})$, same as the ACIs obtained earlier.

In most of the above work we have used the MLE as the CAN estimator to construct the asymptotic pivot or its studentized version to obtain ACI. Recall that by Hodges–LeCam method we can improve the MLE at a specific point. Section 7.7 discussed the $N(\theta, 1)$ case in detail. We also noted there the improvement of MLE at one point in the parameter space is obtained at the cost of higher MSE at infinitely many points in the parameter space. We now show that the ACI obtained by using the Hodgegs–LeCam superefficient estimator could have very poor coverage probability for many points in the parameter space.

For any $\varepsilon > 0$ consider the coverage probability

$$P(\bar{X}, \varepsilon, \theta) = P[\,|\,\bar{X} - \theta\,| \le \varepsilon \mid \theta] = \Phi(\sqrt{n}\,\varepsilon) - \Phi(-\sqrt{n}\,\varepsilon).$$

Observe that for any θ_n and any sequence $\varepsilon_n \to 0$ such that $\sqrt{n}\,\varepsilon_n \to \infty$ we have $P(\bar{x}, \varepsilon_n, \theta_n) \to 1$ as $n \to \infty$. Recall that the superefficient estimator is given by

$$\begin{aligned} T &= \bar{x} \quad \text{if } |\,\bar{x}\,| \ge a_n \\ &= \alpha\bar{x} \quad \text{if } |\,\bar{x}\,| < a_n \end{aligned}$$

where $0 < \alpha < 1$, $a_n \to 0$ and $\sqrt{n}a_n \to \infty$. In Section 7.7 we have obtained $G_n(t, \theta)$ the exact d.f. of T under any θ. Note that

$$G_n(t, \theta) = \Phi[\sqrt{n}(a_n - \theta)],\ \alpha a_n \le t < a_n$$

and $G_n(t, \theta)$ remains constant on the interval $[\alpha a_n, a_n]$. Now consider the coverage probability $P(T, \varepsilon, \theta) = P[\,|\,T - \theta\,| \le \varepsilon \mid \theta]$ and take $\varepsilon_n = ca_n$ and let θ_n and a_n be chosen such that $\alpha a_n \le \theta_n - \varepsilon_n \le \theta_n + \varepsilon_n \le a_n$. Then for such a sequence the coverage probability $P(T, \varepsilon_n, \theta_n) = P[\,|\,T_n - \theta_n\,| \le \varepsilon_n \mid \theta_n] \equiv 0$. Thus whereas coverage probability $P(\bar{x}, \varepsilon, \theta) \to 1$ as $n \to \infty$ for any choice of ε and θ, there exist sequences ε_n and θ_n such that the coverage probability $P(T, \varepsilon_n, \theta_n) \to 0$ as $n \to \infty$. For example, we take $a_n = n^{-1/4}$ and

$$\theta_n = \frac{(1+\alpha)\,n^{-1/4}}{2},\ \varepsilon_n = \frac{(1-\alpha)\,n^{-1/4}}{4}.$$

For more details see Kale (1985).

In the next section we discuss the problem of confidence intervals from the view point of coverage probability.

Exercise 10.4 (1) Let $(X_1, \ldots, X_n)$ be a random sample of size n from an exponential distribution with mean θ. Let $\psi(\theta) = e^{-t_0/\theta}$. Obtain ACI for $e^{-t_0/\theta}$ by using studentized version of asymptotic pivot based on the CAN estimator of $\psi(\theta)$. Obtain ACI for $e^{-t_0/\theta}$ by using its monotone nature and ACIs for θ given by (10.3.3) and (10.3.8). Further consider the variance stabilization transformation $\varphi(\bar{x}) = \log \bar{x}$. Using this obtain ACI's for θ and $\psi(\theta)$. Show that all the ACIs obtained in these diferent ways differ from each other by $0(1/\sqrt{n})$.

(2) Let $(X_1, \ldots, X_n)$ be a random sample from $N(\mu, \sigma^2)$, $\mu \in R_1$, $\sigma^2 > 0$. An asymptotic pivot for $\mu + \xi_p\sigma$ can be based on $\bar{x} + \xi_p \hat{\sigma}$ with its variance estimated. Using this obtain ACI of size $1 - \gamma$ for $\mu + \xi_p\sigma$.

Let $x_{(r)}$ be CAN for $\mu + \xi_p\sigma$ where $r = [np] + 1$. Obtain ACI based on $x_{(r)}$ and its estimated asymptotic variance and using median for estimating μ and interquartile range for estimating σ. For expressions for variance of linear functions of order statistics for large samples from $N(\mu, \sigma^2)$ refer to David (1981).

(3) Let $(X_1, \ldots, X_n)$ be a random sample from $U(\mu - 1, \mu + 1)$. Obtain ACI for μ based on $\bar{X}$, M_n the median, $X_{(1)} + 1$, $X_{(n)} - 1$, and $(X_{(1)} + X_{(n)})/2$.

(4) Let $(X_1, \ldots, X_n)$ be a random sample from $U(0, \theta)$. We have already seen that $u(x_{(n)}, \theta) = \dfrac{n(\theta - x_{(x)})}{\theta} \xrightarrow{d} Z$, where Z is standard exponential. Obtain ACI for θ using the asymptotic pivot $u(x_{(n)}, \theta)$ and compare this ACI with SELCI for θ obtained in Example 10.2.1.

10.4 Unbiased Confidence Intervals

Neyman (1937) introduced a criterion other than the length of the CI based on probability of covering a wrong value of θ. Let $(T_1(x), T_2(x)) = E$ be a CI for θ of size $1 - \gamma$ and consider $C(\theta, \theta') = P[T_1(x) \le \theta' \le T_2(x) \mid \theta]$ i.e. the probability that $(T_1(x), T_2(x))$ covers $\theta' \neq \theta$ which is the underlying true value θ. We call $(T_1(x), T_2(x))$ as unbiased CI of size $1 - \gamma$ if

$$\left.\begin{aligned} C_E(\theta, \theta') &= 1 - \gamma \quad \text{for} \quad \theta = \theta' \\ &\le 1 - \gamma \quad \text{for} \quad \theta \neq \theta' \end{aligned}\right\} \tag{10.4.1}$$

Thus $(T_1(x), T_2(x))$ is an unbiased CI if probability of coverage of true value θ is always larger than the probability of coverage of a wrong value $\theta' \neq \theta$ the true value of the parameter. This is analogous to unbiasedness of a test of level α which requires that the probability of rejecting H_0 when H_1 holds should always be larger than the probability of rejecting H_0 when H_0 holds (i.e. power $\ge$ level). We give two examples one in which SELCI is unbiased and the other in which it is not.

EXAMPLE 10.4.1 Let $(X_1, \ldots, X_n)$ be a random sample of size n from $N(\theta, 1)$, then within the class of CI of type $(\bar{x} + a, \bar{x} + b)$ we have seen that SELCI of size $1 - \gamma$ is given by

$$E = \left(\bar{x} - \frac{a}{\sqrt{n}}, \bar{x} + \frac{a}{\sqrt{n}}\right) \quad \text{where } a = \xi_{1-\gamma/2}.$$

Now $C_E(\theta, \theta') = P\left[\bar{X} - \frac{a}{\sqrt{n}} < \theta' < \bar{X} + \frac{a}{\sqrt{n}} \mid \theta\right]$. By routine calculations we have

$$C_E(\theta, \theta') = \Phi(\delta + a) - \Phi(\delta - a) = C_E(\delta) \tag{10.4.2}$$

where $\delta = \sqrt{n}(\theta' - \theta)$.

Note that $C_E(0) = 1 - \gamma$ and as $\delta \to \pm\infty$, $C_E(\delta) \to 0$.

$$\text{Now } \frac{dC_E(\delta)}{d\delta} = \exp\left\{-\frac{(a^2 + \delta^2)}{2}\right\} \{e^{-\delta a} - e^{\delta a}\} \frac{1}{\sqrt{2\pi}}$$

and as $a > 0$,

$$\frac{dC_E(\delta)}{d\delta} > 0 \quad \text{if} \quad \delta < 0$$

$$= 0 \quad \text{if} \quad \delta = 0$$

$$< 0 \quad \text{if} \quad \delta > 0$$

This implies that $C_E(\delta)$ increases over $(-\infty, 0)$, reaches the maximum value $1 - \gamma$ at $\delta = 0$ and then decreases over $(0, \infty)$. Therefore $C_E(\delta) < 1 - \gamma$ for $\delta \neq 0$ and E is an unbiased CI of size $1 - \gamma$.

It is left to the reader to verify that CI of size $1 - \gamma$ of the type $(\bar{x} + c, \bar{x} + d)$ with $c < d$ is unbiased only when $-c = d = \frac{\xi_{1-\gamma/2}}{\sqrt{n}}$ (Guenther, 1971).

EXAMPLE 10.4.2 Consider $(X_1, \ldots, X_n)$ a random sample of size n from exponential distribution with mean θ. Let $E = \left(\frac{\Sigma x_i}{b}, \frac{\Sigma x_i}{a}\right)$ be CI for θ of size $1 - \gamma$. Then, as seen before, for E to be SELCI, a and b are given by

$$G_n(b) - G_n(a) = 1 - \gamma \text{ and } a^{n+1}e^{-a} = b^{n+1}\, e^{-b} \tag{10.4.3}$$

Now consider

$$C_E(\theta, \theta') = P\left(\frac{\Sigma X_i}{b} < \theta' < \frac{\Sigma X_i}{a} \mid \theta\right) = P\left(\frac{a\theta'}{\theta} < \frac{\Sigma X_i}{\theta} < \frac{\theta'}{\theta} b \mid \theta\right)$$

$$= G_n\left(b\frac{\theta'}{\theta}\right) - G_n\left(a\frac{\theta'}{\theta}\right).$$

Then letting $\frac{\theta'}{\theta} = \lambda$ we have at $\lambda = 1$, $C_E(\theta, \theta') = 1 - \gamma$. Consider $G_n(b\lambda) - G_n(a\lambda) = \psi(\lambda, a, b)$. Then $\psi(\lambda, a, b) \to 0$ as $\lambda \to 0$ and $\lambda \to \infty$. Now CI, E will be unbiased if $\frac{d\psi}{d\lambda} = g_n(b\lambda).\, b - g_n(a\lambda)a = 0$ at $\lambda = 1$ i.e. $a^n e^{-a} = b^n e^{-b}$. Having selected a and b to satisfy $a^n e^{-a} = b^n e^{-b}$, we show that E is an unbiased CI of size $1-\gamma$. This can be proved if we show that $\dfrac{d\psi(\lambda, a, b)}{d\lambda} = 0$ at $\lambda = 1$ and $\left(\dfrac{d^2\psi}{d\lambda^2}\right)_{\lambda=1} < 0$.

Now $$\left[\frac{d^2\psi}{d\lambda^2}\right]_{\lambda=1} = [g_n'(b\lambda)\, b^2 - g_n'\,(a\lambda)\, a^2]_{\lambda=1} = g_n'(b)\, b^2 - g_n'\,(a)\, a^2$$

For $n \geq 2$, $g_n'(u) = \dfrac{1}{\Gamma(n)}\, e^{-u}\,(n-1)u^{n-2} - \dfrac{1}{\Gamma(n)}\, e^{-u}\, u^{n-1}$.

Thus

$$g_n'(b)b^2 - g_n'(a)a^2 = \frac{1}{\Gamma(n)}\,\{e^{-b}(n-1)b^n - e^{-b}b^{n+1} - e^{-a}(n-1)\, a^n + e^{-a}a^{n+1}\}$$

But $e^{-b}\, b^n = e^{-a}a^n$ and therefore

$$g_n'(b)b^2 - g_n'(a)a^2 = \frac{1}{\Gamma(n)}\,\{e^{-a}\, a^n\, a - e^{-b}\, b^n\, b\} \qquad (10.4.4)$$

Thus LHS of (10.4.4) $= \dfrac{1}{\Gamma(n)}\, e^{-a}\, a^n(a-b) < 0$ as $a < b$.

For $n = 1$, we have $G_n(u) = 1 - e^{-u}$ and $g_n(u) = e^{-u}$ and a, b are given by $e^{-a} - e^{-b} = 1-\gamma$ and $e^{-a}\, a = e^{-b}b$. Here $\psi(\lambda, a, b) = e^{-a\lambda} - e^{-b\lambda}$ and we can show that $\left[\dfrac{d^2\psi}{d\lambda^2}\right]_{\lambda=1} < 0$.

We note that SELCI of size $1-\gamma$ for θ has a and b defined by $G_n(b) - G_n(a) = 1-\gamma$ and $e^{-b}\, b^{n+1} = e^{-b}\, a^{n+1}$ whereas for the unbiased CI of same size $1-\gamma$ we must have a and b such that $G_n(b) - G_n(a) = 1-\gamma$ and $e^{-b}\, b^n = e^{-b}\, a^n$ and SELCI is not an unbiased CI.

Neyman (1937) then suggested obtaining optimum CI which is "best" within the class of unbiased CI of size $1-\gamma$, i.e. for which $C_E(\theta, \theta') = 1-\gamma$ for $\theta = \theta'$ and $C_E(\theta, \theta') \leq 1-\gamma$ and is uniformly as small as possible for $\theta \neq \theta'$.

Neyman criterion of unbiasedness is more appropriate in case of one sided CIs of the type $(T_1(x), \infty)$ or $(-\infty, T_2(x))$ where the length of CI is infinite. However the wrong values of θ are to defined as $\theta' < \theta$ in case of CI bounded below by $T_1(x)$ and $\theta' > \theta$ in case of CI bounded above $T_2(x)$. We illustrate this by an example.

EXAMPLE 10.4.2 (Contd.). Suppose we are interested in a CI of the type $(T_1(x), \infty)$ for θ which is natural as θ represents in this case average life of an item under consideration. Then we select the same pivot $\dfrac{\Sigma x_i}{\theta}$ and interval

$E = \left(\frac{\Sigma x_i}{a}, \infty\right)$ where $P\left(\frac{\Sigma X_i}{a} \le \theta \mid \theta\right) = 1 - \gamma$ or a such that $G_n(a) = 1 - \gamma$.

Here the probability of E covering a wrong value $\theta' \ne \theta$ is given by

$$C_E(\theta, \theta') = G_n\left(a\frac{\theta'}{\theta}\right)$$

$$= 1 - \gamma \quad \text{if } \theta = \theta'$$

$$< 1 - \gamma \quad \text{if } \theta' < \theta.$$

Note that $C_E(\theta, \theta') \ge 1 - \gamma$ if $\theta' > \theta$ but this is not objectionable since if $\frac{\Sigma x_i}{a} \le \theta \Rightarrow \frac{\Sigma x_i}{a} \le \theta'$ if $\theta' > \theta$. Thus wrong values here are $\theta' < \theta$, and covering of any such value is less desirable.

Neyman's formulation of the CI problem is similar to his formulation of the problem of tests of hypotheses where, Type I error namely rejecting H_0 when H_0 is true was bounded above by α. This is equivalent to probability of accepting H_0 when H_0 is true being bounded below by $1 - \alpha$. Thus rather than looking at critical region (CR) or test function φ for H_0, in the CI problem we look at the compliment of CR, the acceptance region or the acceptance function $1 - \varphi$. The reader would note that the compliment of the CI of level $1 - \gamma$ for θ provides critical region for testing $H_0 : \theta' = \theta$ vs H_1: $\theta' \ne \theta$ in the two sided case. Similar remark holds for H_0: $\theta' \le \theta$ vs H_1: $\theta' > \theta$ in the case of one sided CI $(T_1(x), \infty)$.

For any $\theta_0 \in \Omega$ suppose there exists a non-randomized test of level γ based on a statistic $s(x)$ which rejects H_0: $\theta = \theta_0$ against H_1: $\theta \ne \theta_0$ when $K_2(\theta_0) \le s(x)$ or $s(x) \le K_1(\theta_0)$ for each $\theta_0 \in \Omega$. Then we have

$$P[K_1(\theta_0) \le s(x) \le K_2(\theta_0) \mid \theta_0] = 1 - \gamma, \ \forall \ \theta_0 \in \Omega. \quad (10.4.5)$$

Suppose $K_1(\theta_0)$ and $K_2(\theta_0)$ as functions of $\theta_0 \in \Omega$ are strictly increasing then (10.4.5) is equivalent to

$$P[K_2^{-1}(s(x)) \le \theta_0 \le K_1^{-1}(s(x)) \mid \theta_0] = 1 - \gamma, \ \forall \ \theta_0 \in \Omega \quad (10.4.6)$$

and $(K_2^{-1}(s(x)), K_1^{-1}(s(x)))$ is a CI for θ of size $1 - \gamma$. If $K_1(\theta_0)$ and $K_2(\theta_0)$ as functions of $\theta_0 \in \Omega$ are strictly decreasing then $(K_1^{-1}(s(x)), K_2^{-1}(s(x)))$ is a CI for θ of size $1 - \gamma$.

Example 10.4.1 For $N(\theta, 1)$ case for testing H_0: $\theta = \theta_0$ vs $\theta \ne \theta_0$ we have $s(x) = \bar{x}$ and $K_1(\theta_0) = \theta_0 - \frac{\xi_{1-\gamma/2}}{\sqrt{n}}$, $K_2(\theta_0) = \theta_0 + \frac{\xi_{1-\gamma/2}}{\sqrt{n}}$. Both $K_1(\theta_0)$ and $K_2(\theta_0)$ are strictly increasing and $K_1^{-1}(\bar{x}) = \bar{x} + \frac{\xi_1 - \gamma/2}{\sqrt{n}}$ and

$K_2^{-1}(\bar{x}) = \bar{x} - \dfrac{\xi_1 - \gamma/2}{\sqrt{n}}$ so that $\left(\bar{x} - \dfrac{\xi_1 - \gamma/2}{\sqrt{n}}, \bar{x} + \dfrac{\xi_1 - \gamma/2}{\sqrt{n}}\right)$ is CI for θ of size $1 - \gamma$.

EXAMPLE 10.4.2 (*Contd.*). Consider one sided problem H_0: $\theta \le \theta_0$ vs H_1: $\theta > \theta_0$ then we have $s(x) = \Sigma\, x_i$ and we reject H_0 if $\Sigma\, x_i > K_\gamma(\theta_0)$ where $K_\gamma(\theta_0) = \theta_0\, G_n^{-1}(1 - \gamma)$. Now $K_\gamma(\theta_0)$ is strictly increasing and therefore $K_\gamma^{-1}(\Sigma\, x_i) = \dfrac{\Sigma x_i}{G_n^{-1}(1-\gamma)}$ and we have $P[K_\gamma^{-1}(\Sigma\, x_i) < \theta_0 \mid \theta_0] = 1 - \gamma, \forall\, \theta_0 \in \Omega$ so that $\left(\dfrac{\Sigma x_i}{G_n^{-1}(1-\gamma)}, \infty\right)$ is a CI of size $1 - \gamma$ for θ the mean of the exponential distribution.

For detailed study of connection between the problem of CI and that of testing of hypotheses, refer Lehmann (1959). We emphasize that view in this chapter is that the limits of CI are random and these cover true value θ of the parameter with probability $1 - \gamma$ specified in advance. Historically Bayesian approach briefly outlined at the end of Chapter 9 viewed the problem of confidence intervals as expressing uncertainity about θ after obtaining data using posterior probability of $\theta \in (a, b)$ given by $\int_a^b p(\theta \mid x)\, d\theta$. On the other hand Fisher was not in favour of using Bayesian approach based on the assumption of a known prior probability distribution particularly when the prior was either subjective or chosen with mathematical convinience in mind. Fisher therefore proposed a solution where uncertainity about θ is expressed by fiducial probability statement. In the next section we will briefly consider Bayesian approach and Fisher's approach to the problem of CI.

10.5 Bayesian and Fiducial Intervals

Let $(x_1, \ldots, x_n)$ be a random sample of size n from $\{f(x, \theta), \theta \in \Omega \subset R_1\}$. Following Bayesian approach, the uncertainity about θ is expressed by a prior probability distribution given by pdf $p(\theta)$ defined on Ω. Then the joint pdf of (X, θ) is given by $L(x, \theta)\, p(\theta)$ and the conditional pdf of θ for fixed x, the posterior distribution of θ, is given by

$$p(\theta \mid x) = \frac{L(x, \theta)\, p(\theta)}{\int L(x, \theta)\, p(\theta)\, d\theta} \tag{10.5.1}$$

For any interval $(a, b) \in \Omega$ we have then

$$P(a < \theta < b) = \int_a^b p(\theta \mid x)\, d\theta \tag{10.5.2}$$

The RHS of (10.5.2) is a function of the observations x. For example as seen at the end of chapter 9, in the model $\{b(1, \theta), 0 < \theta < 1\}$ with uniform prior

$p(\theta) = 1$, for $0 < \theta < 1$, for the data in which t successes are observed in n trials we have

$$P(a < \theta < b) = \int_a^b \theta^t (1 - \theta)^{n-t} / B((t + 1, n - t + 1) \qquad (10.5.3)$$

Observe that RHS of (10.5.3) changes as t changes. With uniform prior over $\Omega = (0, 1)$ the posterior distribution of θ given t is a beta distribution with parameter t and $n - t + 1$. Note that the posterior distribution of θ given x depends on x only through $t = \Sigma x_i$ which is minimal sufficient statistic for θ in the model specified by $\{b(1, \theta), \theta \in (0, 1)\}$. In the joint pdf of observations the data $(x_1, \ldots, x_n)$ are variables and θ is fixed where as in the posterior distribution θ is a variable and x is fixed and t becomes the parameter. Indeed if T is a sufficient statistic for the model $\{L(x, \theta), \theta \in \Omega\}$ and $p(\theta)$ is any prior over Ω, it is easy to see that the posterior distribution of θ depends on x only through $T(x) = t$. To prove this we observe that

$$\begin{aligned} p(\theta \mid x) &= L(x, \theta)\, p(\theta) / \int_\Omega L(x, \theta)\, p(\theta)\, d\theta \\ &= g(T(x), \theta)\, h(x)\, p(\theta) / \int_\Omega g(T(x), \theta) h(x) p(\theta)\, d\theta \\ &= g(t, \theta)\, p(\theta) / \int_\Omega g(t, \theta)\, p(\theta)\, d\theta \qquad (10.5.4) \end{aligned}$$

In fact Kolmogorov (1950) defines a statistic T to be sufficient in the Bayesian framework if for any prior $p(\theta)$ on Ω, the posterior distribution of θ given x depends on x only through $T(x)$. One can show that sufficiency as defined earlier in Ch. 2 is equivalent to Bayesian sufficiency [Raiffa and Schlaifer (1961)].

Since $p(\theta \mid x)$ depends on the prior distribution, $p(\theta)$, for two different priors for the same data x we get different values for $P(a < \theta < b)$. For example if we assume a beta type prior distribution for θ given by

$$p(\theta) = \frac{1}{B(\alpha, \beta)} \theta^{\alpha-1} (1 - \theta)^{\beta-1}, 0 < \theta < 1$$

a beta distribution with α, β known, then

$$P(a < \theta < b) = \int_a^b \theta^{\alpha+t-1} (1 - \theta)^{\beta+n-t-1} / B(\alpha + t, \beta + n - t).$$

Here $p(\theta \mid x)$ is beta distribution with parameters $(\alpha + t, \beta + n - t)$.

In a similar way consider the situation where we have a random sample of size n from $N(\theta, \sigma_0^2)$ and where prior distribution of θ is itself normal with mean θ_0 and variance η_0^2 where θ_0 and η_0^2, are assumed to be known.

Then using (10.6.4) and the fact that $\bar{X} \sim N(\theta_0, \sigma_0^2/n)$ is the sufficient statistic by straight forward but somewhat complicated calculations we can show that the posterior distribution of θ given $\bar{x}$ is normal with mean which is weighted average of θ_0 and $\bar{x}$ with weights inversely proportional to the variances η_0^2 and σ_0^2/n, respectively, or posterior mean of

$$\theta = \frac{\theta_0}{\eta_0^2} + \frac{\bar{x}}{\sigma_0^2/n} \Big/ \frac{1}{\eta_0^2} + \frac{1}{\sigma_0^2/n}$$

or

$$\bar{\theta} = \frac{\sigma_0^2\theta_0 + n\,\eta_0^2\,\bar{x}}{\sigma_0^2 + n\,\eta_0^2}.$$

The variance of the posterior distribution of θ is given by

$$\left\{\frac{1}{\eta_0^2} + \frac{1}{\sigma_0^2/n}\right\}^{-1} = \left\{\frac{\sigma_0^2 + n\,\eta_0^2}{\sigma_0^2\,\eta_0^2}\right\}^{-1} = \frac{\sigma_0^2\,\eta_0^2}{\sigma_0^2 + n\,\eta_0^2}.$$

For more detailed derivation we refer to Appendix A 1.1 of Box and Tiao (1973). Therefore under the above set up

$$P(a < \theta < b) = \Phi\left(\frac{(b - \bar{\theta})}{\sigma\eta_0}(\sigma_0^2 + n\eta_0^2)^{1/2}\right) - \Phi\left(\frac{(a - \bar{\theta})}{\sigma\eta_0}(\sigma_0^2 + n\eta_0^2)^{1/2}\right)$$

As an exercise, the readers should work out the case of a sample of size n form the exponential distribution with mean θ with prior distribution of θ as inverse gamma with pdf $p(\theta) = \dfrac{1}{\Gamma(\lambda_0)} \exp\left\{-\dfrac{1}{\theta}\right\}\left(\dfrac{1}{\theta}\right)^{\lambda_0 - 1} \dfrac{1}{\theta^2}$, $\theta > 0$, λ_0 known.

Note that $\Sigma X_i = T$ is the minimal sufficient statistic for θ with pdf given by

$$g(t, \theta) = \frac{1}{\Gamma(n)} \frac{1}{\theta^n} e^{-t/\theta}\, t^{n-1}, t > 0, \theta > 0.$$

Since $P(a < \theta < b) = P\left(\dfrac{1}{b} < \dfrac{1}{\theta} < \dfrac{1}{a}\right)$ it is easier to work with a reparametrization $\phi = \dfrac{1}{\theta}$ so that joint distribution of t and ϕ is

$$g(t, \phi)\, p(\phi) = \frac{1}{\Gamma(n)} \frac{1}{\Gamma(\lambda_0)} \exp\{-\phi(t + 1)\} t^{n-1} \phi^{\lambda_0 - 1} \ldots > 0, t > 0$$

The marginal distribution of t is beta distribution of second kind with pdf $\dfrac{\Gamma(n + \lambda_0)}{\Gamma(n)\,\Gamma(\lambda_0)} \dfrac{t^{n-1}}{(1 + t)^{n+\lambda_0}}$, $0 < t < \infty$. The posterior distribution of ϕ given t is

$p(\phi \mid t) = \dfrac{1}{\Gamma(n + \lambda_0)} \exp\{-\phi(t + 1)\}\ \phi^{n+\lambda_0-1}\,(t + 1)^{n+\lambda_0}$, $0 < \phi < \infty$ which is a gamma pdf. Hence

$$P(a < \theta < b) = G_{n+\lambda_0}\left(\frac{t+1}{a}\right) - G_{n+\lambda_0}\left(\frac{t+1}{b}\right).$$

The above approach has assumed prior distributions of θ in a such a way that the posterior distribution is easy to work out. Further we have also assumed that the interval (a, b) is specified and therefore $P(a < \theta < b)$ depends on data x, through sufficient statistic t. Thus for different values of t we will get different evaluations although on any data x where value of t remains same, $P(a < \theta < b)$ will remain same.

On the other hand one can select (a, b) such that $P(a < \theta < b) = 1 - \gamma$ with $(b - a)$ as small as possible. Such an interval (a, b) will be generally around the mode of the posterior distribution of θ as the posterior density $p(\theta \mid x)$ is thickest in the interval containing the posterior mode as compared to the other intervals all having the same posterior probability $1 - \gamma$. Such intervals are called as Highest Posterior Density (HPD) intervals. Using the standard techniques of minimization subject to constraint, in the normal case the HPD interval of size $1 - \gamma$ is given by

$$\bar{\theta} \pm \xi_{1-\gamma/2}\left\{\frac{\sigma_0^2\,\eta_0^2}{\sigma_0^2 + n\eta_0^2}\right\}^{1/2} \quad \text{where } \bar{\theta} = \frac{\sigma_0^2\theta_0 + n\eta_0^2\,\bar{x}}{\sigma_0^2 + n\eta_0^2}$$

Note that the above Bayesian interval is close to the SELCI for θ given by $\bar{x} \pm \xi_{1-\gamma/2}\,\sigma_0/\sqrt{n}$ for large n, since

$$\left\{\frac{\sigma_0^2\,\eta_0^2}{\sigma_0^2 + n\eta_0^2}\right\}^{1/2} = \frac{\sigma_0}{\sqrt{n}} + 0\left(\frac{1}{\sqrt{n}}\right) \text{ and } \bar{\theta} = \bar{x} + 0(1/n).$$

On the other hand even for small n but large η_0^2 we have

$$\left\{\frac{\sigma_0^2\eta_0^2}{\sigma_0^2 + n\eta_0^2}\right\}^{1/2} = \frac{\sigma_0}{\sqrt{n}}\left\{1 + \frac{\sigma_0^2}{n\eta_0^2}\right\}^{1/2} = \frac{\sigma_0}{\sqrt{n}} + 0\left(\frac{1}{\eta_0}\right)$$

and $\bar{\theta} = \bar{x} + 0\left(\frac{1}{\eta_0}\right)$. Therefore for large values of η_0, the classical SELCI for θ of size $1 - \gamma$ agrees with HPD interval of same size $1 - \gamma$ upto $0\left(\frac{1}{\eta_0}\right)$.

Similarly in the exponential case the HPD interval for θ would be given by (a, b) such that $\left(\frac{1}{a} - \frac{1}{b}\right)$ is minimized subject to the constraint $G_{n+\lambda_0}\left(\frac{t+1}{a}\right) - G_{n+\lambda_0}\left(\frac{t+1}{b}\right) = 1 - \gamma$. Here (a, b) is determined by size constraint and the equation

$$g_{n+\lambda_0}\left(\frac{t+1}{a}\right)\left(\frac{t+1}{a}\right)^2 = g_{n+\lambda_0}\left(\frac{t+1}{b}\right)\left(\frac{t+1}{b}\right)^2$$

or
$$e^{-(t+1)/a}\,\frac{1}{a^{n+\lambda_0-1}} = e^{-(t+1)/b}\,\frac{1}{b^{n+\lambda_0-1}}$$

For given λ_0 and observed t we can use Tate and Klett (1959) tables or other numerical methods to determine (a, b).

One of the main objections to Bayesian inference based on inverse probability has been its heavy dependence on the assumed prior distribution of the parameter θ. This prior distribution expresses uncertainty about $\theta \in \Omega$. The underlying chance behaviour of data x is expressed by the model given by $\{L(x, \theta), \theta \in \Omega\}$. It is thus possible that using the same model but different prior distributions $p_1(\theta)$ and $p_2(\theta)$, the experimenter or the statistician may arrive at different posterior distributions $p_1(\theta \mid x)$ and $p_2(\theta \mid x)$ for the same data x. This is regarded as loss of objectivity so crucial to scientific method. On the other hand the two experimenters or statisticians based on two different models $\{L_1(x, \theta), \theta \in \Omega\}$ and $\{L_2(x, \theta), \theta \in \Omega\}$ may recommend different estimators of parameter θ based on the same observed data x. This is not regarded as loss of objectivity or violation of scientific method. This diffference in our attitude towards data x and parameter θ is perhaps a consequence of making possibly uncertain statements about the true value of θ in a given model is regarded as a legitimate scientific activity.

The method of inverse probability, initiated by Bayes-Laplace, had great followers such as Gauss, K. Pearson and even Fisher to start with. However slowly due to criticism from Boole (1854), Venn (1866) and due to contradictions that could arise from specification of different priors, Fisher appears to have turned away from the method of inverse probability as tool for inference at least in those situations in which the prior probabiity distribution had no objective basis. Fisher therefore proposed confidence intervals for unknown parameter θ using Fiducial probability distributions as an alternative to using posterior distributions based on a prior $p(\theta)$.

The fiducial intervals as proposed by Fisher are based on pivotal quantities and can be obtained only in special models where there is an inherent symmetry between the observable r.v. x and parameter θ e.g. in $N(\theta, 1)$ model or in general one parameter exponential family in its canonical form. Thus in $N(\theta, 1)$ we have pdf of $\bar{x}$ is given by

$$f(\bar{x}, \theta) = \left\{\frac{n}{2\pi}\right\}^{1/2} \exp\left\{\frac{-n(\bar{x}-\theta)^2}{2}\right\}, \bar{x} \in R_1, \theta \in R_1.$$

Here $\bar{x}$ is r.v. and θ is fixed. But $f(\bar{x}, \theta)$ is symmetric in $\bar{x}$ and θ, and one could interchange their roles or fix $\bar{x}$ and consider θ as a variable. In $(\bar{x}, \theta)$ plane the set $-a \le \bar{x} - \theta \le a$ is such that $P[-a \le \bar{x} \le a \mid \theta] = 2\Phi(\sqrt{n}a) - 1$, $\forall\ \theta \in R_1$ and $-a \le \bar{x} - \theta \le a$ is equivalent to $\bar{x} - a \le \theta \le \bar{x} + a$ or $\theta - a \le \bar{x} \le \theta + a$. Fisher suggested that the probability $2\Phi(\sqrt{n}a) - 1$ can be associated with the interval $(\bar{x} - a, \bar{x} + a)$ and to distinguish it from the probability associated with $(\theta - a, \theta + a)$ assigned by the distribution of $\bar{x}$

called it the fiducial probability of θ and consequently defined fiducial distribution of θ as $N\left(\bar{x}, \frac{1}{n}\right)$ with $\bar{x}$ fixed and viewed as a parameter of the fiducial distribution. In many situations fiducial intervals and CI for θ coincide but its interpretation is quite different. The classical theory due to Neyman treats $(T_1(x), T_2(x))$ as random and θ fixed whereas Fisherian approach treats θ as a variable and $(T_1(x), T_2(x))$ as fixed. Fisher did not want to claim that θ is a random variable but maintained that the uncertainity about θ can be expressed as a probability distribution which he called as fiducial probability distribution to distinguish it from the classical probability distributions used in the model $\{L(x, \theta), \theta \in \Omega\}$. In this Fisher (1930) has followed Bayes (1763) wherein initially the uncertainity about the parameter θ is called as Chance to distinguish it from the Probability of the observed data. Although later in the same paper (1763), Bayes stated that by Chance he means the same thing as Probability, Fisher maintained the distinction between the two. Savage (1954) described this attempt of Fisher as "trying to make an omlette without breaking an egg". The statistical inference based on fiducial approach did not get as much following as Bayesian inference. For example whereas Kendall and Stuart Vol. II (1967) has a chapter (pp. 134–160) on Fiducial Intervals its revised version by Stuart and Ord (1991) has only a few pages (pp. 781–787) devoted to this topic. For fiducial argument we refer to Kendall and Stuart Vol. II (1967) and references contained therein, as well as the articles by Edwards (1983), Buehler (1983) and Stone (1983). For Bayesian inference we refer to some of the old texts such as Lindley (1965), Box and Tiao (1973) as well as some recent ones namely Berger (1985) and O'Hagan (1994) and Bernardo and Smith (1994).

References

Abramowitz, M. and Stegun I.A. (1968). Handbook of Mathematical Functions. National Bureau of Standards. Appl. Math. Ser. 55, Washington.

Apostol, T.M. (1964). *Mathematical Analysis* (4th Ed.). Addison-Wesley Publishing Co. Inc. London, pp. 143–148.

Arnold, B.C. (1968). Parameter estimation for a multivariate exponential distribution. *Jour Amer. Stat. Assn.,* 63, 848–852.

Barnett, V. and Lewis, T. (1984). *Outliers in Statistical Data.* 2nd Ed. John Wiley & Sons, New York.

Bartlett, M.S. (1947). The use of transformations. *Biometrics,* 3, 39–52.

Bayes, T. (1763). An essay towards solving a problem in the doctrine of chances. *Phil. Trans. Roy. Soc.* 53, (Reproduced in *Bayesian Statistics* by S. James Press (1989). John Wiley and Sons, New York).

Berger, J.O. (1985). *Statistical Decision Theory and Bayesian Analysis,* 2nd Edition, Springer Verlag, New York, Berlin, Heidelberg, Tokyo.

Bernardo, J.M. and Smith A.F.M. (1994). *Bayesian Theory.* John Wiley & Sons, New York.

Beuhler, R.J. (1963). Fiducial Inference, *Encyclopedia of Statistical Sciences,* Vol. 3, pp. 76–81. Ed. Johnson N.L. and Kotz, S., John Wiley & Sons, New York.

Bhattacharya, G.K. and Johnson, R.A. (1973). On test of independence in bivariate exponential distribution. *Jour. Amer. Stat. Assn.* 68, 704–706.

Birch, M.W. (1964). A new proof of the Pearson Fisher Theorem. *Ann. Math. Stat.* 35, 92–98.

Boole, G. (1954). An investigations of the Laws of thought on which are founded the Mathematical Theories of Logic and Probabilities. Dover Publications (1951), New York.

Box, G.E.P. and Tiao, G.C. (1973). *Bayesian Inference in Statistical Analysis,* Addison-Wesley, Reading, MA USA.

Box Joan Fisher (1978). *R.A. Fisher Life of a Scientist.* John Wiley and Sons, New York.

Chanda, K.C. (1954). A note on the consistency and maxima of the roots of likelihood equations. *Biometrika,* 41, 56–61.

Chandrasekar, B. (1983). Contributions to the theory of unbiased estimation functions. Ph. D. Thesis, University of Pune, Pune-411007.

Cochran, W.G. (1934). The distribution of quadratic forms in a normal system. *Proc. Camb. Phil. Soc.,* 30, 178–191.

Cramer, H. (1946). *Mathematical Methods of Statistics.* Princeton University Press, Princeton N.J. 11th printing 1966.

Cohen, A.C. (1960). Estimating the parameter in a conditional Poisson Distribution. *Biometrics,* 16, 203–211.

Crichton, M. (1990). *Jurassic Park.* (pp. 160–166). Ballantine Books, New York.

David, H.A. (1981). *Order Statistics.* 2nd Edition, John Wiley and Sons, New York.

Desu, M.M. and Raghavrao, D. (1990). *Sample Size Methodology.* Academic Press, San Diego.

Draper, N.R. and Smith, H. (1981). *Applied Regression Analysis.* 2nd Edition, John Wiley & Sons, New York.

Edwards, A.W.F. (1983). Fidicual Distributions. *Encyclopedia of Statistical Sciences,* Vol. 3, pp. 70–76, Johnson N.L. and Kotz S., John Wiley and Sons, New York.

Elderton, W.P. (1954). *Frequency Curves and Correlation.* 4th Edition, Cambridge University Press, Cambridge.

Feller, W. (1957). *An Introduction to Probability Theory and its Applications.* Vol. I (2nd Edition). Ch. 6 Sec. 7, John Wiley and Sons, New York.

Ferguson, T.S. (1967). *Mathematical Statistics: A Decision Theoretic Approach.* Academic Press, New York.

Fisher, R.A. (1912) On an absolute Criterion for Fitting Frequency Curves. *Messnger of Mathematics,* 41, 155–160.

Fisher, R.A. (1920). A mathematical examination of the method of determining the accuracy of an observation by the mean error and by the mean square error. *Monthly Notices Roy. Astronomical Soc.,* 80, 757–770.

Fisher, R.A. (1922). On the mathematical foundations of theoretical statistics. *Phil. Trans. Roy. Soc.* Ser. A, 222, 309–368.

Fisher, R.A. (1925). Theory of Statistical Estimation. *Proceedings of Cambridge Phil. Soc.,* 22, 700–725.

Fisher, R.A. (1930). Inverse Probability. *Proc. Comb. Phil. Soc.* 26, 528–535.

Fisher, R.A. (1937). Professor Karl Pearson and the method of moments. *Ann. Eug.* 7, 308–317.

Fisher, R.A. (1954). *Statistical methods for research workers* (13th Edition), Oliver and Boyd Ltd. Edinburgh; pp. 299–320.

Fisher, R.A. (1959). *Statistical Methods and Scientific Inference.* 2nd Ed. Hafner Publishing Co. New York.

Fisher, R.A. and Yates, F. (1963). *Statistical Tables for Biological, Agricultural and Medical Research.* 6th Edition, Oliver and Boyd Edinburgh.

Fisz, M. (1963). *Probability Theory and Mathematical Statistics.* John Wiley and Sons, New York.

Foutz, R.V. (1977). On the unique consistent solution of the likelihood equations. *Jour. Amer. Stat. Assn.* 72, 147–148.

Freund, J.E. (1961). A bivariate extension of the exponential distribution. *Jour. Amer. Stat. Assn.* 56, 971–977.

Guenther, W.C. (1969). Shortest confidence intervals. *Amer. Stat.* 23, 22–25.

Guenther, W.C. (1971). Unbiased Confidence intervals. *Amer. Stat.* 25, 51–53.

Hanagal, D.D. and Kale, B.K. (1992). Large sample tests for testing symmetry and independence in some Bivariate Exponential Models. *Comm. Stat. Theory Meth.* 21(9), 2625–2643.

Huzurbazar, V.S. (1948). The likelihood equation, consistency and the maxima of the likelihood function. *Ann. Eugenics,* 14, 183–200.

Huzurbazar, V.S. (1955). Confidence intervals for the parameter of a distribution admitting a sufficient statistic when the range depends on the parameter. *Jour. Roy. Stat. Soc.* B 17, 86–90.

Ibragimov, I.A. and Hasminski, R.S. (1981). *Statistical Estimation, Asymptotic Theory,* Springer, New York, Heidelberg Berlin, p. 91.

Kale, B.K. (1961). On the solution of the likelihood equation by iteration processes. *Biometrika,* 48, 452–456.

Kale, B.K. (1962). On the solution of likelihood equations by iteration processes. The multiparametric case. *Biometrika,* 49, 479–486.

Kale, B.K. and Chandrasekar, B. (1983). On the equivalence of optimality criteria for vector unbiased statistics. *Jour. Ind. Stat. Assn.* 21, 49–55.

Kale, B.K. (1985). A note on the super efficient estimator. *Jour. Stat. Plan. Inference,* 259–263.

Kendall, M.G., Stuart, A. (1958). *Advanced Theory of Statistics,* Vol. I, Charles Griffin and Co. London.

Kendall, M.G. and Stuart, A. (1967). *The Advanced Theory of Statistics,* Vol. II, Inference and Relationship. Hafner Publishing Co., New York.

Kendall, M.G. and Plackett, R.L. (1977). *Studies in the History of Statistics and Probability,* Vol. II, Charles Griffin, London.

Klebanov, L.B., Linik, Yu. V. and Rukhin, A.L. (1971). Unbiased estimation and matrix loss functions. Soviet Math. Dokl. 12, 1526–1528.

Kolmogorov, A.N. (1950). Unbiased estimates. *Amer. Math. Soc. Translation No. 98 (1953)* pp. 155–177 of Izvestia Akad. Nauk. SSSR, Ser. Math. 14 (1950), 303–326.

Koshal, R.S. (1933). Application of the method of maximum likelihood to the improvement of curves fitted by the method of moments. *Jour. Roy. Stat. Soc.* 96, 303–13.

Koshal, R.S. (1935). Application of the method of maximum likelihood to the derivation of efficient statistics for fitting of frequency curves. *Jour. Roy. Stat. Soc.* 98, 128.

Kunte, S. and Gore, A.P. (1992). The paradox of large samples. *Current Science,* 62, 393–395.

Laplace, P.S. (1774). Memoire sur les probabilities in Ouvres Complete de Laplace, 9, 393–485, Gauthier Villars, Paris.

Laplace, P.S. [(1812), (1820)]. Theorie analytique des Probabilities, Courcier, Paris.

Laurent, A.G. (1963). Conditional distribution of order statistics and distribution of the reduced i-th order statistics of the exponential model. *Ann. Math. Stat.* 34, 652–657.

Le Cam, L. (1953). On some asymptotic properties of maximum likelihood estimates and related Bayes estimates. Univ. of California Publications in Stat. 1, 277–330.

Lehmann, E.L. (1959). *Testing of Statistical Hypotheses,* John Wiley & Sons, New York.

Lehmann, E.L. and Scheffe, H. (1950). Completeness, similar regions and unbiased estimation. *Sankhya,* 10, 305–340.

Lindly, D.V. (1965). *Introduction to Probability and Statistics from a Bayesian View Point.* (Parts 1 and 2). Cambridge University Press, Cambridge.

Marshall, A.W. and Olkin, I. (1967). A multivariate exponential distribution. *Jour. Amer. Stat. Assn.* 62, 30–44.

Neyman, J. and Pearson, E.S. (1933). On the problem of the most efficient tests of statistical hypotheses. *Phil. Trans. Roy. Soc.* A–231, 289.

O'Hagan, A. (1994). *Kendall's Advanced Theorky of Statistics.* Vol. 2B, Bayesian Inference, John Wiley, New York.

Pairman, E. (1919). Tables of Digamma and Trigamma Functions, Tracts for Computers 1. Ed. K. Pearson, Cambridge University Press.

Paranjape, S.A. (1994). Moment Estimation of Weibull Shape parameter: Which moment? (private communication)

Pearson E.S. (1978). Lectures by K. Pearson on "The History of Statistics in 17th and 18th centuries". Charles Griffin, London.

Pearson, K. (1900). On the criterion that a given system of deviations from the probable in the case of a correlated system of variables is such that it can be reasonably supposed to have arisen from random sampling. *Phil. Mag. Series* 5, Vol. L, pp. 157–175.

Pearson, K. (1936). Method of moments and method of maximum likelihood. *Biometrika,* 28, 34–59.

Pearson, E.S., Hartley, H.O. (1970). *Biometrika Tables for Statisticians,* Vol. I, Cambridge University Press, Cambridge.

Rao, C.R. (1947). Large sample tests of statistical hypotheses concerning several parameters with applications to problem of estimation. *Proc. Comb. Phil. Soc.* 44. 50–57.

Rao, C.R. (1962). Apparent anamolies and irregularities in maximum likelihood estimation (with discussion), *Sankhya* B, 24, 73–102.

Rao, C.R. (1962). Efficient estimates and optimum inference procedures in large samples (with discussion) *Jour. Roy. Stat. Soc.* B, 24, 46–72.

Rao, C.R. (1963). Criterion of estimation in large samples. *Sankhya* A, 25, 189–206.

Rao, C.R. (1973). *Linear Statistical Inference,* (2nd Edition), Wiley Eastern, New Delhi.

Rao, C.R. (1992). Statistics as a last resort, In Glimpses of India's Statistical Heritage Editors. Ghosh, J.K., Mitra, S.K. and Parthasarathy, K.R., Wiley Eastern, New Delhi, pp. 153–213.

Rohatgi, V.K. (1976). *An Introduction to Probability Theory and Mathematical Statistics.* John Wiley and Sons. New York.

Rudin, Walter (1964). *Principles of Mathematical Analysis,* McGraw-Hill Book Co. New York.

Savage, L.J. (1954). *The Foundation of Statistics,* John Wiley and Sons, New York.

Serfling, R.J. (1980). *Approximation Theorems of Mathematical Statistics.* John Wiley and Sons, New York.

Shewhart, W.A. (1950). *Contributions to Mathematical Statistics- R.A. Fisher*, John Wiley and Sons, New York

Stone, M. (1983). Fiducial Probability. *Encyclopedia of Statistical Sciences,* Vol. 3, pp. 81–86 Ed. Johnson N.L. and Kotz S., John Wiley & Sons, New York.

Stuart, A. and Ord, J.K. (1991). *Kendall's Advanced Theory of Statistics,* Vol. 2, Fifth Edition, published by Edward Arnold, a division of Hodder and Stoughton, London.

Student (1908). The probable error of a mean. *Biometrika,* 6, 1–25.

Tarone, R.E. and Gruenhage, G. (1975). A note on the uniqueness of roots of the likelihood equations for vector valued parameters. *Jour. Amer. Stat. Assn.* 70, 903–904.

Tate, R.F. and Klett, G.W. (1959). Optimum confidence intervals for the variance of a normal distribution. *Jour. Amer. Stat. Assn.,* 54, 674–682.

Venn, J. (1866). *The Logic of Chance. An Essay on the Foundations and Province of the Theory of Probability with Special Reference to its Logical bearings and its applications to Moral and Social Sciences.* 3rd Ed. McMillan, London.

Wald, A. (1943). Tests of statistical hypotheses concerning several parameters when the number of observations is large. *Trans. Amer. Math. Soc.* 54, 426–482.

Wald, A. (1947). *Sequential Analysis*. John Wiley & Sons, New York.

Wald, A. (1950). *Statistical Decision Functions.* John Wiley & Sons, New York.

Wilks, S.S. (1938). The large sample distribution of the likelihood ratio for testing composite hypotheses. *Ann. Math. Stat.* 9, 60–62.

Wolfowitz, J. (1965). Asymptotic efficiency of maximum likelihood estimators. *Theory of Prob. and its Appl.* 10, 247–260.

Index